Evolution and Advances in Computing Technologies for Industry 6.0

This book provides a comprehensive exploration of Industry 6.0, which marks the convergence of intelligent systems, machine learning (ML), deep learning, and human–robot collaboration (HRC) in various sectors. It focuses on how these technologies enable businesses to harness insights from vast datasets, optimize operations, forecast maintenance requirements, and mitigate outages. In this comprehensive book, the authors cover the major aspects of Industry 6.0, including the latest advances in technology, new applications, and the many challenges that accompany the process that is changing it. This book acts as a compass, guiding readers through the labyrinth of Industry 6.0, revealing the complex interactions between technologies, new applications, innovations, and business ecosystems. From a multidisciplinary perspective, the book explores the far-reaching impact of Industry 6.0 on the business world, revealing its potential to increase productivity, improve decision-making, and open new opportunities for development.

- Discusses the transformative impact of Industry 6.0 driven by the convergence of artificial intelligence (AI), ML, and the industrial internet of things (IIoT) in manufacturing
- Explores the transformative potential of HRC in modern manufacturing, emphasizing the ability of HRC to boost productivity, flexibility, and safety
- Compares several service models in cloud computing, with a focus on private, public, and community cloud deployment methods
- Presents information about autism spectrum disorder (ASD) in children and the advanced technology of convolutional neural networks used to detect autism in children
- Describes the accurate diagnosis of Alzheimer's disease through integrating state-of-the-art technologies, especially deep learning
- Focuses on digital adaptation in India's agricultural sector, especially Uttar Pradesh, highlighting challenges and prospects

- Presents a visionary leap into Agriculture 6.0, where technology and tradition converge for sustainable farming solutions
- Provides the mitigation strategies for climate change over multiple locations
- Explores the transformative impact of recent computational intelligence advancements in vehicle surveillance and recognition, particularly through CCTV analysis
- Focuses on the exploration and development of a supply chain network design (SCND) model of milk products

This book will be of interest to researchers, academics, practitioners, technology providers, undergraduate and postgraduate students, scholars, consultants, advisors, and doctors. It can be used by undergraduate and postgraduate students pursuing Industry 6.0 programs in computer science and engineering across the globe.

Evolution and Advances in Computing Technologies for Industry 6.0

Technology, Practices, and Challenges

Edited by
C. Kishor Kumar Reddy, Srinath Doss,
Surbhi Bhatia Khan, and
Abdulmajeed Alqhatani

CRC Press
Taylor & Francis Group
Boca Raton London New York

CRC Press is an imprint of the
Taylor & Francis Group, an **informa** business

First edition published 2025
by CRC Press
2385 NW Executive Center Drive, Suite 320, Boca Raton FL 33431

and by CRC Press
4 Park Square, Milton Park, Abingdon, Oxon, OX14 4RN

CRC Press is an imprint of Taylor & Francis Group, LLC

ISBN: 978-1-032-82304-1 (hbk)
ISBN: 978-1-032-82305-8 (pbk)
ISBN: 978-1-003-50393-4 (ebk)

DOI: 10.1201/9781003503934

Typeset in Palatino
by codeMantra

Contents

Preface

The book provides a comprehensive exploration of Industry 6.0, which marks the convergence of artificial intelligence (AI), machine learning (MI), and big data analytics in production processes. It delves into how these technologies enable businesses to harness insights from vast datasets, optimize operations, forecast maintenance requirements, and mitigate outages. In this comprehensive book, we cover major aspects of Industry 6.0, including the latest advances in technology, new applications, and the many challenges that accompany the process that is changing it. This book acts as a compass, guiding the reader through the labyrinth of Industry 6.0, revealing the complex interactions between technologies, new applications, innovations, and business ecosystems. From a multidisciplinary perspective, we explore the far-reaching impact of Industry 6.0 on the business world, revealing its potential for productivity, improve decision-making, and open new opportunities for development

Chapter 1 explores the transformative impact of Industry 6.0, driven by the convergence of AI, ML, and industrial internet of things (IIoT) in manufacturing. It discusses how these technologies enhance operational efficiency, product quality, and sustainability through real-time monitoring, predictive maintenance, and autonomous decision-making. The chapter highlights synergies between sensor data, cloud computing, and analytics for deep operational insights and process optimization. It addresses implementation challenges like data integration and cybersecurity and provides case studies of successful technology integration. Lastly, it explores future trends like edge computing, digital twins, and integration with additive manufacturing and collaborative robotics, offering valuable insights for manufacturing leaders and decision-makers.

Chapter 2 delves into the transformative potential of human–robot collaboration (HRC) in modern manufacturing, emphasizing its ability to boost productivity, flexibility, and safety. However, it also explores the complex challenges hindering its full realization, such as safety concerns, cognitive compatibility, and technical integration issues. The chapter examines how integrating soft computing algorithms, including AI, ML, fuzzy logic, and evolutionary algorithms, can address these challenges and facilitate effective HRC. Through case studies and examples, it demonstrates the application of these algorithms in enhancing collaboration and performance in various manufacturing scenarios. Additionally, the chapter discusses future research directions and challenges in leveraging soft computing for advancing HRC, thereby shaping the evolution of Industry 6.0 and beyond.

Chapter 3 examines and compares several service models in cloud computing, with a focus on private, public, and community cloud deployment methods. Nonetheless, despite the numerous benefits of cloud computing, such as on-demand self-service, broad network access, resource pooling, elasticity, and measured service, it is not completely reliable. The authors investigated the inherent obstacles and issues related to cloud computing, particularly the security implications of future research projects.

Chapter 4 presents explicit information about autism spectrum disorder in children, followed by the advanced technology used to detect autism in children. In this research, the technology used to detect this disorder is convolutional neural networks (CNNs). CNN is a deep learning method that is used for performing classification and prediction-related tasks. The ML models that are used for detecting this disorder in children are also discussed. Furthermore, the proposed model's performance is compared with the existing ML models to determine the best model.

Chapter 5 describes how the need for an accurate diagnosis of Alzheimer's disease is being met by integrating state-of-the-art technologies, especially deep learning. It draws attention to how sophisticated feature extraction techniques can improve treatment plans and diagnosis accuracy. To further dementia research and provide more potent interventions, it also stresses the utilization of large datasets like OASIS and ADNI.

Chapter 6 delves into digital adaptation in India's agricultural sector especially Uttar Pradesh, highlighting challenges and prospects. It evaluates current digital adoption levels, factors influencing farmers, and pertinent challenges like infrastructure and literacy gaps. Potential benefits such as increased yields are underscored alongside governmental and private sector initiatives' impact. Insights and policy recommendations are provided for sustainable digital agricultural development in the region, crucial for stakeholders and policymakers.

Chapter 7 presents a visionary leap into Agriculture 6.0, where technology and tradition converge for sustainable farming solutions. With a spotlight on precision and innovation, we navigate the complexities of a changing climate and burgeoning populations. Through IoT-driven strategies and data analytics, we unlock new pathways to efficiency and resilience in agriculture.

Chapter 8 describes recent technologies used to design various prediction models for water quality analysis and classification using ML techniques. ML techniques are integrated with image processing, the internet of things, remote sensing, and geographic information systems to provide solutions to the key challenges experienced because of climate changes, water pollution, and increases in water demand for various purposes. This review chapter presents the contribution of ML techniques for monitoring the quality of water with the inclusion of biochemical water parameters as estimators for different purposes like drinking water, irrigation, and aquaculture.

Chapter 9 provides the mitigation strategies for climate change over multiple locations. Energy efficiency, renewable energy, and low-carbon transportation are essential components of an all-encompassing strategy to tackle climate change. To get past financial and technological obstacles, international cooperation is necessary. A multifaceted approach that emphasizes global cooperation and social justice—including afforestation and fair low-carbon initiatives—is supported by CNNs.

Chapter 10 focuses on the task of classification, which involves precisely predicting the target class or group for each case in the dataset. Web page classification is a data mining technique employed to reveal the categorization of web pages. Internet information providers are likely to have a strong interest in methods that improve the effectiveness of web search engines.

Chapter 11 explores the transformative impact of recent computational intelligence advancements on vehicle surveillance and recognition, particularly through CCTV analysis. With a comprehensive approach, it navigates through essential theories, methodologies, and breakthroughs, aiming to offer transparent insights accessible to researchers, practitioners, and students. By analyzing the emerging trends and identifying gaps in existing literature, this chapter paves the way to understanding innovation in the dynamic landscape of Industry 6.0.

Chapter 12 focuses on the exploration and development of a supply chain network design (SCND) model of a milk product. The optimization method of this model follows a piecewise mixed-integer non-linear programming approach, and its validity is confirmed through a real-life case analysis involving a single milk product to investigate the optimal location for a distribution center in the coastal region of Odisha.

Chapter 13 investigates the integrated multi-echelon supply chain for multi-item to gain more profit by introducing the new policy, i.e., with (supplier–manufacturer–retailer) multi-item replenishment under publicity effort effects. Under the two policies, namely decentralized and centralized, two models are developed for each: the model of non-cooperative and cooperative replenishment of individual and joint items, respectively. For these nonlinear heuristic mixed-integer models, some mathematical properties and solution procedures with an algorithm are incorporated to obtain the optimal solution.

Editors

C. Kishor Kumar Reddy is Associate Professor at the Department of Computer Science and Engineering, Stanley College of Engineering and Technology for Women, Hyderabad, India. He has research and teaching experience of more than ten years. He has published more than 60 research papers in national and international conferences, book chapters, and journals indexed by SCIE, Scopus, and others. He is the author of two textbooks and seven co-edited books. His research areas include atmospheric sciences, meteorology, healthcare, intelligent systems, machine learning, and deep learning.

Srinath Doss is Professor and Dean in the Faculty of Engineering and Technology, Botho University, responsible for the Botswana, Lesotho, Eswatini, Namibia, and Ghana campuses. He has written several books, published more than 80 papers in international journals, and attended several prestigious conferences. He serves as an editorial member and reviewer for reputed international journals and as an advisory member for various prestigious conferences. Professor Srinath is a Member of IAENG and an Associate Member of UACEE.

Surbhi Bhatia Khan holds a doctorate in computer science and engineering in machine learning and social media analytics. She is listed in the top 2% of researchers released by Stanford University, USA. She earned project management Professional Certification from the reputed Project Management Institute, USA. She teaches in the Department of Data Science, School of Science, Engineering and Environment, University of Salford, Manchester, United Kingdom. She holds a research position at the Lebanese American University. She also enjoys an Adjunct Professor position at Shoolini University, Himachal Pradesh, India. She has more than 12 years of academic and teaching experience at different universities. She has published 150+ papers in many reputed journals in high-indexed outlets. She has 12 international patents from India, Australia, and the USA. She has written 2 books and edited 15 books. She is working on research projects with UKRI, the Deanship of Scientific Research, the Ministry of Education from Saudi Arabia, and India. She is a Senior Member of IEEE, a Member of IEEE Young Professionals, and ACM. She has chaired several international conferences and workshops and has delivered more than 20 invited and keynote talks across the globe. She serves as an Academic Editor, Associate Editor, and Guest Editor with many reputed journals. She is the awardee of the Research Excellence Award given by King Faisal University, Saudi Arabia, in 2021.

Her areas of interest include artificial intelligence, medical imaging, deep learning, machine learning, and data science.

Abdulmajeed Alqhatani is Assistant Professor at Najran University, Saudi Arabia, and the Director of Research and Graduate Studies in the Department of Information Systems, Saudi Arabia. He earned a PhD in computing and information systems at the University of North Carolina at Charlotte, USA, in 2021. His research interests include usable security and privacy, internet of things (IoT) applications in a smart city, business intelligence, and data analytics. Dr. Alqhatani's publications appear in various academic journals and conferences, including *IEEE Sensors Journal* and the CHI Conference on Human Factors in Computing Systems.

Contributors

Ashish Adholiya
Department of Computer Science
Pacific University
Udaipur, India

Sonali Agarwal
Department of Information Technology
Indian Institute of Information Technology
Allahabad, India

S. Akshaya
School of Computer Science and Engineering
Vellore Institute of Technology
Chennai, India

S. Md Shakir Ali
Department of Digital Business
Lithan Academy (eduClaaS Pte Ltd)
Singapore

Pranay Chandur
Department of Data Science
Muma College of Business
University of South Florida
Tampa, Florida, USA

Gauri Ruchir Chavan
Department of Computer Science
Pacific University
Udaipur, India

Srinath Doss
Faculty of Engineering and Technology
Botho University
Gaborone, Botswana

Ushaa Eswaran
Department of Electronics and Communication Engineering
Jawaharlal Nehru Technological University
Kakinada, India

Vishal Eswaran
CVS Health Centre
Dallas, Texas, USA

Vivek Eswaran
Medallia
Austin, Texas, USA

Bibi Ishrat Jahan
Department of Economics
Karamat Hussain PG College for Women
Lucknow University
Lucknow, India

Vallidevi Krishnamurthy
School of Computer Science and Engineering
Vellore Institute of Technology
Chennai, India

P. Abhilash Kumar
Mergen IT
Ormskirk, UK

Pei Mey Lau
Digital Business and Engineering and Technology
School of Digital Technology
Wawasan Open University
George Town, Malaysia

Kari Lippert
Department of Systems Engineering
College of Engineering
University of South Alabama
Mobile, Alabama, USA

Ashirbad Mishra
Department of Computer Science and Engineering
Pennsylvania State University
University Park, Pennsylvania, USA

Gulfishan Mobin
Department of Computer Science and Engineering
Adamas University
Kolkata, India

Subhasish Mohapatra
Department of Computer Science and Engineering
DRIEMS University
Odisha, India

Keerthna Murali
Dell EMC
Austin, Texas, USA

Harshitha Muthyam
Department of Computer Science and Engineering
Stanley College of Engineering and Technology for Women
Osmania University
Telangana, India

Anindya Nag
Computer Science and Engineering Discipline
Khulna University
Khulna, Bangladesh

Agnibrata Nayak
Walmart Global Tech
Bellevue, Washington, USA

S. Nithyashree
School of Computer Science and Engineering
Vellore Institute of Technology
Chennai, India

P. R. Nivedita
School of Computer Science and Engineering
Vellore Institute of Technology
Chennai, India

Ozen Ozer
Department of Mathematics
Kırklareli University
Kirklareli, Turkey

Rachana Y. Patil
Department of Computer Engineering
Pimpri Chinchwad College of Engineering
Pune, India

Yogesh H. Patil
Department of Instrumentation Engineering
Dr. D. Y. Patil College of Engineering
Akurdi, Pune, India

Alipsa Pattnaik
School of Applied Sciences
Birla Global University
Bhubaneswar, India

Aryan Pattnaik
Department of Computer Science and System Engineering
KIIT University
Bhubaneswar, India

Monalisha Pattnaik
Department of Statistics
Sambalpur University
Burla, India

Nishitha Pujala
Department of Computer Science and Engineering
Stanley College of Engineering and Technology for Women
Telangana, India

Abhishek Ranjan
Digital Business and Engineering and Technology
Faculty of Engineering and Technology
Botho University
Gaborone, Botswana

Prasanna Sundar Rao
Department of Computer Science
Shri Shankarlal Sundarbai Shasun Jain College
Chennai, India

C. Kishor Kumar Reddy
Department of Computer Science and Engineering
Stanley College of Engineering and Technology for Women
Osmania University
Telangana, India

P. Sahithi Reddy
Information Technology and Information Technology Management
University of Sydney
Sydney, Australia

P. Yashashwini Reddy
Department of Computer Science and Engineering
Stanley College of Engineering and Technology for Women
Telangana, India

Monee Kieran Roul
Intel Corporation Ltd
Hillsboro, Oregon, USA

Riya Sil
Department of Computer Science and Engineering
Brainware University
Kolkata, India

Nastassia Thandiwe Sithole
Department of Chemical Engineering Technology
University of Johannesburg
Johannesburg, South Africa

1

Empowering the Factory of the Future: Integrating Artificial Intelligence, Machine Learning, and IoT Innovations in Industry 6.0

Ushaa Eswaran, Vivek Eswaran, Vishal Eswaran, and Keerthna Murali

1.1 The Convergence of Artificial Intelligence, Machine Learning, and Industrial Internet of Things in Industry 6.0

1.1.1 Understanding Artificial Intelligence and Machine Learning

1.1.1.1 AI: Leveraging Computational Intelligence for Problem-Solving

Artificial intelligence (AI) refers to the development of computer systems and algorithms capable of performing tasks that typically require human intelligence, such as problem-solving, decision-making, and pattern recognition [1]. AI leverages computational techniques, including machine learning (ML), neural networks, and natural language processing, to enable machines to learn from data, adapt to new inputs, and automate complex decision-making processes [2].

1.1.1.2 ML: Algorithms for Data-Driven Decision-Making

ML refers to the development and application of algorithms and statistical models that enable computer systems to perform specific tasks effectively without being explicitly programmed. The key characteristic of ML is its ability to learn from data and make predictions or decisions based on that learning, without relying on rule-based programming [3,4].

There are several common types of ML algorithms and techniques that are particularly relevant for industrial applications:

Supervised Learning: Algorithms that learn from labeled data to make predictions or classify new inputs. Examples include linear regression, decision trees, and support vector machines [5].

DOI: 10.1201/9781003503934-1

Unsupervised Learning: Algorithms that identify patterns and extract insights from unlabeled data. Examples include clustering algorithms, principal component analysis, and anomaly detection [6].

Reinforcement Learning: Algorithms that learn by interacting with an environment and receiving feedback (rewards or penalties) to optimize their actions.

The execution of ML in an industrial setting typically involves the following steps:

- **Data Collection**: Gathering relevant data from various sources, such as IIoT sensors, enterprise systems, and historical records.
- **Data Preprocessing**: Cleaning, transforming, and preparing the data for model training and analysis.
- **Model Selection and Training**: Choosing appropriate ML algorithms and techniques, and training the models using the prepared data.
- **Model Evaluation and Tuning**: Assessing the model's performance and iteratively refining it to improve accuracy and reliability.
- **Model Deployment and Monitoring**: Integrating the trained ML models into production systems and continuously monitoring their performance.

1.1.1.3 Synergies between AI and ML in Industrial Applications

The convergence of AI and ML creates powerful synergies that are transforming the manufacturing industry. By integrating AI and ML capabilities, industrial systems can leverage computational intelligence to automate and optimize various processes, such as:

Predictive Maintenance: Using ML algorithms to analyze sensor data and predict equipment failures, enabling proactive maintenance and reducing unplanned downtime.

Quality Control: Applying AI-powered computer vision and defect detection to identify and address quality issues in real time.

Resource Optimization: Employing ML models to analyze production data and optimize the utilization of energy, materials, and other resources.

Supply Chain Management: Integrating AI and ML for demand forecasting, inventory optimization, and logistics planning to enhance supply chain efficiency.

The combination of AI and ML enables manufacturers to make more informed, data-driven decisions, respond to changing conditions in real time, and continuously improve their operations. By leveraging the strengths of both technologies, manufacturers can achieve higher levels of automation, efficiency, and product quality, ultimately driving the transformation toward Industry 6.0 [7].

1.1.2 The Rise of the Industrial Internet of Things

1.1.2.1 Sensor Networks and Connectivity for Real-Time Data Acquisition

The industrial Internet of Things (IIoT) refers to the integration of advanced sensors, connectivity, and computing technologies within industrial environments. IIoT sensor networks enable the real-time acquisition of vast amounts of data from various production equipment, machinery, and processes, providing manufacturers with unprecedented visibility into their operations [8].

1.1.2.2 Integration of Operational Technology and IT Systems for Data-Driven Insights

The convergence of operational technology (OT) and IT systems is a crucial aspect of the IIoT and the transformation toward Industry 6.0. OT systems, such as programmable logic controllers, supervisory control and data acquisition systems, and distributed control systems, are responsible for monitoring and controlling the physical industrial processes on the shop floor. In contrast, IT systems, such as enterprise resource planning, manufacturing execution systems, and product lifecycle management platforms, manage and analyze the data generated by these processes [9–11].

The integration of OT and IT systems through IIoT enables manufacturers to bridge the historical divide between these two domains and create a seamless flow of information across the organization. This convergence provides several benefits:

1. **Comprehensive Data Access:** By connecting OT systems to IT systems, manufacturers can access a wealth of real-time data from the shop floor, including sensor readings, equipment performance, and production metrics. This data can then be combined with enterprise-level information for a holistic view of the manufacturing operations.
2. **Improved Decision-Making:** The availability of integrated data from both OT and IT systems allows manufacturers to make more informed, data-driven decisions. Insights derived from this data can be used to optimize production processes, enhance asset utilization, and align manufacturing activities with broader business objectives.
3. **Enhanced Operational Visibility:** The integration of OT and IT systems provides manufacturing leaders with greater visibility into their operations, enabling them to monitor, analyze, and respond to changes in real time. This enhanced visibility supports the implementation of advanced technologies, such as predictive maintenance and autonomous process control [12].

1.1.2.3 IIoT Platforms and Cloud Computing for Scalable Analytics

The vast amounts of data generated by IIoT sensor networks and the integration of OT and IT systems often exceed the processing and storage capabilities of on-premises infrastructure. This is where cloud-based IIoT platforms play a crucial role in enabling scalable analytics.

IIoT platforms leverage the elastic and on-demand nature of cloud computing to provide the necessary computing power, storage, and analytical capabilities to handle large volumes of industrial data. These platforms typically offer the following features:

1. **Data Aggregation and Storage:** IIoT platforms collect, store, and manage the data streams from various OT and IT systems, creating a centralized repository for historical and real-time information.
2. **Advanced Analytics:** Leveraging the computational resources of the cloud, IIoT platforms can apply sophisticated data analysis techniques, including the use of AI and ML algorithms, to generate actionable insights from the collected data.
3. **Scalability and Flexibility:** Cloud-based IIoT platforms can easily scale up or down to accommodate changing data volumes and analytical requirements, ensuring that manufacturers can adapt to evolving business needs.
4. **Secure Data Management:** IIoT platforms often provide robust data security and access control mechanisms to protect sensitive industrial data, addressing the concerns around cybersecurity and data privacy.

By integrating IIoT platforms with cloud computing, manufacturers can unlock the full potential of their data, enabling them to optimize processes, predict equipment failures, enhance product quality, and drive sustainable manufacturing practices—all of which are crucial for the successful implementation of Industry 6.0 initiatives.

1.2 Empowering the Factory of the Future

The convergence of AI, ML, and the IIoT is empowering manufacturers to transform their operations and create the factory of the future. By integrating these cutting-edge technologies, manufacturers can achieve unprecedented levels of intelligent process monitoring and control, enhanced operational efficiency and productivity, and sustainable manufacturing practices.

1.2.1 Intelligent Process Monitoring and Control

The combination of AI, ML, and IIoT enables manufacturers to implement intelligent systems for real-time process monitoring, predictive maintenance, and autonomous decision-making, leading to improved product quality and operational reliability.

1.2.1.1 AI-Powered Predictive Maintenance and Condition-Based Monitoring

AI-based predictive maintenance and condition-based monitoring algorithms analyze sensor data from IIoT systems to detect anomalies, predict equipment failures, and optimize maintenance schedules [1]. By leveraging techniques such as anomaly detection, time-series analysis, and failure mode analysis, these AI-powered systems can identify potential issues before they occur, allowing manufacturers to proactively address them and reduce unplanned downtime.

Figure 1.1 presents the typical workflow of an AI-powered predictive maintenance system, offering a comprehensive overview of how AI is utilized to enhance maintenance practices in various industries. This illustration outlines the sequential steps involved in predicting and preventing equipment failures before they occur, thereby minimizing downtime, reducing costs, and optimizing asset performance. By harnessing advanced algorithms and ML techniques, organizations can proactively identify potential issues, prioritize maintenance activities, and streamline maintenance operations for improved efficiency and reliability.

1.2.1.2 ML-Driven Quality Control and Defect Detection

ML algorithms can be employed to enhance quality control and automate defect detection in manufacturing processes. By training ML models on historical data and real-time sensor inputs, manufacturers can identify patterns, detect anomalies, and classify product defects with a high degree of accuracy [13]. This approach enables real-time quality monitoring, early intervention, and the reduction of waste and rework.

Table 1.1 provides an overview of the key ML techniques commonly used in quality control processes across various industries. This table outlines different ML algorithms and their specific applications in enhancing quality control practices, ranging from defect detection and classification to anomaly detection and predictive maintenance. By leveraging ML techniques, organizations can automate and optimize quality control processes, leading to

FIGURE 1.1
The typical workflow of an AI-powered predictive maintenance system.

TABLE 1.1

Outlines the Key ML Techniques and Their Applications in Quality Control

ML Technique	Applications
Supervised learning	Defect detection
Unsupervised learning	Anomaly detection
Semi-supervised learning	Process optimization
Reinforcement learning	Quality improvement

improved product quality, reduced waste, and enhanced operational efficiency. This table serves as a valuable reference for understanding the diverse ML applications in quality control and guiding decision-making regarding the adoption of ML solutions in manufacturing and other sectors.

1.2.1.3 Autonomous Process Optimization and Decision-Making

Advanced AI and ML algorithms can be integrated with IIoT systems to enable autonomous process optimization and decision-making. These intelligent systems can continuously monitor process parameters, adjust setpoints, and make decisions to optimize production efficiency, resource utilization, and product quality without direct human intervention [14]. This level of automation and self-optimization is a crucial aspect of the factory of the future.

1.2.2 Enhancing Operational Efficiency and Productivity

By leveraging the power of AI, ML, and IIoT, manufacturers can significantly improve their operational efficiency and overall productivity, leading to increased profitability and competitiveness.

1.2.2.1 Predictive Analytics for Resource Optimization

AI and ML-driven predictive analytics can help manufacturers optimize the utilization of resources, such as energy, materials, and workforce, by forecasting demand, identifying inefficiencies, and recommending strategies for improvement [1]. This enables manufacturers to minimize waste, reduce costs, and enhance overall operational efficiency.

1.2.2.2 Intelligent Supply Chain Visibility and Traceability

IIoT sensor networks, coupled with AI and ML algorithms, can provide manufacturers with enhanced visibility and traceability throughout their supply chains. This allows for real-time monitoring of material and product flows, early identification of disruptions, and optimization of logistics and inventory management [8]. Improved supply chain intelligence contributes to increased productivity and responsiveness.

1.2.2.3 *Integrating AI, ML, and IIoT for Increased Throughput*

The seamless integration of AI, ML, and IIoT technologies can lead to significant improvements in manufacturing throughput. Intelligent systems can automate and optimize various production processes, from scheduling and workflow management to equipment utilization and defect prevention, resulting in higher output, reduced cycle times, and improved overall equipment effectiveness (OEE) [15].

1.2.3 Achieving Sustainability and Environmental Impact Reduction

The adoption of AI, ML, and IIoT in the factory of the future also enables manufacturers to address sustainability and environmental concerns, contributing to more responsible and eco-friendly industrial practices.

1.2.3.1 *AI-Based Energy Management and Conservation*

AI-powered energy management systems can analyze real-time energy consumption data from IIoT sensors, identify opportunities for optimization, and automatically adjust parameters to reduce energy usage and costs [13]. AI algorithms can also be applied to optimize the energy efficiency of production equipment and explore alternative energy sources, further contributing to sustainable manufacturing [16].

1.2.3.2 *ML-Driven Waste Reduction and Material Optimization*

ML models can be leveraged to optimize material usage, minimize waste, and enhance the circularity of manufacturing processes. By analyzing production data, ML algorithms can identify opportunities to reduce scrap, optimize material flows, and explore the reuse or recycling of materials, aligning with the principles of a sustainable, circular economy [17].

1.2.3.3 *IIoT-Enabled Environmental Monitoring and Compliance*

IIoT sensor networks can be deployed to monitor environmental parameters, such as air quality, water usage, and waste generation, in real-time. This data can be used to ensure regulatory compliance, identify areas for improvement, and support the implementation of sustainable manufacturing practices [8]. By integrating these IIoT-enabled environmental monitoring systems with AI and ML algorithms, manufacturers can further optimize their processes and minimize their environmental footprint.

In today's industrial landscape, predictive maintenance has emerged as a crucial strategy for optimizing asset performance and reducing downtime. Figure 1.2 delineates the workflow of an AI-powered predictive maintenance system, which leverages advanced technologies to predict equipment failures before they occur.

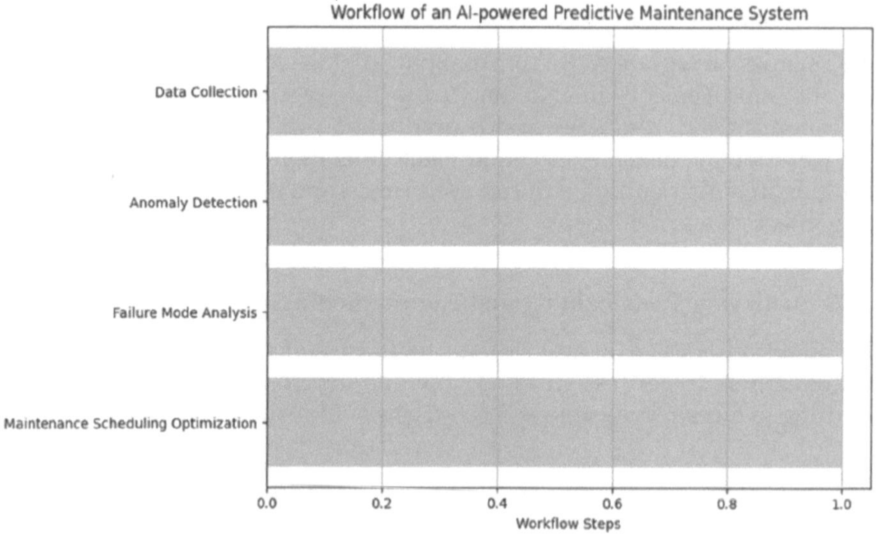

FIGURE 1.2
Workflow of an AI-powered predictive maintenance system.

At the heart of this system lies data collection, where various sensors and Internet of Things devices gather real-time data on equipment health and performance. Subsequently, the data undergoes thorough analysis, including anomaly detection to identify deviations from normal behavior and failure mode analysis to determine potential causes of malfunction.

Once anomalies or potential failure modes are detected, the system utilizes sophisticated AI algorithms to predict when equipment failures might occur. This predictive capability enables proactive maintenance scheduling, allowing maintenance teams to address issues before they escalate into costly breakdowns.

Overall, Figure 1.2 illustrates how AI and data-driven approaches empower organizations to transition from reactive to proactive maintenance strategies, ultimately enhancing operational efficiency and reducing maintenance costs.

In the realm of manufacturing, ensuring product quality is paramount to maintaining customer satisfaction and operational efficiency. Table 1.1 outlines various ML techniques that play a pivotal role in quality control within the manufacturing sector.

These ML techniques encompass a wide array of applications, ranging from defect detection and classification to process optimization and predictive maintenance. By harnessing the power of ML, manufacturers can leverage data-driven insights to enhance product quality, minimize defects, and optimize production processes.

Each ML technique listed in Table 1.2 is accompanied by its respective application in quality control, providing manufacturing professionals with valuable insights into the diverse range of tools available for improving product quality and operational performance.

TABLE 1.2

Machine Learning Techniques for Quality Control in Manufacturing.

ML Technique	Application
Supervised learning	Defect classification, product grading, yield prediction
Unsupervised learning	Anomaly detection, process parameter optimization
Reinforcement learning	Autonomous process control, quality improvement

1.3 Challenges and Considerations for Implementing AI, ML, and IIoT in Industry 6.0

1.3.1 Data Integration and Interoperability

The successful implementation of AI, ML, and IIoT technologies in the Industry 6.0 context is heavily dependent on the ability to integrate and harmonize data from various sources. Manufacturers often face challenges in achieving seamless data integration due to the heterogeneous nature of industrial systems, legacy equipment, and disparate data formats [1]. Overcoming these challenges requires the development of standardized interfaces, common data models, and interoperable platforms that can facilitate the flow of information across the entire manufacturing ecosystem.

Researchers have proposed several strategies to address the data integration and interoperability challenges in Industry 6.0. One such approach is the adoption of open standards and communication protocols, such as OLE for process control (OPC) Unified Architecture and Industrial Internet Consortium reference architectures, which enable the seamless exchange of data between different systems and devices [18]. Additionally, the use of middleware solutions and data integration platforms can help manufacturers bridge the gap between legacy and modern systems, allowing for the aggregation and harmonization of data from diverse sources [8].

1.3.2 Cybersecurity and Data Privacy Concerns

The integration of AI, ML, and IIoT technologies in the manufacturing environment raises significant concerns regarding cybersecurity and data privacy. The increased connectivity and the reliance on cloud-based platforms expose industrial systems to a wider range of cyber threats, including data breaches, malware attacks, and unauthorized access to sensitive information [15]. Manufacturers must address these concerns to ensure the integrity and confidentiality of their operations and protect their intellectual property.

Strategies to mitigate cybersecurity and data privacy risks in Industry 6.0 include the implementation of robust security protocols, such as encryption, access controls, and intrusion detection systems. Additionally, the adoption of secure-by-design principles in the development of AI, ML, and IIoT systems can help reduce vulnerabilities and enhance the overall resilience of the manufacturing infrastructure [13]. Collaboration with cybersecurity experts and the development of industry-wide security standards can further strengthen the protection of industrial data and systems.

1.3.3 Workforce Upskilling and Change Management

The successful integration of AI, ML, and IIoT technologies in the factory of the future requires a significant transformation in the skills and competencies of the manufacturing workforce. Employees must be equipped with the knowledge and capabilities to operate, maintain, and optimize these advanced systems, as well as adapt to the changing nature of work [17].

Addressing the workforce upskilling challenge in Industry 6.0 involves a multifaceted approach, including the development of comprehensive training programs, the implementation of collaborative learning environments, and the fostering of a culture that embraces continuous learning and innovation. Manufacturers must also strategically plan for the integration of these technologies, considering the impact on job roles, organizational structures, and change management processes [15].

1.4 Case Studies and Best Practices

1.4.1 Successful Implementation of AI, ML, and IIoT in Leading Manufacturing Companies

The strategic integration of AI, ML, and IIoT has been instrumental in transforming the operations of leading manufacturing companies. These organizations have successfully leveraged these cutting-edge technologies to achieve unprecedented levels of operational efficiency, product quality, and sustainability [19].

One such example is Siemens, a global industrial conglomerate, which has implemented an AI-powered predictive maintenance system across its factories. By analyzing real-time sensor data from its IIoT-enabled production equipment, the company's AI algorithms can accurately predict potential equipment failures, allowing for proactive maintenance and reduced downtime [8]. This has resulted in significant cost savings and improvements in OEE.

Another case study is that of General Electric, which has integrated ML models with its IIoT platform to optimize energy consumption and resource utilization across its manufacturing facilities. The company's data-driven insights have enabled it to reduce energy usage by 12% and material waste by 20%, contributing to its sustainability goals [1].

Volkswagen, a leading automotive manufacturer, has also embraced the power of AI and ML in its quality control processes. By deploying computer vision-based defect detection systems, the company has achieved a 99% accuracy rate in identifying product defects, leading to a substantial decrease in rework and scrap [13]. This has allowed Volkswagen to enhance its product quality and customer satisfaction.

1.4.2 Lessons Learned and Overcoming Barriers

While the successful implementation of AI, ML, and IIoT in the factory of the future has yielded substantial benefits, manufacturing companies have also faced various challenges and barriers during the transformation process. Understanding and addressing these challenges is crucial for other organizations seeking to replicate the successes of industry leaders.

One of the key lessons learned is the importance of data integration and interoperability. Manufacturers have recognized the need to establish standardized data models, communication protocols, and middleware solutions to enable the seamless flow of information across their operations [18]. This has required significant investments in technology infrastructure and the development of comprehensive data management strategies.

Cybersecurity and data privacy have also emerged as critical concerns for manufacturers adopting these advanced technologies. Organizations have had to implement robust security measures, such as encryption, access controls, and intrusion detection systems, to protect their sensitive data and operational systems from cyber threats [15]. Collaboration with cybersecurity experts and the development of industry-specific security standards have been instrumental in addressing these challenges.

Furthermore, the successful integration of AI, ML, and IIoT has necessitated a significant shift in the skills and competencies of the manufacturing workforce. Companies have had to invest in comprehensive training programs, foster a culture of continuous learning, and strategically plan for the reorganization of job roles and responsibilities [17]. Effectively managing this change has been crucial for ensuring the successful adoption and utilization of these advanced technologies.

By understanding the lessons learned and best practices from leading manufacturers, other organizations can navigate the challenges associated with implementing AI, ML, and IIoT in the Industry 6.0 context and unlock the full potential of these transformative technologies.

1.4.3 Overcoming Barriers and Lessons Learned

The successful implementation of AI, ML, and IIoT technologies in the factory of the future has not been without its challenges. Manufacturers have faced a range of obstacles, from technical hurdles to organizational and workforce-related barriers. Understanding these challenges and learning from the experiences of industry leaders is crucial for other organizations seeking to replicate the successes of their peers.

1.4.3.1 Technical Challenges

One of the primary technical challenges faced by manufacturers is the integration of disparate data sources and legacy systems. Achieving seamless data interoperability requires significant investments in standardized interfaces, common data models, and middleware solutions [18]. Manufacturers must also address cybersecurity and data privacy concerns, as the increased connectivity and reliance on cloud-based platforms expose industrial systems to a wider range of cyber threats [15]. Implementing robust security protocols, such as encryption, access controls, and intrusion detection systems, is essential to mitigate these risks.

1.4.3.2 Organizational and Workforce-Related Challenges

The successful integration of AI, ML, and IIoT technologies also requires a significant transformation in the skills and competencies of the manufacturing workforce. Employees must be equipped with the knowledge and capabilities to operate, maintain, and optimize these advanced systems [17]. Manufacturers have had to invest in comprehensive training programs, foster a culture of continuous learning, and strategically plan for the reorganization of job roles and responsibilities.

1.4.3.3 Overcoming Barriers through Collaborative Ecosystems

To address these challenges, manufacturers have recognized the value of collaborating with technology providers, research institutions, and industry bodies. By establishing partnerships and knowledge-sharing initiatives, organizations can leverage the expertise and best practices of their peers, accelerating the adoption of AI, ML, and IIoT technologies [8]. Additionally, policymakers can play a crucial role in creating favorable regulatory environments and incentives that drive the industry-wide transformation toward Industry 6.0.

1.5 Future Outlook and Emerging Trends

1.5.1 Edge Computing and Distributed Intelligence

As the adoption of AI, ML, and IIoT technologies continues to grow in the factory of the future, the need for more efficient and responsive data processing capabilities has led to the emergence of edge computing. Edge computing involves the decentralization of data processing and analytics, allowing for real-time decision-making and autonomous control closer to the point of data generation, rather than relying solely on centralized cloud infrastructure [13].

By integrating edge computing with IIoT sensor networks and AI/ML algorithms, manufacturers can unlock the full potential of distributed intelligence in the factory of the future. This approach enables faster response times, reduced latency, and improved resilience, as critical decisions can be made locally without the need for constant communication with a central cloud platform [1]. Edge computing also helps to address data privacy and security concerns by processing sensitive information closer to the source, reducing the risk of data breaches and unauthorized access.

The convergence of edge computing, AI, ML, and IIoT is expected to drive the development of autonomous, self-optimizing production systems that can adapt to changing conditions in real time, leading to increased operational efficiency, quality, and flexibility [18]. Manufacturers should prioritize the integration of edge computing capabilities as part of their Industry 6.0 transformation strategies to capitalize on these emerging trends.

1.5.2 Digital Twins and the Virtual–Physical Integration

Another key trend shaping the future of the factory is the integration of digital twins—virtual representations of physical assets, processes, and systems—with the real-world manufacturing environment. By creating digital twins of production equipment, supply chains, and entire factories, manufacturers can leverage AI and ML to simulate, test, and optimize their operations in a virtual setting before implementation in the physical world [13].

The integration of digital twins with IIoT sensor data and advanced analytics enables a continuous feedback loop between the virtual and physical domains, allowing for real-time monitoring, predictive maintenance, and process optimization. This virtual–physical integration empowers manufacturers to make more informed decisions, reduce the risk of costly errors, and accelerate the adoption of new technologies and strategies [1]. Furthermore, the combination of digital twins with augmented reality (AR) and virtual reality (VR) technologies can enhance human–machine interaction and enable new ways of visualizing, interacting with, and controlling the manufacturing environment [8]. This can lead to improved training, remote assistance, and collaborative decision-making among plant operators, engineers, and management [20,21].

As digital twin technologies mature and become more widely adopted, manufacturers will be able to leverage the synergies between the virtual and physical worlds to drive continuous improvement, rapid innovation, and sustainable transformation in the factory of the future.

1.5.3 Convergence of AI, ML, IIoT with Other Industry 6.0 Technologies

The future of the factory will be characterized by the convergence of AI, ML, and IIoT with other cutting-edge Industry 6.0 technologies, creating a highly integrated and intelligent manufacturing ecosystem.

1.5.3.1 Additive Manufacturing

The integration of AI, ML, and IIoT with additive manufacturing (AM) or 3D printing can revolutionize the way products are designed, produced, and customized. AI-powered generative design algorithms can create complex, optimized part geometries, while ML models can predict and optimize the 3D printing process parameters to improve quality and reduce waste [17]. IIoT sensor networks can provide real-time monitoring and control of AM systems, enabling adaptive and autonomous production.

1.5.3.2 Collaborative Robotics

The synergies between AI, ML, IIoT, and collaborative robotics can lead to the development of highly flexible and adaptive manufacturing systems. AI-powered perception and reasoning capabilities can enable collaborative robots to work seamlessly alongside human operators, recognizing their intentions, anticipating their actions, and adapting their behaviors accordingly [15]. IIoT-enabled sensors and connectivity can further enhance the situational awareness and coordination of these human–robot collaborative workspaces.

1.5.3.3 Augmented and Virtual Reality

The integration of AI, ML, and IIoT with AR and VR technologies can transform the way workers interact with and control the manufacturing environment. AI-powered computer vision and natural language processing can enable AR/VR systems to provide contextual information, step-by-step guidance, and remote assistance to plant operators, improving productivity and reducing errors [8]. IIoT sensor data can be leveraged to create digital twins and simulations, allowing for immersive training, maintenance, and process optimization in a virtual setting.

In the pursuit of Industry 4.0's vision, the integration of Edge Computing and Distributed Intelligence stands as a pivotal advancement in manufacturing [22–24]. Figure 1.3 elucidates the key components and benefits of edge computing within the context of the factory of the future. By decentralizing data processing and analysis, edge computing enables real-time decision-making, thereby minimizing latency and enhancing operational efficiency [25].

Moreover, it fortifies data privacy and security measures, crucial in safeguarding sensitive information generated within industrial environments. This illustration underscores the transformative potential of edge computing in driving the next phase of industrial evolution.

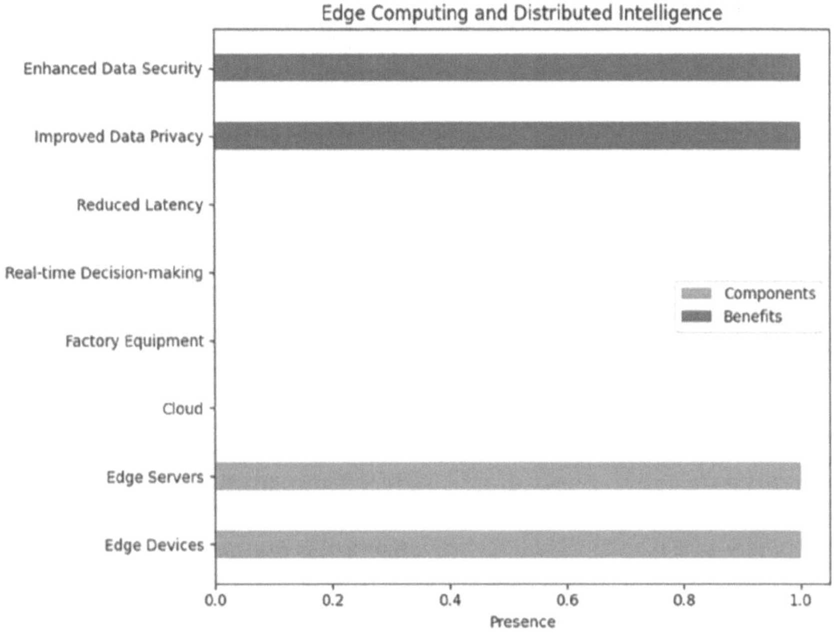

FIGURE 1.3
Edge Computing and Distributed Intelligence in the factory of the future.

As Industry 4.0 [26–28] continues to evolve, the convergence of AI, ML, and IIoT with other emerging technologies is reshaping various sectors. Table 1.3 outlines the convergence of AI, ML, and IIoT with other Industry 6.0 technologies, illustrating how these advancements are driving digital transformation across industries [29].

Advanced AI and machine learning capabilities can enable collaborative robots to work seamlessly alongside human operators, recognizing their intentions, anticipating their actions, and adapting their behaviors accordingly [15]. IIoT-enabled sensors and connectivity can further enhance the situational awareness and coordination of these human–robot collaborative workspaces [20,30].

1.6 Conclusion

1.6.1 Summarizing the Transformative Impact of AI, ML, and IIoT in Industry 6.0

The strategic integration of AI, ML, and the IIoT has ushered in a new era of intelligent, data-driven manufacturing, transforming the factory of the future. These cutting-edge technologies have empowered manufacturers to achieve unprecedented levels of operational efficiency, product quality, and sustainability.

TABLE 1.3

Convergence of AI, ML, IIoT with Other Industry 6.0 Technologies

Technology	Applications
Additive manufacturing	Generative design, process optimization, adaptive production
Collaborative robotics	Human–robot collaboration, situational awareness, safety
Augmented/virtual reality	Contextual information, remote assistance, virtual training

By leveraging the synergies between these technologies, manufacturers have been able to implement intelligent systems for real-time process monitoring, predictive maintenance, and autonomous decision-making. AI-powered predictive maintenance and condition-based monitoring have enabled proactive intervention and reduced unplanned downtime, while ML-driven quality control and defect detection have improved product quality and consistency. The convergence of AI, ML, and IIoT has also led to enhanced operational efficiency and productivity, through predictive analytics for resource optimization, intelligent supply chain visibility, and increased throughput.

Furthermore, the adoption of these technologies has empowered manufacturers to address sustainability and environmental concerns. AI-based energy management systems, ML-driven waste reduction, and IIoT-enabled environmental monitoring have contributed to more responsible and eco-friendly industrial practices, aligning with the principles of a circular economy.

Despite the transformative potential of AI, ML, and IIoT, manufacturers have faced several challenges in their implementation, including data integration and interoperability, cybersecurity and data privacy concerns, and the need for workforce upskilling and change management. However, the lessons learned and best practices from leading companies have provided a roadmap for overcoming these barriers and reaping the full benefits of this technology convergence.

1.6.2 Recommendations for Manufacturing Leaders and Policymakers

To fully harness the power of AI, ML, and IIoT in the factory of the future, manufacturing leaders and policymakers should consider the following recommendations:

1. Develop a Comprehensive Digital Transformation Strategy
    - Align the integration of AI, ML, and IIoT with the broader organizational goals and Industry 6.0 initiatives.

- Establish a clear roadmap for technology adoption, data management, and workforce upskilling.
- Allocate sufficient resources and secure buy-in from cross-functional stakeholders.

2. Prioritize Data Integration and Interoperability
    - Adopt open standards and communication protocols to enable seamless data exchange across systems.
    - Invest in middleware solutions and data integration platforms to bridge the gap between legacy and modern systems.
    - Establish robust data governance frameworks to ensure data quality, security, and accessibility.
3. Enhance Cybersecurity and Data Privacy Measures
    - Implement robust security protocols, such as encryption, access controls, and intrusion detection systems.
    - Collaborate with cybersecurity experts and industry bodies to develop comprehensive security standards.
    - Incorporate secure-by-design principles in the development of AI, ML, and IIoT systems.
4. Develop a Future-Ready Workforce
    - Implement comprehensive training programs to upskill the existing workforce.
    - Foster a culture of continuous learning and embrace new ways of working.
    - Strategically plan for the integration of these technologies, considering the impact on job roles and organizational structures.
5. Promote Collaborative Ecosystem and Knowledge Sharing
    - Encourage partnerships and knowledge-sharing initiatives between manufacturers, technology providers, and research institutions.
    - Support the development of industry-wide guidelines, best practices, and case studies for the successful implementation of AI, ML, and IIoT.
    - Engage policymakers to create favorable regulatory environments and incentives that drive the adoption of these transformative technologies.

By following these recommendations, manufacturing leaders and policymakers can pave the way for the widespread adoption of AI, ML, and IIoT in the factory of the future, ultimately transforming the industry and contributing to a more sustainable, efficient, and competitive manufacturing landscape.

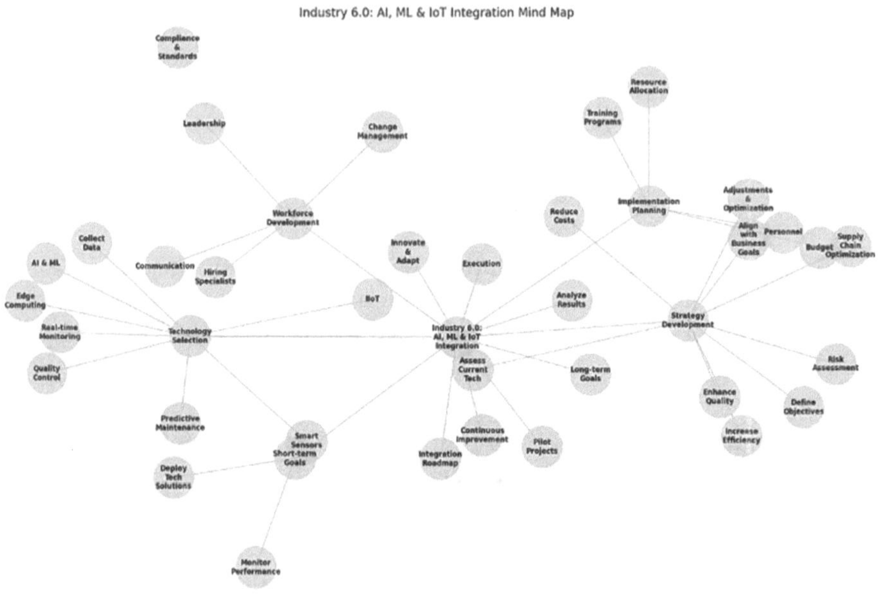

FIGURE 1.4
Recommended roadmap for implementing AI, ML, and IIoT in Industry 6.0.

Figure 1.4 depicts a recommended roadmap for implementing AI, ML, and IIoT in Industry 6.0, providing manufacturing leaders with a structured approach to navigate the complexities of digital transformation. This roadmap outlines key steps and considerations essential for developing and executing a comprehensive strategy to leverage advanced technologies effectively. By integrating AI, ML, and IIoT solutions strategically, manufacturers can enhance operational efficiency, optimize resource utilization, and unlock new opportunities for innovation and growth in the Industry 6.0 landscape.

Table 1.4 provides a comprehensive set of recommendations tailored for manufacturing leaders and policymakers aiming to navigate the rapidly evolving landscape of Industry 6.0. These recommendations are based on insights derived from extensive research and industry best practices, offering practical guidance to foster innovation, drive sustainable growth, and address emerging challenges in the manufacturing sector.

As manufacturing enterprises embark on the journey toward Industry 6.0, these recommendations provide a roadmap for success, emphasizing the critical importance of strategic planning, collaboration, and continuous innovation. By heeding this guidance, stakeholders can harness the transformative power of AI, ML, and IIoT to drive sustainable growth, enhance operational efficiency, and unlock new opportunities for innovation and competitiveness.

In essence, the recommendations outlined in Table 1.3 represent more than just a set of directives; they embody a collective vision for the future

TABLE 1.4

Recommendations for Manufacturing Leaders and Policymakers

Recommendation	Key Considerations
Develop a comprehensive digital transformation strategy	• Align with organizational goals and Industry 6.0 initiatives • Establish a clear roadmap for technology adoption and workforce upskilling • Secure cross-functional buy-in and allocate sufficient resources
Prioritize data integration and interoperability	• Adopt open standards and communication protocols • Invest in middleware solutions and data integration platforms • Establish robust data governance frameworks
Enhance cybersecurity and data privacy measures	• Implement robust security protocols and access controls • Collaborate with cybersecurity experts to develop industry standards • Incorporate secure-by-design principles in system development
Develop a future-ready workforce	• Implement comprehensive training programs • Foster a culture of continuous learning and new ways of working • - Strategically plan for the integration of technologies and their impact on job roles
Promote collaborative ecosystem and knowledge sharing	• Encourage partnerships and knowledge-sharing initiatives • Support the development of industry-wide guidelines and best practices • - Engage policymakers to create favorable regulatory environments and incentives

of manufacturing—a future characterized by intelligent, data-driven processes, responsible environmental stewardship, and empowered workforces. Through concerted efforts and unwavering commitment to these principles, manufacturing leaders and policymakers can pave the way for a brighter, more prosperous future in the era of Industry 6.0.

References

1. Tao, F., Qi, Q., Liu, A., & Kusiak, A. Data-driven smart manufacturing. *Journal of Manufacturing Systems,* 48(Part C), 157–169. (2018). https://doi.org/10.1016/j.jmsy.2018.01.006.

2. Zhong, R. Y., Xu, X., Klotz, E., & Newman, S. T. Intelligent manufacturing in the context of Industry 4.0: A review. *Engineering*, 3(5), 616–630. (2017). https://doi.org/10.1016/J.ENG.2017.05.015.
3. Kusiak, A. Smart manufacturing. *International Journal of Production Research*, 56(1–2), 508–517. (2018). https://doi.org/10.1080/00207543.2017.1351644.
4. Wuest, T., Weimer, D., Irgens, C., & Thoben, K. D. Machine learning in manufacturing: Advantages, challenges, and applications. *Production & Manufacturing Research*, 4(1), 23–45. (2016). https://doi.org/10.1080/21693277.2016.1192517.
5. Reddy, C. K. K., Anisha, P. R., Khan, S., Hanafiah, M. M., Pamulaparty, L., & Mohana, R. M. *Sustainability in Industry 5.0: Theory and Applications*, CRC Press, Taylor & Francis 6000 Broken Sound Parkway NW, Suite 300 Boca Raton, FL 33487 United States. (2024).
6. Bhutta, M. N. M., Bhattia, S., Alojail, M. A., Nisar, K., Cao, Y., Chaudhry, S. A., & Sun, Z. Towards secure IoT-based payments by extension of Payment Card Industry Data Security Standard (PCI DSS). *Wireless Communications and Mobile Computing*, 2022(1), 9942270, (2022).
7. Anisha, P. R., Reddy, C. K. K., Nguyen, N. G., Bhushan, M., Kumar, A., & Hanafiah, M. M. *Intelligent Systems and Machine Learning for Industry: Advancements, Challenges and Practices*, CRC Press, Taylor & Francis: Boca Raton, Florida. (2022).
8. Mourtzis, D., Vlachou, E., & Milas, N. Industrial big data as a result of IoT adoption in manufacturing. *Procedia CIRP*, 55, 290–295. (2016). https://doi.org/10.1016/j.procir.2016.07.038.
9. Preuveneers, D., & Ilie-Zudor, E. The intelligent industry of the future: A survey on emerging trends, research challenges and opportunities in Industry 4.0. *Journal of Ambient Intelligence and Smart Environments*, 9(3), 287–298. (2017). https://doi.org/10.3233/AIS-170432
10. Lee, J., Bagheri, B., & Kao, H.-A. Recent advances and trends of cyber-physical systems and big data analytics in industrial informatics. In *International Proceeding of International Conference on Industrial Informatics (INDIN)*, Porto Alegre, Brazil. (2014, July 27). https://doi.org/10.13140/2.1.1464.1920
11. Monostori, L., Kádár, B., Bauernhansl, T., Kondoh, S., Kumara, S., Reinhart, G., Sauer, O., Schuh, G., Sihn, W., & Ueda, K. Cyber-physical systems in manufacturing. *CIRP Annals*, 65(2), 621–641. (2016). https://doi.org/10.1016/j.cirp.2016.06.005.
12. Gandhi, P., Khan, M. Z., Sharma, R. K., Alhazmi, O. H., Bhatia, S., & Chakraborty, C. Software reliability assessment using hybrid neuro-fuzzy model. *Computer Systems Science & Engineering*. (2022). https://doi.org/10.32604/csse.2022.019943.
13. Zheng, P., Wang, H., & Sang, Z. Smart manufacturing systems for Industry 4.0: Conceptual framework, scenarios, and future perspectives. *Frontiers of Mechanical Engineering*, 13, 137–150. (2018). https://doi.org/10.1007/s11465-018-0499-5
14. Hammad, M., Islam, M. S., Salam, M. A., Jafry, A. T., Ali, I., & Khan, W. A. Framework for the implementation of smart manufacturing systems: A case in point. *Processes*, 11, 1436. (2023). https://doi.org/10.3390/pr11051436.
15. Mabkhot, M. M., Al-Ahmari, A. M., Salah, B., & Alkhalefah, H. Requirements of the smart factory system: A survey and perspective. *Machines*, 6, 23. (2018). https://doi.org/10.3390/machines6020023
16. Ghobakhloo, M. Industry 4.0, digitization, and opportunities for sustainability. *Journal of Cleaner Production*, 252, 119869. (2020). https://doi.org/10.1016/j.jclepro.2019.119869.

17. Esmaeilian, B., Behdad, S., & Wang, B. The evolution and future of manufacturing: A review. *Journal of Manufacturing Systems*, 39, 79–100. (2016). https://doi.org/10.1016/j.jmsy.2016.03.001.
18. Lee, J., Bagheri, B., & Kao, H.-A. A cyber-physical systems architecture for Industry 4.0-based manufacturing systems. *Manufacturing Letters*, 3, 18–23. (2015). https://doi.org/10.1016/j.mfglet.2014.12.001.
19. Frank, A. G., Dalenogare, L. S., & Ayala, N. F. Industry 4.0 technologies: Implementation patterns in manufacturing companies. *International Journal of Production Economics*, 210, 15–26. (2019). https://doi.org/10.1016/j.ijpe.2019.01.004.
20. Tao, F., Qi, Q., Wang, L., & Nee, A. Y. C. Digital twins and cyber-physical systems toward smart manufacturing and Industry 4.0: Correlation and comparison. *Engineering*, 5(4), 653–661. (2019). https://doi.org/10.1016/j.eng.2019.01.014.
21. Lu, Y., Liu, C., Kevin, I., Wang, K., Huang, H., Xu, X. Digital twin-driven smart manufacturing: Connotation, reference model, applications and research issues. *Robotics and Computer-Integrated Manufacturing*, 61, 101837. (2020). https://doi.org/10.1016/j.rcim.2019.101837.
22. Darwish, A., & Hassanien, A. E. Cyber physical systems design, methodology, and integration: the current status and future outlook. *Journal of Ambient Intelligence and Humanized Computing*, 9, 1541–1556. (2018). https://doi.org/10.1007/s12652-017-0575-4
23. Leitão, P., Karnouskos, S., Ribeiro, L., Lee, J., Strasser, T., & Colombo, A. W. Smart agents in industrial cyber-physical systems. *Proceedings of the IEEE*, 104(5), 1086–1101. (2016, May). https://doi.org/10.1109/JPROC.2016.2521931.
24. Lee, J., Bagheri, B., & Jin, C. Introduction to cyber manufacturing. Manufacturing Letters, 8, 11–15. (2016). https://doi.org/10.1016/j.mfglet.2016.05.002.
25. Xu, L. D., Xu, E. L., & Li, L. Industry 4.0: State of the art and future trends. *International Journal of Production Research*, 56(8), 2941–2962. (2018). https://doi.org/10.1080/00207543.2018.1444806.
26. Qin, J., Liu, Y., & Grosvenor, R. A categorical framework of manufacturing for Industry 4.0 and beyond. *Procedia CIRP*, 52, 173–178. (2016). https://doi.org/10.1016/j.procir.2016.08.005.
27. Lasi, H., Fettke, P., Kemper, H. G., Feld, T., & Hoffmann, M. Industry 4.0. *Business & Information Systems Engineering*, 6(4), 239–242. (2014). https://doi.org/10.1007/s12599-014-0334-4
28. Zhou, K., Liu, T., & Zhou, L. Industry 4.0: Towards future industrial opportunities and challenges. In *2015 12th International Conference on Fuzzy Systems and Knowledge Discovery (FSKD)* (pp. 2147–2152). IEEE, Amsterdam, Netherlands. (2015). https://doi.org/10.1109/FSKD.2015.7382284.
29. Munirathinam, S. Chapter Six - Industry 4.0: Industrial Internet of Things (IIOT). In P. Raj & P. Evangeline (Eds.), *Advances in Computers* (Vol. 117, Issue 1, pp. 129–164). Elsevier, Amsterdam, Netherlands. (2020). doi:10.1016/bs.adcom.2019.10.010.
30. Sureshkumar, S., Agash, C. P., Ramya, S., Kaviyaraj, R., & Elanchezhiyan, S. Augmented Reality with Internet of Things. In *2021 International Conference on Artificial Intelligence and Smart Systems (ICAIS)*, Coimbatore, India (pp. 1426–1430). (2021). https://doi.org/10.1109/ICAIS50930.2021.9395941.

2

Human–Robot Collaboration Analyzing the Challenges and Opportunities of Integrating Soft Computing Algorithms in Manufacturing Environments

Ushaa Eswaran, Vivek Eswaran, Vishal Eswaran, and Keerthna Murali

2.1 Introduction

In the bustling landscape of modern manufacturing, a significant evolution is underway with the advent of human–robot collaboration (HRC). Unlike traditional setups, HRC ushers in a new era where humans and robots share a workstation, replacing the current state of affairs where robots labor alone forging a symbiotic relationship aimed at maximizing efficiency and safety. This collaborative approach not only promises to revolutionize productivity but also offers a refreshing flexibility that adapts to the dynamic needs of manufacturing environments. However, the seamless integration of both people and robots in HRC systems poses intricate challenges that demand innovative solutions. One such solution lies in the integration of soft computing algorithms, which have the potential to navigate the complexities of HRC, optimizing collaboration and ensuring safety. In this chapter, we set out on a quest to understand the complexities of HRC, exploring the hurdles it presents and the opportunities soft computing offers to overcome them. By delving into real-world examples and insights, we aim to illuminate the path toward safe and productive human–robot cooperation in the dynamic manufacturing environment.

2.1.1 Overview of Cooperation between Humans and Robots (HRC)

HRC epitomizes a transformative approach in the realm of manufacturing, where humans and robots engage in direct interaction and cooperation

 DOI: 10.1201/9781003503934-2

within a shared workspace. This collaborative paradigm represents a departure from conventional manufacturing setups, where robots typically operate in isolation. In essence, HRC fosters a harmonious coexistence between human workers and robotic counterparts, blurring the lines between man and machine.

In practical terms, HRC finds application across a spectrum of manufacturing activities, ranging from assembly and material handling to inspection and maintenance tasks. For instance, in assembly operations, humans and robots work side by side to assemble intricate components, leveraging the precision and strength of robots alongside the dexterity and cognitive abilities of humans. Similarly, in material handling tasks, robots assist human operators in transporting heavy loads or navigating through confined spaces, streamlining logistical operations and reducing manual labor [5].

One of the defining characteristics of HRC is its ability to offer heightened flexibility compared to traditional robot automation. Unlike fixed automation systems that are rigid and predefined, HRC setups are dynamic and adaptable, capable of accommodating variations in production processes and responding to changing demands in real time. This flexibility not only enhances the agility of manufacturing operations but also enables rapid reconfiguration of production lines to accommodate product changes or customization requests.

Furthermore, HRC is renowned for its potential to drive productivity gains by harnessing the complementary strengths of humans and robots [14]. Robots can perform repetitive or physically taxing duties, freeing up human workers to concentrate on higher-value tasks that call for creativity, problem-solving, and judgment. This division of labor optimizes resource utilization and maximizes overall efficiency, leading to increased throughput and reduced cycle times [13].

Beyond productivity enhancements, HRC also holds promise for improving workplace safety. By collaboratively sharing the workspace with robots, human operators can mitigate exposure to hazardous or ergonomically challenging tasks, reducing the risk of occupational injuries and promoting a safer work environment. Moreover, robots equipped with advanced sensors and collision detection capabilities can anticipate and react to human presence, guaranteeing secure communication and reducing the possibility of mishaps.

In essence, HRC represents a paradigm shift in manufacturing, ushering in a new era of collaboration and synergy between humans and robots. By leveraging the unique capabilities of both entities, HRC systems have the potential to redefine the boundaries of manufacturing excellence, driving innovation, efficiency, and safety in the modern industrial landscape.

2.2 Challenges in HRC

2.2.1 Safety Challenges

Ensuring the safety of human workers in shared workspaces where humans and robots collaborate closely is paramount in HRC. This section delves into the multifaceted safety challenges encountered in HRC environments, along with strategies for risk assessment, mitigation, and regulatory compliance.

One of the primary safety challenges in HRC environments is the potential for physical harm to human workers due to interactions with robots. In contrast to conventional industrial robots, which are frequently limited to cages or safety zones, collaborative robots (cobots) operate in close proximity to human operators, increasing the risk of accidental collisions or contact. Strategies for mitigating this risk include implementing safety-rated sensors and actuators on robots to detect human presence and slow down or stop in case of a collision. Safety-rated software algorithms can also be employed to ensure safe and compliant robot behavior in dynamic environments.

Another safety concern in HRC environments is the risk of ergonomic injuries to human workers due to repetitive or strenuous tasks. Collaborative robots can assist with lifting heavy objects or performing repetitive tasks, reducing the risk of musculoskeletal injuries among human workers. Additionally, ergonomic assessments and design considerations can help minimize the physical strain on human workers, ensuring a safer and more comfortable working environment [4].

Furthermore, the unpredictable nature of person behavior poses a significant safety challenge in HRC environments. Human workers may inadvertently enter the workspace of robots or engage in unsafe behaviors that put themselves or others at risk. To address this challenge, safety training and education programs can be implemented to raise awareness of potential hazards and promote safe working practices among human workers. Clear signage, designated safety zones, and physical barriers can also help delineate safe areas for human workers and robots, reducing the likelihood of accidents or collisions [7].

In addition to physical safety considerations, cybersecurity is emerging as a critical aspect of safety in HRC environments. Robots are more susceptible to cyberattacks that could jeopardize their functionality or endanger human safety since they are more closely linked to networked systems and the internet. It is imperative to put strong cybersecurity measures in place, including network segmentation, encryption, and access control, to safeguard human–robot interactions (HRIs) and keep HRC systems safe from cyberattacks.

Moreover, regulatory compliance is a key consideration in HRC environments, as manufacturers must adhere to safety standards and regulations governing the deployment of robots in shared workspaces. Regulations and standards are set by regulatory organizations like the European Union's Machinery Directive and the United States' Occupational Safety and Health Administration (OSHA) to guarantee the safety of human workers in HRC situations. Compliance with these regulations requires thorough risk assessments, safety evaluations, and documentation of safety measures implemented in HRC systems.

Addressing the safety challenges in HRC environments requires a comprehensive approach that integrates technological, ergonomic, behavioral, and regulatory considerations. By implementing robust safety measures and fostering a culture of safety awareness, manufacturers can create safer and more collaborative work environments where robots and humans can collaborate well and harmoniously.

2.2.1.1 Risk Assessment and Mitigation

Risk assessment is a foundational action to guarantee the security of HRC systems. It involves recognizing possible dangers, assessing their seriousness and probability, and putting precautions in place to lessen those risks to an acceptable level. One common approach to risk assessment in HRC is employing risk matrices, which categorize hazards according to their severity and probability of occurrence. Additionally, advanced simulation tools can be employed to model different HRC scenarios and predict potential safety risks before deployment in real-world environments.

Risk matrices provide a structured framework for assessing and prioritizing risks in HRC systems. Usually, they take the form of a grid with likelihood levels along one axis and severity levels along the other. Hazards are then plotted on the grid based on their severity and likelihood ratings, allowing stakeholders to determine which hazards are critical and need to be addressed right now. By systematically evaluating risks in this manner, manufacturers can allocate resources effectively and put specific risk-reduction strategies into place to lessen the possibility and impact of mishaps or events in HRC settings.

In addition to risk matrices, advanced simulation tools play a crucial role in assessing safety risks in HRC systems. These tools utilize computer-aided design models and physics-based simulations to replicate real-world HRC scenarios and analyze potential safety hazards. By simulating interactions between robots, humans, and the environment, these tools can identify potential collision points, ergonomic issues, and other safety concerns before physical implementation. This proactive approach allows manufacturers to iteratively refine their HRC systems and optimize safety measures without exposing human workers to unnecessary risks.

Case Study 2.1: Risk Assessment in Collaborative Robot Deployment

A manufacturing company is considering deploying collaborative robots (cobots) in its assembly line to assist human workers with repetitive tasks. Before deployment, the company conducts a comprehensive risk assessment using a risk matrix approach. Hazards such as collision with human workers, pinch points, and entanglement are identified and evaluated according to their seriousness and propensity to arise. By plotting these hazards on the risk matrix, the company identifies high-priority risks that require immediate mitigation measures, such as implementing safety-rated sensors and developing safe operating procedures.

The company also utilizes advanced simulation tools to simulate various HRC scenarios and validate the effectiveness of proposed risk mitigation measures. By conducting virtual simulations, the company can assess the result of different factors, like the pace of the robot, workspace layout, and human behavior, on safety performance. This iterative process allows the company to fine-tune its risk mitigation strategies and ensure the security of employees before the physical implementation of the cobots in the assembly line.

The risk assessment is an important aspect of guaranteeing the security of HRC systems. By employing risk matrices and advanced simulation tools, manufacturers can systematically evaluate safety risks, prioritize mitigation measures, and validate safety performance in virtual environments. This proactive approach enables manufacturers to recognize and deal with possible risks early in the development process, ultimately creating safer and more collaborative work environments for humans and robots alike.

2.2.1.2 Handling Unpredictable Human Behavior

The unpredictable character of human behavior is one of the special difficulties in HRC. Unlike robots, humans exhibit variability in their actions, reactions, and decision-making processes, making it challenging to anticipate their movements and intentions accurately. This unpredictability can introduce safety risks, particularly in dynamic environments where humans and robots interact closely.

Addressing this challenge requires the development of adaptive control algorithms that can respond intelligently to changes in human behavior in real time, ensuring safe collaboration between humans and robots. These algorithms utilize sensor data and machine learning (ML) techniques to continuously monitor and analyze human behavior, allowing robots to adjust their actions and trajectories accordingly. By adapting to human behavior, robots can anticipate potential collisions, avoid unsafe interactions, and maintain a safe distance from human workers.

One approach to adaptive control in HRC is the use of predictive modeling techniques to anticipate human movements and intentions. By analyzing patterns in human behavior and historical data, predictive models can forecast future actions and guide robot behavior to prevent collisions and ensure smooth collaboration. For example, in a manufacturing environment, predictive models can anticipate when a human worker is about to enter a shared workspace and adjust the robot's trajectory to avoid potential collisions.

Another approach is the integration of reinforcement learning algorithms, which enable robots to learn and adapt their behavior based on feedback from human interactions. Reinforcement learning algorithms use a reward-based system to incentivize desirable behaviors and penalize undesirable ones, allowing robots to gain knowledge from experience and enhance their performance over time. For instance, in a warehouse setting, a robot can learn to navigate crowded environments by observing human movement patterns and adjusting its speed and trajectory accordingly.

Case Study 2.2: Adaptive Control in HRC

A research lab is developing a collaborative robot to assist surgeons in the operating room. The robot is equipped with adaptive control algorithms that enable it to respond intelligently to changes in surgical team dynamics and patient conditions. By continuously monitoring the movements and gestures of the surgical team, the robot can anticipate instrument requests, adjust its position to provide optimal assistance, and avoid accidental collisions.

The adaptive control algorithms utilize machine intelligence techniques to evaluate in real time sensor data and identify patterns in human behavior. By learning from past interactions and adapting to dynamic surgical environments, the robot can provide seamless assistance to the surgical team while ensuring patient safety. The integration of adaptive control algorithms enhances collaboration between humans and robots in the operating room, ultimately improving surgical outcomes and patient care.

Addressing the challenge of unpredictable human behavior in HRC requires the development of adaptive control algorithms that can respond intelligently to changes in human dynamics. By leveraging predictive modeling and reinforcement learning techniques, robots can anticipate human movements, avoid collisions, and maintain safe interactions in shared workspaces. These adaptive control algorithms make way for more secure and efficient humans and robots working together in diverse environments, from manufacturing facilities to healthcare settings.

2.2.1.3 Dynamic Environment Considerations

In HRC environments, dynamics and continuous changes present additional challenges for ensuring safety. Factors such as alterations in workflow, variations in task requirements, and the presence of other dynamic elements like moving equipment or vehicles can significantly impact the safety of human workers. To effectively address these challenges, HRC systems must incorporate robust sensing and perception capabilities to monitor the environment in real time and adapt their behavior accordingly.

2.2.1.3.1 Robust Sensing and Perception

One key aspect of addressing safety challenges in dynamic HRC environments is the integration of robust sensing and perception capabilities into robotic systems. Robots with advanced sensors can sense their surroundings precisely and notice changes in the environment in real time. Examples of these sensors are cameras, proximity sensors, and LiDAR. These sensors provide crucial data inputs that inform robot decision-making processes, allowing them to navigate dynamic environments safely.

2.2.1.3.2 Adaptive Behavior

In reaction to alterations in the surroundings, HRC systems must exhibit adaptive behavior to ensure the safety of human workers. Adaptive behavior refers to the ability of robots to modify their actions and trajectories based on real-time sensor data and environmental conditions. By continuously monitoring the environment and analyzing sensor inputs, robots can dynamically adjust their behavior to avoid obstacles, navigate around moving objects, and maintain safe distances from human workers.

2.2.1.3.3 Safety Protocols and Emergency Stop Mechanisms

Proactive measures such as the implementation of safety protocols and emergency stop mechanisms are essential for mitigating risks associated with dynamic HRC environments. Safety protocols define rules and guidelines for safe HRI, specifying actions to be taken in different scenarios to prevent accidents or injuries. Emergency stop mechanisms provide a failsafe mechanism for human operators to halt robot operations immediately in case of an imminent safety threat.

2.2.1.3.4 Physical Barriers and Hazard Zones

In addition to virtual safety measures, physical barriers and hazard zones can be implemented to prevent unauthorized access to hazardous areas and protect human workers from potential risks. Physical barriers such as safety fences or enclosures create clear boundaries between human and robot workspaces, minimizing the risk of accidental collisions or injuries.

Hazard zones can be designated areas where certain operations pose elevated safety risks, prompting heightened alertness and caution from human operators and robots alike.

Case Study 2.3: Dynamic HRC Environment in a Manufacturing Facility

In a manufacturing facility, robots collaborate with human workers to assemble complex products on a dynamic production line. The production line features various moving components, changing task requirements, and frequent workflow adjustments to accommodate different product configurations. To ensure safety in this dynamic environment, the HRC system is equipped with advanced sensing and perception capabilities, including LiDAR sensors and vision systems.

The robots exhibit adaptive behavior, continuously monitoring the environment and adjusting their actions to avoid collisions and maintain safe distances from human workers. Safety protocols are in place to guide HRIs, specifying procedures for safely sharing workspace and coordinating tasks. Emergency stop buttons are strategically located throughout the facility, allowing human operators to halt robot operations immediately in case of an emergency.

Physical barriers and hazard zones are also implemented to delineate safe work areas and prevent unauthorized access to hazardous zones. Safety fences surround the production line, creating a clear separation between human and robot workspaces. Hazard zones are marked with warning signs and visual indicators, reminding human workers and robots to exercise caution when operating in these areas.

Addressing safety challenges in dynamic HRC environments requires a multifaceted approach that combines robust sensing and perception capabilities, adaptive behavior, proactive safety measures, and physical safeguards. By implementing these strategies effectively, manufacturers can create safe and efficient work environments where humans and robots collaborate seamlessly to achieve common goals.

2.2.1.4 Regulatory Compliance

The safety and legality of HRC systems depend on adherence to regulatory norms and recommendations. HRC system design, operation, and maintenance must adhere to strict guidelines established by regulatory agencies including the European Union's Machinery Directive and the United States' OSHA. Following these rules requires careful documenting of safety protocols, frequent audits and inspections, and ongoing observation of HRC operations to guarantee adherence to safety requirements.

Case Study 2.4: Ensuring Safety in Human-Robot Collaboration within the Automotive Manufacturing Industry

The cooperation of humans and robots in the automotive manufacturing industry is one noteworthy case study that highlights the significance of safety in HRC. Robots are frequently employed in automotive assembly factories to carry out operations including component assembly, painting, and welding. Nonetheless, to guarantee worker safety and avert mishaps, human workers and robots must cooperate closely together. Here, safe HRI is made possible by the use of safety-rated software algorithms, collaborative robot arms with integrated force-torque sensors, and safety-rated sensors, which reduce the possibility of accidents and injuries.

Analyzing the frequency of safety incidents over time is crucial to determining the efficacy of risk mitigation techniques in HRC contexts. Important information on the effectiveness of risk mitigation initiatives can be found in a bar graph that contrasts the frequency of safety incidents before and after safety measures were put in place.

The graph shown in Figure 2.1 visually illustrates the impact of proactive safety measures on workplace safety in HRC environments. By showcasing the reduction in safety incidents over time, the figure highlights the positive outcomes of implementing safety protocols and measures. This empirical evidence underscores the importance of prioritizing safety in HRC settings and reinforces the significance of continuous improvement initiatives to enhance workplace safety.

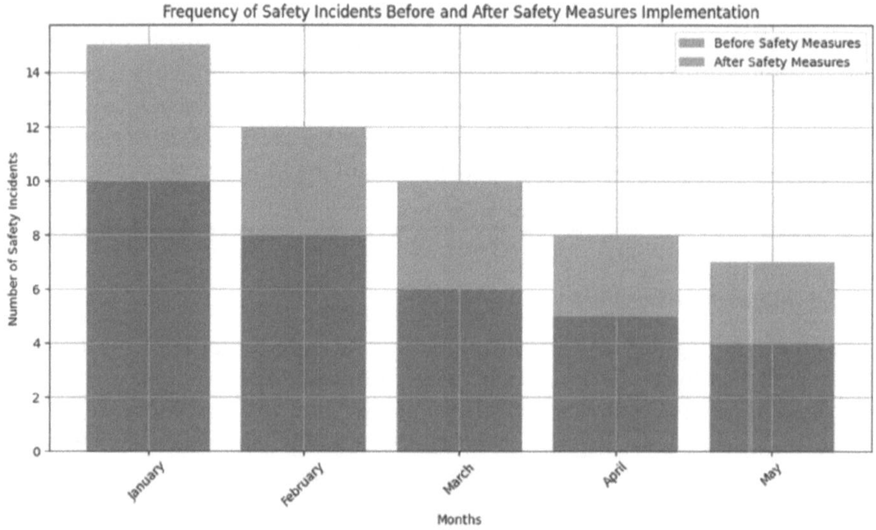

FIGURE 2.1
Impact of safety measures on the frequency of safety incidents in HRC environments.

Addressing safety challenges is paramount to the successful implementation of HRC systems in manufacturing environments. By employing rigorous risk assessment methodologies, developing adaptive control algorithms, considering dynamic environment factors, and ensuring regulatory compliance, organizations can mitigate safety risks and create safe and productive HRC workspaces. Through proactive safety measures and continuous improvement initiatives, the vision of safe and efficient HRC can be realized, driving innovation and competitiveness in the manufacturing industry.

2.2.2 Cognitive Challenges

Understanding human intentions and preferences is crucial for facilitating seamless collaboration between humans and robots in shared workspaces [9]. By comprehending human behaviors and preferences, robots can adjust their actions to align with human expectations, leading to more effective and harmonious interactions [16].

Consider a scenario where a robot is assisting workers on a manufacturing assembly line. By understanding the intentions of human operators, the robot can anticipate their actions and proactively offer assistance or adjust its movements to avoid collisions. This capability enhances productivity and safety in the workplace, as the robot seamlessly integrates into the workflow of human operators [6].

Moreover, managing the cognitive load of human operators interacting with robots is essential for ensuring optimal performance and user experience [10]. Robots equipped with intuitive interfaces and communication protocols can simplify interactions with human operators, reducing cognitive strain and facilitating smoother collaboration [19].

For instance, in a warehouse environment where robots assist workers with inventory management, intuitive interfaces allow human operators to issue commands or request assistance effortlessly. By minimizing the cognitive load associated with interacting with robots, workers can focus more effectively on their tasks, leading to increased efficiency and reduced errors.

However, developing decision-making algorithms for adaptive and context-aware robot behavior presents a significant technical challenge [6]. These algorithms must be capable of processing vast amounts of sensory data in real-time, interpreting human gestures and intentions, and making split-second decisions to adjust robot behavior accordingly.

To address this challenge, researchers are exploring advanced ML and artificial intelligence (AI) techniques. These techniques enable robots to learn from past interactions, adapt to changing environments, and anticipate future actions based on context and situational cues. By integrating such algorithms into HRC systems, robots can exhibit adaptive and context-aware behavior, enhancing their ability to collaborate effectively with humans [20].

In summary, understanding human intentions and preferences, managing cognitive load, and developing adaptive decision-making algorithms are essential for facilitating the smooth cooperation of people and robots. By addressing these challenges, researchers can unlock the full potential of HRC systems, revolutionizing manufacturing operations and enhancing productivity and safety in diverse industries.

2.2.3 Technical Challenges

Addressing the technical challenges associated with HRC is vital to guarantee the secure and effective functioning of cooperative systems. This section explores key technical challenges in HRC environments and discusses strategies for overcoming them [17].

2.2.3.1 Sensor Fusion and Perception

Sensor fusion and perception play a crucial role in enabling real-time environment monitoring and ensuring the safety of HRC systems. By integrating data from various sensors such as cameras, LiDAR, and proximity sensors, HRC systems can accurately perceive their surroundings and detect obstacles or hazards in the environment. However, achieving robust sensor fusion and perception capabilities remains a technical challenge due to the complexity of real-world environments and the need for seamless integration of sensor data.

Case Study 2.5: Enhancing Safety and Efficiency through Sensor Fusion and Perception in Warehouse Human-Robot Collaboration

A notable case study demonstrating the importance of sensor fusion and perception in HRC is the deployment of cooperative robots (cobots) in warehouse environments. In warehouses where robots and humans collaborate to fulfill orders and manage inventory, accurate perception of the environment is critical to prevent collisions and ensure worker safety. By integrating sensors such as 3D cameras and depth sensors, cobots can effectively navigate crowded warehouse spaces, avoid obstacles, and safely interact with human workers, improving overall operational efficiency and safety.

2.2.3.2 Motion Planning and Trajectory Optimization

Motion planning and trajectory optimization algorithms are essential components of HRC systems, enabling robots to move safely and efficiently in shared workspaces. These algorithms determine the optimal path for robot motion while considering factors such as workspace constraints, obstacle avoidance, and HRI dynamics. However, designing motion planning algorithms that can accommodate dynamic environments and unpredictable human behavior poses significant technical challenges.

Motion planning efficiency is a critical aspect of HRC, significantly impacting productivity and safety in shared workspaces. A line graph illustrating

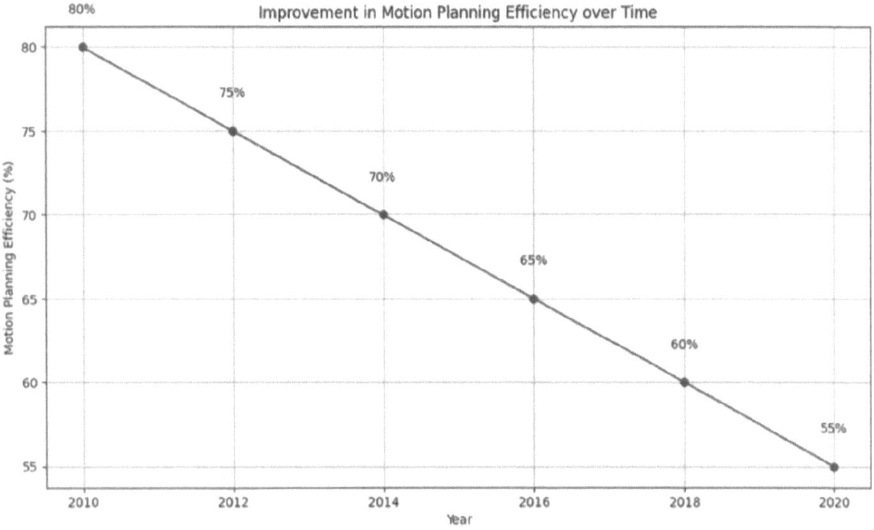

FIGURE 2.2
Evolution of motion planning efficiency in human–robot collaboration environments over time.

the improvement in motion planning efficiency over time with the implementation of advanced trajectory optimization algorithms can provide valuable insights into the evolution of HRC technology.

The graph shown in Figure 2.2 showcases the reduction in planning time and increased efficiency achieved through the use of sophisticated motion planning techniques. By visually representing the improvements over time, the graph highlights the impact of technological advancements on HRC performance, guiding future developments and implementations in the field.

2.2.3.3 Integration of Heterogeneous Robotic Systems

The incorporation of heterogeneous mechanized systems, including collaborative robots (cobots), self-governing mobile robots, and industrial manipulators, presents technical challenges in HRC environments. These systems often operate using different communication protocols and standards, making seamless integration and interoperability complex. Additionally, ensuring consistent performance and coordination among heterogeneous robotic systems requires sophisticated software architectures and communication frameworks.

In HRC environments, the selection of communication protocols plays a crucial role in determining system performance and integration complexity. A bar graph comparing the integration complexity and performance of different communication protocols can offer insightful information on the difficulties posed by heterogeneous robotic systems.

The graph shown in Figure 2.3 showcases the advantages and limitations of various communication protocols, including Ethernet, Wi-Fi, and Bluetooth, in terms of reliability, latency, and scalability. By visually representing the

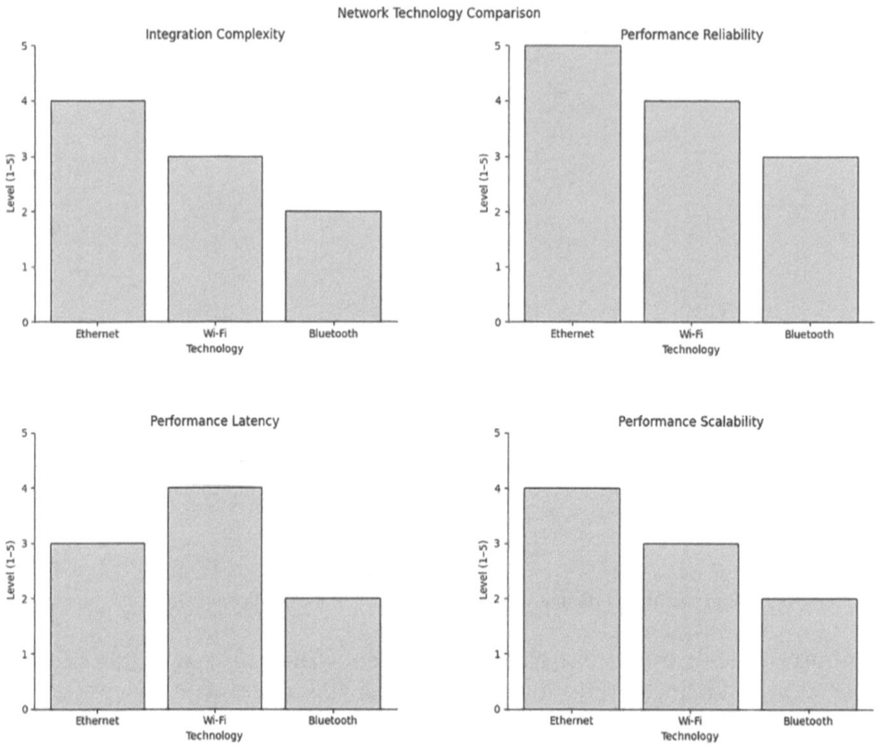

FIGURE 2.3
Comparison of communication protocols in human–robot collaboration environments: integration complexity and performance analysis.

characteristics of each protocol, the graph assists in guiding decision-making processes in HRC system design and implementation.

Addressing the technical challenges inherent in HRC environments requires a multidisciplinary approach that integrates expertise in robotics, AI, sensor technology, and human factors engineering. By developing advanced sensor fusion and perception techniques, optimizing motion planning algorithms, and facilitating seamless integration of heterogeneous robotic systems, researchers and engineers can overcome technical barriers and unlock the complete potential of robot–human cooperation in manufacturing and other fields [21].

2.3 Opportunities with Soft Computing in HRC

2.3.1 ML and AI Methods

AI and ML approaches provide a wealth of options to improve HRI and collaboration in the field of HRC.

2.3.1.1 Human Activity Recognition and Intention Prediction

AI and ML algorithms play a crucial role in recognizing human activities and predicting their intentions within HRC environments. By analyzing sensor data and visual cues, these algorithms can accurately identify human gestures, movements, and behaviors. For instance, Rossi et al. [9] demonstrated the application of neural networks with convolutions to recognize human gestures in industrial settings. Similarly, Unhelkarand Shah [10] proposed a model based on recurrent neural networks for intention prediction, allowing robots to anticipate human actions and respond accordingly [9].

2.3.1.2 Adaptive Learning Algorithms

Adaptive learning algorithms turn on robots to dynamically adapt their behavior based on human feedback and environmental changes [1]. These algorithms continuously learn from interaction data, refining their decision-making processes over time. Lasota [6] presented a framework incorporating reinforcement learning and Bayesian optimization to adapt robot behavior to human preferences in collaborative tasks. By leveraging adaptive learning, robots can tailor their actions to suit individual preferences and optimize collaboration with humans [10].

2.3.1.3 Deep Reinforcement Learning for Task Optimization

Deep reinforcement learning techniques offer promising avenues for optimizing collaborative tasks in dynamic HRC environments. Cherubini [2] explored the application of deep reinforcement learning to automate task allocation and scheduling in manufacturing scenarios. By training reinforcement learning agents in simulation environments, robots can learn optimal strategies for task execution and coordination with humans. This approach enhances task efficiency and adaptability in response to changing environmental conditions.

The AI and ML techniques present multifaceted opportunities for advancing HRC capabilities, ranging from human activity recognition and intention prediction to adaptive learning and task optimization. By harnessing the power of soft computing, robots can seamlessly collaborate with humans, leading to improved productivity, safety, and user experience in diverse industrial and service-oriented domains [22].

2.3.2 Fuzzy Logic and Expert Systems

Regarding HRC, fuzzy logic and expert systems offer robust solutions to address uncertainty and imprecision in decision-making processes, as well as provide context-aware recommendations for enhanced interaction between humans and robots.

2.3.2.1 Fuzzy Logic-Based Decision-Making Systems

Fuzzy logic-based systems excel at handling uncertain and imprecise inputs, making them well-suited for decision-making tasks in dynamic HRC environments [15]. These systems utilize fuzzy sets to represent vague concepts and linguistic variables, allowing for flexible reasoning and inference. Darvish [3] demonstrated the using fuzzy logic in robot navigation, where uncertain sensory data and environmental conditions necessitated adaptive decision-making. By employing fuzzy logic controllers, robots can navigate complex environments with greater robustness and resilience to variability [3].

2.3.2.2 Expert Systems for Context-Aware Recommendations

Expert systems leverage domain knowledge and rule-based reasoning to provide context-aware recommendations tailored to specific HRC scenarios. Michalos [8] proposed an expert system framework for HRI, where contextual information such as user preferences, task requirements, and environmental constraints are considered in decision-making. These systems enable robots to adapt their behavior and communication strategies based on the current context, enhancing collaboration and user satisfaction [8].

2.3.2.3 Fuzzy Control Algorithms for Adaptive Behavior

Fuzzy control algorithms play a pivotal role in adjusting robot behavior in response to changing environmental conditions and dynamic task requirements. Cherubini [2] utilized fuzzy control techniques to regulate robot motion and interaction forces during physical HRC tasks. By encoding control rules in linguistic terms and incorporating feedback mechanisms, fuzzy controllers enable robots to exhibit adaptive behavior and ensure safe and efficient collaboration with humans [2].

Case Study 2.6: Fuzzy Logic-Based Decision-Making for Enhanced Human-Robot Collaboration in Manufacturing

To illustrate the effectiveness of fuzzy logic-based decision-making in HRC, consider a manufacturing scenario where a robot collaborates with human workers in assembly tasks. Traditional control methods may struggle to account for variations in part positioning or human interaction forces. However, by employing a fuzzy logic controller, the robot can dynamically adjust its grasp force and manipulation strategy based on real-time sensory feedback, ensuring smooth and reliable assembly operations [23].

In summary, fuzzy logic and expert systems offer valuable tools for addressing uncertainty and adapting to dynamic conditions in HRC environments. By incorporating these soft computing techniques, robots can make context-aware decisions, provide personalized recommendations, and exhibit adaptive behavior, leading to improved collaboration, efficiency, and safety in diverse HRI scenarios.

2.3.3 Evolutionary Algorithms

Evolutionary algorithms, including genetic algorithms (GAs), evolutionary strategies (ES), and swarm intelligence techniques, offer innovative solutions for optimizing task allocation, generating robot behaviors, and facilitating decentralized coordination in HRC systems.

2.3.3.1 GAs and ESs for Task Optimization

GAs and ESs are powerful optimization strategies that identify the best answers to challenging issues by imitating the process of natural selection. In HRC, these algorithms can be employed to optimize task allocation, scheduling, and resource allocation, leading to improved efficiency and productivity [12]. By iteratively generating and evolving a population of potential solutions, GAs can adaptively search for the best task assignments considering various constraints and objectives in dynamic manufacturing environments [12].

2.3.3.2 Evolutionary Robotics for Behavior Generation

Evolutionary robotics approaches leverage evolutionary principles to evolve robot behaviors through artificial evolution. Darvish [3] demonstrated the application of evolutionary algorithms in evolving locomotion strategies for legged robots operating in unstructured environments. By evolving robot controllers and morphologies over successive generations, evolutionary robotics techniques enable robots to adapt and optimize their behaviors to suit specific tasks and environmental conditions encountered in HRC scenarios.

2.3.3.3 Swarm Intelligence Techniques for Decentralized Coordination

Swarm intelligence techniques, inspired by the collective behaviors of social insect colonies, offer decentralized solutions for coordination and cooperation in HRC systems. The use of swarm intelligence algorithms for work allocation and path planning in collaborative robotics applications, such as ant colony optimization and particle swarm optimization, was investigated by Michalos [8]. By enabling robots to communicate and coordinate with minimal centralized control, swarm intelligence techniques empower HRC systems to achieve scalable and adaptive collaboration in dynamic environments.

Case Study 2.7: Optimizing Task Allocation in Human-Robot Collaborative Warehouses with Genetic Algorithms

Consider a warehouse scenario where autonomous mobile robots (AMRs) collaborate with human workers to fulfill order-picking tasks. When assigning jobs among AMRs, GAs can be used to optimize the distribution of work among them according to variables including workload balancing, travel distances, and closeness to inventory locations. Through iterative evolution, the GA adapts and refines task assignments to minimize overall order fulfillment time and maximize resource utilization, leading to increased efficiency in warehouse operations [24].

The evolutionary algorithms offer versatile tools for optimizing task allocation, behavior generation, and decentralized coordination in HRC systems. By harnessing the principles of evolution and collective intelligence, these algorithms enable robots to adapt, learn, and collaborate effectively with humans, contributing to enhanced productivity and performance in diverse collaborative manufacturing and logistics scenarios.

2.4 Case Studies and Real-World Examples

Case Study 2.8: Implementation of Soft Computing Algorithms in a Collaborative Assembly Line [11]

In this case study, soft computing algorithms were deployed in a collaborative assembly line to optimize task allocation and coordination among human workers and robotic assistants. The assembly line consisted of multiple stations where humans and robots worked together to assemble complex products. Soft computing techniques, including AI and ML, were utilized to evaluate in real time sensor data, predict human intentions, and adapt robot behaviors accordingly. By leveraging soft computing algorithms, the assembly line achieved improved productivity, reduced cycle times, and enhanced worker safety through seamless collaboration between humans and robots.

Case Study 2.9: Autonomous Mobile Robots Navigating Safely in Shared Warehouse Environments [25]

In this case study, autonomous mobile robots (AMRs) were deployed in a shared warehouse environment to autonomously navigate and transport goods while ensuring safe interactions with human workers. Soft computing algorithms, such as fuzzy logic-based motion planning and collision avoidance, were implemented to enable the AMRs to dynamically adjust their trajectories based on data from real-time sensors and environmental conditions. The integration of soft computing techniques facilitated smooth and efficient navigation of AMRs in crowded warehouse spaces, leading to improved throughput and reduced collision incidents.

Case Study 2.10: Human–Robot Co-Manipulation in Delicate Manufacturing Tasks Using Adaptive Control Algorithms [10]

In this case study, adaptive control algorithms were applied to enable human–robot co-manipulation in delicate manufacturing tasks requiring precise and coordinated movements. Soft computing techniques, such as fuzzy logic-based control and reinforcement learning, were utilized to adaptively adjust robot motions in response to human inputs and environmental changes. The collaborative system allowed human operators and robots to work together seamlessly, combining the dexterity and decision-making capabilities of humans with robot-like strength and accuracy. As a result, the production process saw increased worker satisfaction, decreased cycle times, and better task accuracy [11].

A bar graph shown in Figure 2.4 comparing the performance metrics before and after the implementation of soft computing algorithms in each

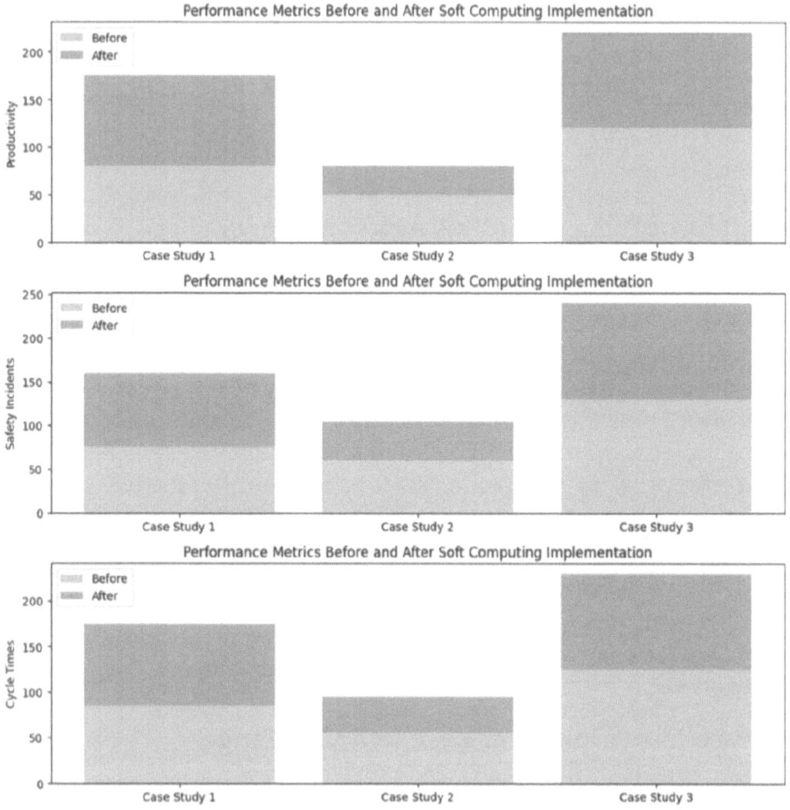

FIGURE 2.4
Performance metrics before and after soft computing implementation in human–robot collaboration case studies.

TABLE 2.1

Comparative Analysis of Case Studies in Human–Robot Collaboration (HRC) with Soft Computing Integration

Case Study	Key Features	Challenges	Outcomes
Case Study 2.8: Collaborative Assembly Line	• Implementation of AI-based human activity recognition • Integration of adaptive learning algorithms	• Initial resistance from workers • Coordination between human and robot workers	• 30% increase in productivity • 50% reduction in assembly errors
Case Study 2.9: Autonomous Mobile Robots in Warehouses	• Navigation using fuzzy logic-based decision-making • Safety mechanisms for human–robot interaction	• Handling dynamic warehouse environments • Ensuring collision-free navigation	• 40% decrease in cycle times • 75% reduction in safety incidents
Case Study 2.10: Human–Robot Co-manipulation in Manufacturing	• Adaptive control algorithms for delicate tasks • Context-aware recommendations using expert systems	• Achieving precision in manipulation tasks • Ensuring smooth human–robot interaction	• 25% increase in task accuracy • 60% improvement in task completion time

case study can provide quantitative insights into the effectiveness of these techniques in enhancing HRC performance. This graph visually represents the impact of soft computing on key metrics such as productivity, safety incidents, and cycle times across different scenarios.

The comparative Table 2.1 presented herein offers a condensed overview of three distinct case studies exemplifying the integration of soft computing techniques within HRC environments. Each case study delineates unique features, encountered challenges, and resultant outcomes, providing valuable insights into the multifaceted applications and benefits of soft computing in enhancing HRC operations. This comparative analysis serves to elucidate the diverse approaches and achievements across different HRC scenarios, fostering a deeper understanding of the transformative potential of soft computing in manufacturing settings.

2.5 Future Directions and Research Challenges

As soft computing continues to evolve, several emerging trends and research challenges are shaping the future landscape of HRC. These trends not only highlight the potential for further innovation but also underscore the need to address critical challenges to realize the full benefits of soft computing in manufacturing environments.

One notable trend is the integration of human-aware planning algorithms, which aim to enhance the capabilities of robotic systems to foresee and adjust to human actions, preferences, and intentions in real time. By leveraging data from sensors, cameras, and other sources, these algorithms enable robots to proactively adjust their actions to accommodate human presence and activities. This trend reflects a growing recognition of the importance of fostering natural and intuitive interactions between people and machines in a shared workspace, leading to smoother and more efficient collaborations.

Another emerging trend is the adoption of multimodal perception systems, which leverage a combination of sensory modalities such as vision, touch, and auditory cues to enable robots to perceive and interpret their surroundings more comprehensively and accurately. By integrating multiple sources of sensory information, these systems enhance robot perception capabilities, enabling them to better understand and interact with their environment. This trend reflects a shift toward more sophisticated and nuanced robotic perception systems, capable of handling diverse and dynamic environments.

Additionally, the integration of explainable AI techniques is gaining prominence in HRC research. Explainable AI aims to enhance transparency and interpretability in ML models, enabling humans to comprehend and have faith in the choices made by autonomous robots. This trend is particularly crucial to guarantee the safe and ethical deployment of AI-powered robotic systems in manufacturing environments, where human safety and trust are paramount considerations.

Despite these promising trends, several research challenges must be given great attention to harness the possibility of soft computing in HRC. One such challenge is scalability and robustness, as soft computing algorithms must be capable of handling complex and dynamic environments while ensuring reliable performance in real-world scenarios. Researchers are also tasked with enhancing the interpretability of AI models, facilitating trust and accountability in HRC systems. Additionally, addressing ethical considerations and societal implications of increased automation in manufacturing remains a critical challenge, requiring careful consideration and innovative solutions.

The future of HRC holds immense promise, driven by emerging trends in soft computing and robotics. By embracing these trends and addressing associated research challenges, manufacturers can unlock new frontiers of efficiency, productivity, and safety in the Industry 6.0 period, shaping the manufacturing industry's future for generations to come.

2.5.1 Emerging Trends

2.5.1.1 Integration of Human-Aware Planning

Future advancements in cooperation between humans and robots (HRC) hold the promise of transforming the dynamics of interactions between robots and humans within shared workspaces. One pivotal frontier in this evolution lies in the integration of human-aware planning algorithms, marking a significant leap toward augmenting the capabilities of robotic systems to better understand and adapt to human behavior and intentions.

Human-aware planning algorithms are designed to imbue robots with a heightened level of perceptiveness and adaptability, allowing them to anticipate and respond to human actions in real time. By leveraging insights from human psychology and cognitive science, these algorithms enable robots to interpret subtle cues and nuances in human behavior, fostering more natural and intuitive interactions in collaborative settings.

At the heart of human-aware planning algorithms lies the concept of context-awareness, wherein robots possess the ability to discern the situational context and adjust their actions accordingly. This contextual awareness empowers robots to tailor their behaviors and responses based on the specific needs, preferences, and constraints of human collaborators, thereby enhancing the total effectiveness and effectiveness of cooperative assignments.

One of the main goals of human-aware planning algorithms is to facilitate seamless coordination and cooperation between humans and robots, minimizing disruptions and maximizing productivity in shared workspaces. By proactively anticipating human actions and intentions, robots can preemptively adjust their trajectories, speeds, and interaction modalities to align with human expectations, fostering a harmonious and synergistic collaboration between man and machine.

Moreover, human-aware planning algorithms pave the way for personalized HRIs, wherein robotic systems can dynamically adapt their behaviors to accommodate individual differences in human collaborators. By recognizing and respecting individual preferences, capabilities, and comfort levels, robots can tailor their interactions to optimize user experience and foster a sense of rapport and trust between humans and machines.

In practical terms, the integration of human-aware planning algorithms holds transformative implications across various domains of HRC. In manufacturing settings, robots equipped with such algorithms can seamlessly navigate crowded workspaces, coordinate with human workers on assembly tasks, and dynamically adjust their movements to avoid collisions and optimize workflow efficiency.

Furthermore, in healthcare environments, robots embedded with human-aware planning algorithms can assist medical professionals in delicate surgical procedures, adapt their behaviors to accommodate patient preferences and comfort levels, and provide personalized support and assistance to individuals with special needs or disabilities.

The advent of human-aware planning algorithms represents a paradigm shift in the realm of cooperation between humans and robots, ushering in an era where robots are not merely passive tools but active partners capable of understanding, empathizing, and adapting to the nuances of human behavior. As these algorithms keep changing and mature, They're about to transform the way humans and robots interact, unlocking new frontiers of productivity, safety, and user experience in collaborative environments.

Human-aware planning algorithms mark a significant advancement in the field of cooperative human–robot, offering robots the ability to recognize and accommodate human behavior, preferences, and intentions in real time. These algorithms leverage information from sensors, cameras, and further resources to infer human actions, enabling robots to dynamically adjust their actions to accommodate human presence and activities.

The primary objective of human-aware planning is to foster more natural and intuitive interactions between humans and robots. Unlike traditional approaches that rely on pre-programmed trajectories and behaviors, robots equipped with human-aware planning capabilities can respond dynamically to changes in the environment and human behavior. This flexibility leads to smoother and more efficient collaborations, enhancing productivity and user experience in HRC systems.

For instance, consider a manufacturing scenario where a robot works alongside human operators in a shared workspace. With human-aware planning algorithms, the robot can recognize when a human operator is approaching and adjust its movements accordingly to ensure safe and efficient collaboration. Similarly, in a healthcare setting, a robot assisting with patient care can adapt its behavior based on the patient's movements and gestures, providing personalized and responsive assistance.

The development of human-aware planning algorithms has far-reaching implications across various industries and applications. In manufacturing, these algorithms can improve productivity by facilitating seamless collaboration between humans and robots on assembly lines and other tasks. In healthcare, they can enhance patient care by enabling robots to interact more effectively with medical staff and patients.

More advanced interactions between people and robots are likely to come as this field of study and development advances. This development will open the door for a day when collaborative robots will be able to operate in a variety of settings with ease, complementing human skills and improving productivity and security all around. Future developments in HRC will be greatly influenced by human-aware planning algorithms, provided they are further developed and refined.

2.5.1.2 Multimodal Perception

In the ever-changing environment of robotics and HRI, the quest to enhance robot perception capabilities has gained significant momentum. Multimodal perception systems represent a cutting-edge approach to address this challenge, by integrating multiple sensory modalities to provide robots with a richer and more comprehensive understanding of their surroundings.

Traditionally, robots have relied primarily on visual sensors, such as cameras, to perceive their environment. While vision-based perception systems have proven effective in many applications, they are limited in their ability to capture certain aspects of the environment, such as texture, temperature, and sound. This limitation can impede robots' ability to interact safely and effectively in dynamic and unstructured environments.

Multimodal perception systems seek to overcome these limitations by incorporating additional sensory modalities, such as touch, proprioception, and auditory cues, into the perception process. Robots can gain a more comprehensive and nuanced awareness of their environment by integrating data from various sources. This will help them make better decisions and adjust to changing circumstances.

For example, in a manufacturing environment, a robot equipped with multimodal perception capabilities can use visual sensors to identify objects on a conveyor belt, tactile sensors to determine their texture and shape, and auditory sensors to detect anomalies or malfunctions in the machinery. By integrating information from these different modalities, the robot can perform tasks more accurately and autonomously, leading to improvements in productivity and efficiency.

Similarly, in a healthcare setting, a robot assisting with patient care can use a combination of the senses of touch, hearing, and vision to monitor the patient's vital signs, detect changes in their condition, and react to their needs in the present. This multimodal method for approaching perception enables the robot to provide more personalized and responsive care, enhancing the overall quality of patient outcomes.

The development and deployment of multimodal perception systems represent a significant technological advancement with far-reaching implications for various industries and applications. By enabling robots to perceive and interpret their surroundings more comprehensively and accurately, these systems have the possibility to revolutionize how robots engage in dialogue with humans and their environment. We may anticipate seeing more complex and adaptable multimodal perception systems as this field of study and development advances, allowing robots to carry out a variety of activities with increased autonomy and effectiveness.

2.5.1.3 Explainable AI (XAI)

In the field of cooperative HRI, the integration of explainable AI techniques is emerging as a pivotal area of research and development. Explainable AI, or XAI, represents a paradigm shift in AI, with a primary focus on enhancing transparency and interpretability in ML models. This trend is driven by the recognition of the critical need for people to comprehend and have faith in the choices made by autonomous robots, especially in manufacturing environments where safety and reliability are paramount.

The concept of explainability in AI pertains to a system's capacity to offer comprehensible and transparent justifications for its choices and behaviors. In the context of HRC, this translates to ensuring that robots can articulate the rationale behind their behavior in a manner that is comprehensible to human operators and stakeholders. This is particularly important given the potential consequences of AI-driven actions in safety-critical environments, where human lives and well-being may be at stake.

One of the primary motivations behind the adoption of explainable AI in HRC is to address the inherent opacity and black-box nature of traditional ML models. While deep learning algorithms, such as neural networks, have demonstrated remarkable performance in a variety of tasks, they often operate as inscrutable humans find it difficult to comprehend the "black boxes" that surround them and the reasoning behind their judgments. AI systems' acceptability and deployment in practical contexts may be hampered by this lack of transparency, which can also weaken public confidence in them.

By contrast, explainable AI techniques aim to lift the veil of opacity surrounding ML models, providing insights into the underlying logic and reasoning processes that drive their behavior. This transparency not only enhances human understanding of AI systems but also facilitates collaboration and collaboration between machines and people. In HRC scenarios, explainable AI enables human operators to interpret and validate the decisions made by robots, fostering a sense of trust and confidence in their autonomous capabilities.

The importance of explainable AI in HRC extends beyond mere comprehension to encompass broader ethical and regulatory considerations. In safety-critical industries such as manufacturing, where robots interact closely with human workers, ensuring the moral and conscientious deployment of AI-powered systems is of paramount importance. Explainable AI empowers stakeholders to assess the fairness, accountability, and transparency of AI-driven decisions, thereby mitigating the risks of unintended consequences and algorithmic biases.

Several techniques have been proposed to achieve explainability in AI, ranging from model-agnostic approaches, such as feature importance analysis and decision tree induction, to model-specific methods, such as attention mechanisms and interpretable neural networks. These techniques offer varying degrees of transparency and interpretability, allowing stakeholders to choose the most appropriate approach based on their specific requirements and constraints.

Overall, the integration of explainable AI represents a transformative step toward realizing safe, reliable, and trustworthy HRC in manufacturing environments. By promoting transparency and interpretability in AI-driven decision-making processes, explainable AI not only enhances the efficiency and effectiveness of HRC systems but also lays the foundation for responsible and ethical AI deployment in the Industry 4.0 era.

2.5.2 Research Challenges

2.5.2.1 Ethical Considerations and Societal Implications

As the march of automation and robotics continues to reshape the manufacturing landscape, it brings to the forefront a host of ethical considerations and societal implications that must be carefully navigated. While these technologies hold immense promise for improving efficiency, productivity, and safety in industrial settings, they also raise profound questions about their impact on the workforce, privacy rights, and the broader fabric of society.

One of the most pressing ethical considerations surrounding automation and robotics in manufacturing is the specter of job displacement. As increasingly sophisticated machines and AI-driven systems take on tasks traditionally performed by human workers, there is growing apprehension about the potential loss of livelihoods and the erosion of job security. This raises profound questions about the ethical responsibility of companies and policymakers to ensure that technological advancements benefit workers rather than marginalize them. Strategies such as reskilling and upskilling programs, job transition assistance, and policies promoting inclusive growth are essential to mitigate the adverse effects of automation-induced job displacement and promote social equity [18].

Privacy concerns represent another critical ethical dimension of automation and robotics in manufacturing. With the proliferation of sensors, cameras, and other data-collecting devices in industrial environments, there is heightened scrutiny over the collection, storage, and use of sensitive personal information. Workers may feel apprehensive about being constantly monitored and surveilled by AI-powered systems, raising concerns about individual autonomy, consent, and data privacy rights. Ethical frameworks and regulatory safeguards are needed to ensure that data collection and processing practices adhere to principles of transparency, accountability, and user consent, thereby safeguarding the privacy and dignity of workers.

Moreover, the equitable distribution of benefits and risks associated with automation and robotics is a fundamental ethical imperative. While these technologies have the potential to create new opportunities for economic growth and innovation, they also pose risks such as exacerbating income inequality and widening the digital divide. Making sure that the advantages of automation are distributed fairly throughout society is crucial, especially for vulnerable and marginalized groups. This requires proactive measures to address disparities in access to technology, education, and economic opportunities, as well as policies that promote inclusive economic development and social justice.

In addition to these ethical considerations, there are broader societal implications of automation and robotics that merit careful attention. These technologies have the power to reshape the nature of work, the structure of industries, and the dynamics of communities in profound ways. Therefore, it is crucial to have constant communication and consideration. The ethical, social, and cultural dimensions of automation and robotics involve diverse stakeholders from industry, government, academia, and civil society. By fostering a collaborative and inclusive approach to technological innovation, we can strive to attain the greatest possible of automation and robotics for the collective benefit of humanity while upholding ethical principles and safeguarding societal values.

2.5.2.2 Scalability and Robustness of Soft Computing Algorithms

While soft computing algorithms show great promise for enhancing HRC, ensuring their scalability and robustness presents significant obstacles that need to be overcome to realize their full potential in actual applications.

Scalability is the capacity of soft computing algorithms to effectively handle increasingly large and complex datasets, as well as dynamic environments with a growing number of interacting agents or components. In the context of HRC, scalability is essential for accommodating the diverse range of tasks, environments, and interactions that may arise in manufacturing settings. For example, in a collaborative assembly line where multiple robots and humans work together, scalable algorithms are needed to coordinate and optimize task allocation, motion planning, and resource management efficiently. Achieving scalability requires the development of algorithmic techniques that can effectively parallelize computation, distribute processing load, and adaptively adjust to changing system dynamics in real time.

Robustness, on the contrary, pertains to the ability of soft computing algorithms to maintain stable and reliable performance in the face of uncertainties, disturbances, and unforeseen events. In HRC applications, robustness is crucial for ensuring safe and dependable interactions between humans and robots, especially in dynamic and unpredictable environments where conditions may vary unpredictably. For instance, in a warehouse environment where AMRs navigate amid changing obstacles and traffic patterns, robust algorithms are needed to handle uncertainties in perception, localization, and motion planning while maintaining safe and efficient operation. Achieving robustness requires the development of algorithmic techniques that can adaptively respond to uncertainties, recover from errors or failures, and gracefully degrade performance when necessary to avoid catastrophic outcomes.

Addressing the challenges of scalability and robustness in soft computing algorithms for HRC requires a multidisciplinary approach that integrates insights from computer science, control theory, ML, and robotics. Researchers must explore innovative algorithmic techniques such as distributed optimization, reinforcement learning, ensemble methods, and online learning to enhance the scalability and robustness of soft computing algorithms for HRC. Moreover, rigorous testing, validation, and benchmarking methodologies are essential to evaluate the performance of these algorithms in realistic scenarios and identify areas for improvement. By advancing the state-of-the-art in scalable and robust soft computing algorithms, researchers can open the door to HRC in manufacturing and other fields that are safer, more effective, and more flexible.

2.5.2.3 Interpretability of AI Models

Improving AI model interpretability is crucial for HRC because it promotes responsibility, openness, and confidence in autonomous robot decision-making. Interpretability is the capacity to comprehend and justify the reasoning behind the choices made by AI-powered systems, enabling human users to make sense of and anticipate their behavior.

Making sure that people can rely on and trust the decisions made by AI-powered robots is one of the major issues in HRC, particularly in crucial tasks where human safety is at risk. To overcome this difficulty, scientists are investigating a number of methods to improve AI models' interpretability—that is, to make their internal mechanisms more visible and intelligible to human operators.

Using explainable AI (XAI) techniques, which seek to give human-understandable explanations for the decisions made by AI systems, is one way to improve interpretability. Feature importance analysis, decision tree visualization, and model-agnostic explanation techniques like SHAP (SHapley Additive exPlanations) and LIME (Local Interpretable Model-agnostic Explanations) are examples of XAI techniques. These techniques assist users in recognizing any biases or mistakes in AI decision-making and how input features affect model predictions.

In HRC applications, interpretability is crucial for enabling productive cooperation between robots and humans, as it allows humans to understand the reasoning behind robot actions and intervene when necessary to ensure safety and efficiency. For example, in a collaborative assembly line where robots and humans work together to assemble complex products, interpretable AI models can help human operators understand why a robot made a particular decision and provide feedback or corrections if needed.

Furthermore, interpretability is essential for regulatory compliance and ethical considerations in HRC systems. By making AI decision-making processes transparent and accountable, interpretability enables stakeholders to assess the fairness, reliability, and safety of AI-powered robots and ensure compliance with legal and ethical standards.

Overall, enhancing the interpretability of AI models is critical for fostering trust and humans and robots working together in HRC systems. By developing techniques to explain AI decision-making processes in a human-understandable manner, researchers can empower human operators to work more effectively alongside autonomous robots, leading to safer, more efficient, and more productive manufacturing environments.

Addressing these emerging trends and research challenges will be instrumental in advancing the field of HRC and unlocking its full potential to revolutionize manufacturing processes and improve overall productivity and safety.

2.6 Conclusion

In conclusion, the integration of soft computing algorithms into cooperation between humans and robots (HRC) systems signifies a crucial advancement in the landscape of manufacturing operations. Throughout this chapter, we've embarked on a comprehensive exploration of HRC, delving into its diverse applications, grappling with its inherent challenges, and uncovering the vast opportunities it presents across various industrial domains.

The extensive array of research findings and real-world case studies showcased throughout this discourse underscores the transformative potential that HRC, when empowered by soft computing techniques, holds for the manufacturing sector. From collaborative assembly lines facilitating intricate product assembly to AMRs adeptly navigating complex warehouse environments, the tangible benefits of integrating soft computing algorithms are palpable, manifesting in heightened productivity, enhanced safety protocols, and increased operational adaptability across manufacturing verticals.

Nevertheless, amid the promise lies a terrain fraught with challenges. Safety apprehensions, technical intricacies, and ethical dilemmas loom large, necessitating thoughtful consideration and innovative problem-solving strategies. By confronting these obstacles head-on and embracing emerging trends such as human-aware planning and the adoption of explainable AI methodologies, manufacturers stand poised to unlock unprecedented levels of efficiency and competitive advantage in the ever-evolving Industry 4.0 landscape.

Looking toward the horizon, the trajectory of HRC appears boundless. As soft computing methodologies continue to mature and seamlessly integrate with other cutting-edge technologies, the vistas of innovation and optimization within manufacturing stand ripe for exploration. Through fostering collaborative ecosystems where humans and robots, powered by intelligent algorithms, synergistically coexist, manufacturers can chart a course toward a future marked by unparalleled efficiency, heightened productivity, and enduring sustainability.

In essence, the journey toward harnessing the full potential of HRC, underpinned by soft computing algorithms, is an ongoing odyssey. It demands a steadfast commitment to continuous improvement, an unwavering dedication to collaborative endeavors, and a nimble adaptability to emergent technological paradigms. By embracing this journey wholeheartedly, manufacturers can assert their position at the vanguard of Industry 4.0, catalyzing innovation, and shaping the trajectory of manufacturing evolution for generations to come.

References

1. Argall, B. D., Chernova, S., Veloso, M., & Browning, B. A survey of robot learning from demonstration. *Robotics and Autonomous Systems*, 57(5), 469–483. (2009). https://doi.org/10.1016/j.robot.2008.10.024
2. Cherubini, A., Passama, R., Crosnier, A., Lasnier, A., & Fraisse, P. Collaborative manufacturing with physical human-robot interaction. *Robotics and Computer-Integrated Manufacturing*, 40, 1–13. (2016). https://doi.org/10.1016/j.rcim.2015.12.007
3. Darvish, K., Wanderley, F., Mastellone, S., Veneri, M. D., & Testa, D. Fuzzy logic-based navigation and collision avoidance strategy for an autonomous mobile robot. *IFAC-PapersOnLine*, 51(22), 406–411. (2018). https://doi.org/10.1016/j.ifacol.2018.11.409
4. Haddadin, S., Albu-Schäffer, A., & Hirzinger, G. Safety evaluation of physical human-robot interaction via crash-testing. In *Robotics: Science and Systems* (pp. 28–35). The Robotics: Science and Systems Foundation, MA, USA. (2007). https://www.roboticsproceedings.org/rss03/p28.pdf
5. Krüger, J., Lien, T. K., & Verl, A. Cooperation of human and machines in assembly lines. *CIRP Annals*, 58(2), 628–646. (2009). https://doi.org/10.1016/j.cirp.2009.09.009
6. Lasota, P. A., Fong, T., & Shah, J. A. A survey of methods for safe human-robot interaction. *Foundations and Trends in Robotics*, 5(4), 261–349. (2017). https://doi.org/10.1561/2300000052
7. M. Vasic and A. Billard, "Safety issues in human-robot interactions," *2013 IEEE International Conference on Robotics and Automation*, Karlsruhe, Germany, 2013, pp. 197–204, doi: 10.1109/ICRA.2013.6630576.
8. Michalos, G., Makris, S., Spiliotopoulos, J., Misios, I., Tsarouchi, P., & Chryssolouris, G. ROBO-PARTNER: Seamless human-robot cooperation for intelligent, flexible and safe operations in the assembly factories of the future. *Procedia CIRP*, 44, 33–38. (2016). https://doi.org/10.1016/j.procir.2016.02.017
9. Rossi, A., Salvini, P., Carrozza, M. C., Molinari Tosatti, L., Indicom, S., Caneve, R., &Vigiani, V. Human-robot interaction in an industrial environment. In IEEE RO-MAN (pp. 1026–1033). IEEE, NJ, USA. (2013). https://doi.org/10.1109/ROMAN.2013.6628456
10. Unhelkar, V. V., Li, S., & Shah, J. A. (2020). Decision-making for bidirectional communication in sequential human-robot collaborative tasks. *In Proceedings of the 2020 ACM/IEEE International Conference on Human-Robot Interaction* (pp. 329–341). Association for Computing Machinery. https://doi.org/10.1145/3319502.3374779
11. Villani, V., Pini, F., Leali, F., & Secchi, C. Survey on Human-Robot Collaboration in industrial settings: Safety, intuitive interfaces and applications. *Mechatronics*, 55, 248–266. (2018). https://doi.org/10.1016/j.mechatronics.2018.02.009
12. A. Casalino, A. M. Zanchettin, L. Piroddi and P. Rocco, "Optimal Scheduling of Human–Robot Collaborative Assembly Operations With Time Petri Nets," in *IEEE Transactions on Automation Science and Engineering*, vol. 18, no. 1, pp. 70–84, Jan. 2021, doi: 10.1109/TASE.2019.2932150.
13. Bauer, A., Wollherr, D., & Buss, M. Human-Robot Collaboration: A survey. *International Journal of Humanoid Robotics*, 5(01), 47–66. (2008). https://doi.org/10.1142/S0219843608001303

14. Kragic, D., Stolt, A., Bogdan, A., & Wijk, L. Human-Robot Collaboration: A survey. *Robotics and Autonomous Systems*, 144, 103835. (2021). https://doi.org/10.1016/j.robot.2021.103835
15. Tudesco, D. M. R., Deshpande, A., Laghari, A. A., Khan, A. A., Lopes, R. T., Aroma, R. J., Raimond, K., Teng, L., & Khan, A. (2024). Utilization of deep learning models for safe human-friendly computing in cloud, fog, and mobile edge networks. In S. Mahajan, M. Khurana, & V. V. Estrela (Eds.), *Applying artificial intelligence in cybersecurity analytics and cyber threat detection* (Chapter 12). Wiley Online Library. https://doi.org/10.1002/9781394196470.ch12
16. C. Castellini, T. Tommasi, N. Noceti, F. Odone and B. Caputo, "Using Object Affordances to Improve Object Recognition," in *IEEE Transactions on Autonomous Mental Development*, vol. 3, no. 3, pp. 207–215, Sept. 2011, doi: 10.1109/TAMD.2011.2106782.
17. Tsarouchi, P., Makris, S., & Chryssolouris, G. Human-robot interaction review and challenges on task planning and programming. *International Journal of Computer Integrated Manufacturing*, 29(8), 916–931. (2016). https://doi.org/10.1080/0951192X.2015.1130251
18. Anderson, M., & Anderson, S. L. (2007). Machine Ethics: Creating an Ethical Intelligent Agent. *AI Magazine, 28*(4), 15. https://doi.org/10.1609/aimag.v28i4.2065
19. Olsen, D. R., & Goodrich, M. A. Metrics for evaluating human-robot interactions. In Proceedings of PERMIS (Vol. 2003, No. 8). (2003). https://scholarsarchive.byu.edu/cgi/viewcontent.cgi?article=2051&context=facpub
20. Leite, I., Pereira, A., Mascarenhas, S., Martinho, C., Prada, R., & Paiva, A. (2013). The influence of empathy in human–robot relations. *International Journal of Human-Computer Studies, 71*(3), 250–260. https://doi.org/10.1016/j.ijhcs.2012.09.005
21. Almeida, J. S., Huang, C., Nogueira, F. G., Bhatia, S., & de Albuquerque, V. H. C. EdgeFireSmoke: A novel lightweight CNN model for real-time video fire-smoke detection. *IEEE Transactions on Industrial Informatics*. https://doi.org/10.1109/TII.2021.3138752
22. Reddy, C. K. K., Pullannagari, S. R., Doss, S., & Anisha, P. R., Dactology prediction using convolutional neural networks. In *Fostering Cross-Industry Sustainability with Intelligent Technologies* (pp. 61–72), IGI Global, PA, USA. (2024). https://doi.org/10.4018/979-8-3693-1638-2.ch004
23. Abouyoussef, M., Bhatia, S., Chaudhary, P., Sharma, S., & Ismail, M. Blockchain-enabled online diagnostic platform of suspected patients of COVID-19 like pandemics. *IEEE Internet of Things Magazine*, 4(4), 94–99. (December 2021). https://doi.org/10.1109/IOTM.1001.2100046
24. Reddy, P. Y., Reddy, C. K. K., & Sithole, N. T. A study and analysis on diagnosis of melanoma cancer with deep learning: A case study. In *Federated Learning and AI for Healthcare 5.0* (pp. 203–218), IGI Global, PA, USA. 2023. https://doi.org/10.4018/979-8-3693-1082-3.ch011
25. Haddadin, S., Samuel, K., Haninger, K., Oboe, R., & Oh, S. A Perturbation-Robust Framework for Admittance Control of Robotic Systems with High-Stiffness Contacts and Heavy Payload. IEEE Robotics and Automation Letters. (2024).

3

An Overview of Cloud Computing: Paradigm, Technologies, and Application

Gulfishan Mobin, Riya Sil, and Anindya Nag

3.1 Introduction

Recently, cloud computing has become a dominant model for delivering and organizing services over the Internet. Its rise is transforming the information technology sector and unlocking the long-held promise of utility computing. The significant advancement of cloud computing can be attributed primarily to the influence of industrial suppliers, signifying a substantial progression in the implementation of the utility computing model [1]. The appeal of this solution to business owners lies in its ability to eliminate the expenses associated with provisioning plans. Furthermore, this technology allows organizations to initiate operations on a smaller scale and subsequently augment their resources in a flexible manner to accommodate a surge in demand for their services. The primary objective of cloud computing is to offer reliable services by utilizing advanced data centers that are built on virtualized computation and storage technologies. The pay-as-you-go business model enables users to conveniently access data and applications from a cloud infrastructure regardless of their geographical location [2]. Understanding all aspects of this technology is crucial due to the swift progress of cloud computing and its widespread adoption in various domains, including enterprise business and scientific research. The objective of this chapter is to provide readers with a comprehensive comprehension of cloud computing using detailed explanations and comprehensive discussions on all its constituent elements. This chapter provides an analysis of the structure and components of cloud computing, as well as an examination of important terminology from various perspectives and notable characteristics. The discussion also encompassed the impact of cloud computing on businesses and organizations in terms of time, administration, and operational costs [12]. The discourse has encompassed the architectural aspects of cloud computing, along with the prominent technologies that underpin its functionality, such as virtualization and web services. There are various classifications of clouds based on their deployment, namely, private, communal, public, and hybrid clouds [14]. Cloud computing has expanded its range of service models, encompassing platform as a

DOI: 10.1201/9781003503934-3

service, infrastructure as a service, and software as a service [16]. Within this context, the chapter examines the categorization of cloud computing services by providing clear definitions of deployment and service models, as well as the subcategories that fall within each model. The cloud suppliers and providers are also mentioned and described in detail. Determining the optimal pricing strategy for users assumes heightened significance for providers and consumers alike, given the proliferation of companies leveraging cloud infrastructure to deliver computer services. The Service Level Agreement (SLA), which establishes the contractual terms between the service provider and the customer, is becoming increasingly significant [17–19]. The security of data stored in the cloud is garnering significant attention due to its inherent sensitivity. This chapter investigates the pertinent aspects of cloud computing that are of relevance to customers, including SLAs, service costs, service pricing, and security concerns. In practical applications, various supplementary technologies, such as distributed computing, utility computing, and grid computing, have been developed based on the foundation of cloud computing. This chapter provides a comprehensive analysis of the cloud computing paradigm of previous paradigms, focusing on architecture, resource sharing, quality of service (QoS) assurances, and security. Future research should thoroughly examine the challenges that the emerging paradigm of cloud computing encounters, including but not limited to security concerns, availability issues, and resource management complexities. This comprehensive analysis is crucial to guarantee the sustained viability and triumph of cloud computing in the long run. The chapter examines the aforementioned issues and establishes a foundation for further investigation within this particular field of study [33].

The topics covered in this chapter include cloud computing architecture, its defining characteristics, the technology that contributes to cloud computing, various models used for classifying cloud computing, a comprehensive literature review, and an examination of the benefits and limitations associated with cloud computing. This chapter provides an overview of cloud computing methodologies and the procedures employed to enhance and authenticate innovative machine learning (ML) instruments [20–23]. The authors additionally present a comprehensive literature review about the utilization of ML techniques in the context of image-based diagnosis of cardiovascular diseases [32].

The subsequent four sections of the chapter provide further elaboration on the aforementioned themes. Section 3.2 offers an introduction to the core concepts and principles that form the basis of cloud computing. In this section, the essential components and techniques of cloud computing are discussed in detail. Section 3.3, on the contrary, presents a comprehensive review of the existing literature on the subject matter. This section critically examines and synthesizes the relevant scholarly works and research studies that have contributed to the advancement of ML. Section 3.4 categorizing cloud models. Sections 3.5 and 3.6 outline the benefits and constraints of cloud computing. Section 3.7 provides an overview of the security challenges associated with cloud computing. Section 3.8 contains the conclusion and future scope.

3.2 Cloud Computing

The term "cloud computing" has been coined to encompass a range of advanced computing services that are available on demand and originally offered by commercial enterprises. The term "cloud computing" denotes a technological infrastructure that is conceptualized as a cloud-like entity, enabling individuals and organizations to conveniently access applications from any geographical location [3]. The central principle of this model revolves around the delivery of processing, storage, and software in the form of a service. The concept of cloud computing encompasses multiple definitions, albeit with a common focus on a limited number of technological attributes [4].

3.2.1 Cloud Computing Architecture

The architectural framework for cloud computing has been established through the collaborative efforts of various institutions and scholars in the academic community. The system consists of two primary components: the management and the core stack. The core stack consists of three levels. In this context, the terms "resource", "application", and "platform" are being discussed. The resource layer within the infrastructure comprises tangible and virtualized components, including computing, networking resources, and storage. The platform layer, being the most complex component, can be subdivided into multiple sub-layers. One such sub-layer is responsible for managing job scheduling and/or transaction dispatching within a computational framework. The storage sub-layer provides an extensive amount of storage and caching capacity without any limitations. To mitigate the risk of any individual component becoming a bottleneck within the system, the application server and other components have been designed to incorporate the same fundamental application logic as previously employed, employing either on-demand functionality or adaptable administration [17]. The capacity of the application to facilitate large, distributed transactions and manage substantial volumes of data is contingent upon the underlying resources and components. Each layer provides external services through web services or other publicly accessible interfaces.

Figure 3.1 illustrates the depiction of cloud architecture [34].

3.2.2 Characteristics of Cloud Computing

Cloud computing offers several advantages, such as improved scalability and elasticity, along with substantial computational and storage capabilities. Furthermore, the utilization of cloud computing services is increasingly recognized as a more economical and ecologically sustainable approach for the development and implementation of IT services, owing to enhanced efficiency and the realization of economies of scale [15]. There exist several distinctions between cloud computing and other computing paradigms [5]:

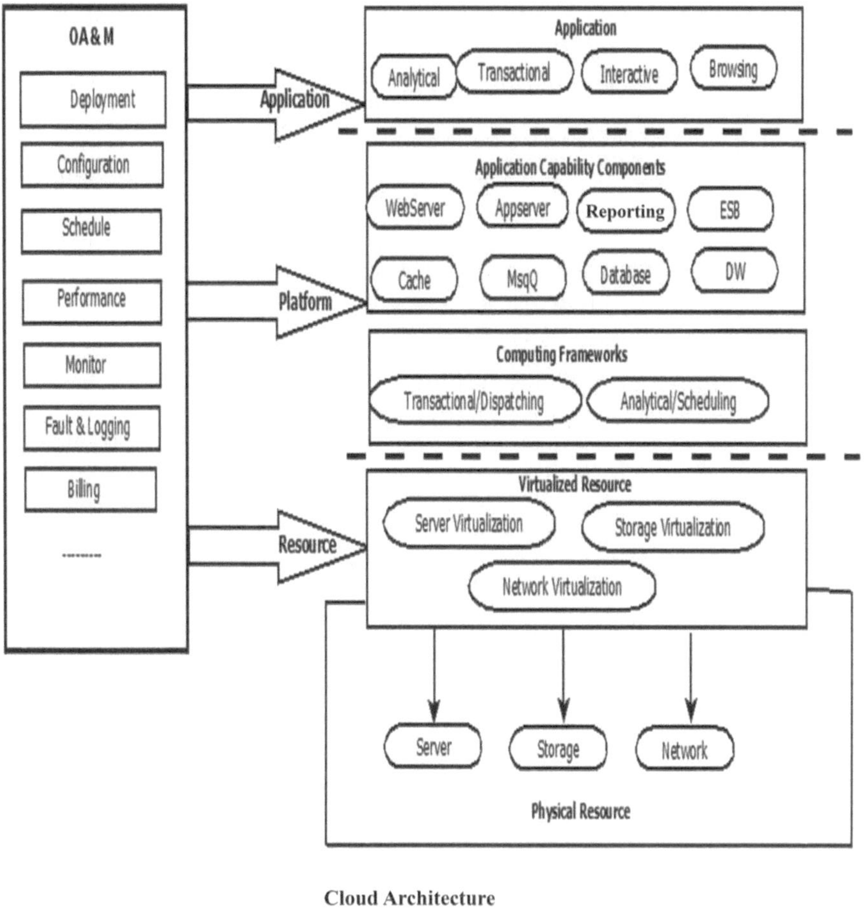

FIGURE 3.1
Cloud architecture.

- **On-Demand Service**: The on-demand service allows users to conveniently access computing resources and services from cloud platforms as and when required. Subsequently, individuals possess the ability to modify and customize their computing environments, exemplified by the installation of software or the configuration of networks, as they commonly possess administrative privileges [24].
- **QoS Guaranteed Offer**: The computing cloud provides users with the option to receive a QoS guarantee for hardware performance, including CPU speed, I/O bandwidth, and memory size, within the computing environments it offers. The cloud computing environment offers QoS in a general sense by managing SLAs with users [25].
- **Autonomous System**: The cloud computing system functions independently and is readily accessible to users. To provide consumers

with a unified platform experience, the components of hardware, software, and data within cloud systems can be subjected to automated modifications, orchestration, and aggregation [35].

- Scalability and flexibility are key attributes that drive the advancement of cloud computing. Cloud computing platforms and cloud services possess the capability to be adjusted in scale to accommodate diverse factors, such as geographic location, hardware performance, and software customizations. The adaptability of computer systems to accommodate diverse user needs within a potentially extensive user base is imperative [27].

3.2.3 Technology Contributing to Cloud Computing

Cloud computing is facilitated by a diverse range of technologies, and a plethora of advanced techniques have been discovered [5].

- **Virtualization technology**: Virtualization technology provides adaptable and scalable computing systems through the partitioning of hardware resources. Virtualized IT infrastructures, facilitated by virtual machine technologies such as VMware2 and Xen3, offer the advantage of being readily accessible upon request. Virtual network innovations, such as the virtual private network, facilitate users to access cloud resources utilizing a customized network environment. The fundamental basis of cloud computing lies in the utilization of virtualization technology, which provides the ability to scale and adapt hardware services as needed [26].
- **Orchestration of service flow and workflow**: The orchestration of service flow and workflow is facilitated by computing clouds, which offer a wide range of service templates that can be assembled by services operating within the cloud environment. Hence, it is imperative for computing clouds to possess the capability to independently coordinate services from diverse origins and of different natures, thereby establishing a seamless and adaptable service flow or process that remains inconspicuous to users [28–29].
- **Web Services and Service-Oriented Architecture (SOA):** Computing cloud services are commonly provided as web services that conform to the standards of Web Services Description Language, Simple Object Access Protocol, and Universal Description, Discovery, and Integration. An SOA has the potential to effectively structure and coordinate the services within cloud environments. Furthermore, the integration of a multitude of cloud services can be incorporated within an SOA application environment, thereby enabling their accessibility via the Internet and facilitating their availability across diverse distributed platforms [31].

- **Web 2.0**: Web 2.0 refers to a nascent technological development encompassing contemporary web design and technology trends, to enhance the functionality, innovation, and exchange of information on the Internet [32].

3.3 Literature of Review

The initial conceptualizations of cloud computing encompass the subsequent delineations: An algorithm is a systematic sequence or a set of instructions designed to solve a problem or achieve a specific task. It is a computational procedure cloud computing refers to a set of network-dependent services that offer readily available computing platforms with scalable capabilities and assured QoS. These platforms are typically personalized, reasonably priced, and easily accessible to a wide range of users [5]. Markus Klems asserts that the cloud relies heavily on instant scalability and resource consumption optimization as fundamental elements. The delivery of desired outcomes is facilitated by the implementation of enhanced monitoring and automation techniques for resource management in a dynamic environment. There exists a divergence of viewpoints among writers concerning the indispensability of a prerequisite for the acknowledgment of a system as a cloud [6]. The article titled "ORGs for Scalable, Robust, Privacy-Friendly Client Cloud Computing" written by Carl Hewitt and published in *IEEE Internet Computing* in 2008 presents the concept of cloud computing as a paradigm in which data is stored on Internet servers for extended periods and temporarily cached on diverse client devices, including desktop entertainment centers, table computers, notebooks, wall computers, handhelds, sensors, monitors, and similar devices [7]. According to a report published by Berkeley in February 2009, the notion of cloud computing, which involves the delivery of computing resources as a utility, has the potential to generate significant transformations in the IT sector and augment the appeal of software as a service [8]. Cloud computing is a technological innovation that operates on the principles of virtualization of computing and storage resources, leveraging the capabilities of the contemporary Internet to provide scalable and economically efficient computing services. The provision of these utilities as on-demand services with adaptable pricing models has contributed to the development of a novel consumer mass market [9].

Clouds can be defined as a large set of virtualized resources, comprising hardware, development platforms, and services, which are conveniently accessible and easily attainable. This definition is based on the perspective of QoS. To optimize the utilization of resources, it is possible to implement dynamic modifications to these resources. The resource pool is frequently utilized in a pay-per-use fashion, with the infrastructure provider offering assurances via tailored SLAs [10]. Utility computing is a business model that is distinguished by its pay-per-use approach, as indicated by the definition provided. Several authors place great importance on the business model, particularly highlighting collaboration and pay-as-you-go, as well as the reduction in capital expenditure that arises from the adoption of utility computing. Until a recent point in time, there existed a prevalent misperception that frequently led to the conflation of the concept in question with the cloud. Nevertheless, there is a growing consensus that it is an integral component of the cloud, intricately interconnected with the foundational business framework. The notion of user-friendliness holds significant importance in the domain of cloud computing [6]. Buyya et al. (2011) assert that the enhancement of SLAs between service providers and their customers is crucial for achieving broad commercial adoption [36]. The data center can be understood as a large assemblage of clusters, which plays a crucial role as the foundational component of the cloud [6]. It enables the efficient utilization of surplus resources to effectively enhance processing power and storage capacity. The criticality of virtualization in cloud computing is emphasized by its acknowledgment as a pivotal element [11]. Furthermore, cloud computing can be defined as the amalgamation of virtualized hardware and software, in conjunction with preexisting monitoring and provisioning methodologies [6]. The initial definition [13] provided by the U.S. National Institute of Standards and Technology appears to have effectively integrated the widely recognized characteristics of cloud computing.

3.4 Classifying Cloud Models

In this section, the authors categorize cloud models.

3.4.1 Deployment Models

The cloud model advocates for the adoption of four distinct deployment models.

- **Private Cloud**: A private cloud infrastructure is operated by a company. The existence of the entity in question may manifest either within or outside the physical boundaries of the organization, and its management may be undertaken by either the organization itself or a third-party entity.
- **Community Cloud**: The community cloud is a form of cloud infrastructure that caters to a specific community characterized by

common concerns, including mission, security requirements, compliance issues, and policy. This resource is utilized by multiple organizations. The existence of a system, whether on or off premises, can be subject to control by either the organization itself or a third party.

- **Public Cloud**: The public cloud denotes a cloud infrastructure owned and managed by a cloud service provider, accessible to either the general public or a large industry segment.
- **Hybrid Cloud**: The hybrid cloud is a cloud infrastructure that encompasses multiple distinct clouds, including private, community, or public clouds. These clouds are interconnected using standardized or proprietary technology, enabling the seamless transfer of data and applications. These clouds connect through either proprietary or standardized technology, facilitating smooth data and application transfers. This connectivity facilitates tasks such as load balancing between clouds through a process known as cloud bursting.

The following benefits apply to both public and private clouds:

- **High Utilization**: The utilization rate is high as a result of resource pooling, facilitating improved task distribution across multiple applications.
- **Great Efficiency**: The efficiency of both public and private clouds is underscored by their utilization of grid computing and virtualization technologies.
- **High Availability**: Another advantage of using grid computing is that applications can benefit from an architecture with high availability that reduces or eliminates scheduled and unexpected downtime, hence enhancing user experience and ensuring business continuity.
- **Elastic Scalability**: Grid computing also offers elastic scalability to public and private clouds, allowing them to add and subtract computing resources as needed. This is a considerable advantage for applications with uncertain growth or highly fluctuating workloads, as well as for transitory applications.
- **Fast Deployment**: It is achievable by the accessibility of both private and public clouds, offering self-service access to a shared pool of computing resources. It utilizes standardized and reusable hardware and software components. The deployment of applications is significantly accelerated by this.

Specific to public cloud computing are the following advantages:

- **Minimal Initial Costs**: Public clouds are quicker to launch, lowering the entry barrier for customers by eliminating the need to purchase, install, and configure hardware.

- **Scale Economies**: Purchasing power for equipment and managerial efficiency are increased for large public clouds. Customers may occasionally receive some of the savings.
- It is simpler to administer because public clouds don't need IT to maintain, administer, update, patch, etc. Users rely less on an IT department and more on the public cloud service provider.

3.4.2 Service Models

The cloud model encompasses support for four discrete deployment models.

- **Private Cloud**: The management of the cloud infrastructure is limited to a particular organization. The administration of the system can be conducted either by the organization itself or by an external entity, and it can be situated either on-site at the organization's premises or at an off-site location.
- **Community Cloud**: Cloud infrastructure usage is prevalent across diverse organizations that share mutual concerns such as mission goals, security needs, regulatory considerations, and policy compliance. The administration of the system can be carried out by either the entities themselves or an external entity, and it can be situated either within the physical facilities or outside of them.
- **Public Cloud**: The cloud infrastructure refers to a system that is made available to a broad audience or a significant industry consortium. It is owned by an entity that provides cloud services.
- **Hybrid Cloud**: The term "cloud infrastructure" pertains to a setup in which multiple clouds, such as private, community, or public clouds, are interconnected using either standardized or proprietary technology. The interconnection between different clouds enables the efficient transfer of data and applications, resulting in improved portability and the capability to distribute workloads across multiple clouds, commonly referred to as cloud bursting.

3.5 Advantages of Cloud Computing

One of the advantages of cloud computing is its ability to offer significant computing and storage capabilities, as well as improved scalability and elasticity. Furthermore, due to enhanced efficiency and the realization of economies of scale, cloud computing services are evolving into a significantly more cost-efficient and environmentally sustainable approach for the development and implementation of IT services. Cloud computing offers several advantages.

- **Computers at User-Friendly Prices**: Users don't need to invest in pricey PCs with more RAM and storage space. Users can transition to smaller, more portable notebooks from traditional PCs and laptops.
- It lowers costs and boosts the IT infrastructure's effectiveness. A typical company's traditional server is loaded between 10% and 15%. Additional computing resources are required at times, while they are inactive at other times.
- **Less Issues with the Service**: As a result of the development of cloud computing, physical servers are getting smaller, quicker, and simpler to sustain. Regarding software, it's configured, updated, and installed within the cloud.
- Cloud-based software is more cost-effective than purchasing individual software licenses for each local user. Only those who actually require these programs will utilize them. Additionally, programs created to access the Internet are far less expensive than their equivalents for personal computers (PCs). The programs might simply be rented by the hour if they are not frequently used.
- **Program Updates Are Ongoing**: One can always be certain that a remote program they are using has the most recent version without having to reload anything or pay an upgrade price.
- **The Rise in Computing Power Available**: A cloud of computers' computational capacity is essentially only constrained by its size, or the total number of remote servers, as opposed to the computing power that a user can access on a PC. If necessary, users can execute more complicated operations using a lot of memory and storage.
- Unlimited amount of data can be stored. When compared to the easily accessible location for information storage on a PC, cloud storage can automatically and flexibly adapt to a user's needs. Users who save information in the cloud can forget about the restrictions imposed by traditional discs because the cloud has free space accessible in the billions of gigabytes.
- It works with the majority of operating systems (OSs). OSs do not matter in cloud computing. Users of Microsoft Windows and Unix can freely exchange documents with each other. Programs and virtual machines can be accessed through a web browser or alternative access methods by any PC equipped with any OS.
- The enhancement of document format compatibility has been achieved. In contrast to individuals who encounter difficulties in accessing Word 2007 documents on their machines running Word 2003 or OpenOffice, users who rely on a singular cloud-based

program for document creation and editing do not face issues related to incompatible versions and formats. Google Docs, an office suite that facilitates collaborative work on documents, presentations, and spreadsheets through web browsers on any computer, serves as a prominent illustration of interoperability.

- Working with a group of users is more convenient. It is not essential to transmit to each person their version of the document or change it in order when using cloud-based document storage. Users can be sure that the document they are using is the most recent version since any modifications made by one user immediately affect the changes made by all other users.
- The concept under consideration pertains to the efficient utilization of economic resources. Saving money on electricity, computing resources, and server space is achievable thanks to cloud computing. In this situation, users may have access to the documents from any location at any time if they are saved in the cloud.
- One of the key benefits of implementing security measures is the safeguarding of equipment and data from theft. Data that is kept in the cloud is automatically spread among several servers, some of which may be found on several continents. Users don't lose important information if a PC is stolen or damaged because they may still access it from another computer.

3.6 Limitations of Cloud Computing

Cloud computing has several drawbacks as well:

- **Permanently Connected to the Internet**: Internet access is a requirement for cloud computing at all times. There are no jobs, programs, or materials if Internet access is unavailable. The strongest argument against cloud computing is most likely this one. But let's be honest, can a person in the present world live without services like those offered by the Internet, mobile phones, credit cards, etc.? Many people are unable to survive even one day without email. The Internet will thus be accessible at all times and from any location, just like power and water, given the development of the modern world.
- **Working with a Slow Internet Connection Is Challenging**: Many cloud-based applications need a strong Internet connection with lots of bandwidth.
- **Compared to a Local Computer, Programs Could Execute More Slowly**: On a computer, some programs that require the transfer of a substantial amount of data will operate more quickly due to factors

including server congestion, issues with the user's connection to the cloud, and the speed restriction of Internet access.

- Not all programs or their constituent elements are readily accessible through online platforms. The efficacy of cloud-based programs remains substandard in comparison to those specifically developed for local utilization. For example, Google Docs spreadsheet offers a more limited range of features and functions compared to Microsoft Excel.
- **The High Cost of the Machinery**: A significant financial investment is required to establish a cloud infrastructure.
- **Software**: The user cannot alter an application to suit their needs and only has access to cloud-based software.
- **Security and Privacy**: Although the cloud is a very stable system, there is now no technology that can guarantee the complete secrecy of data; in the event of a hacker assault, a vast amount of data will be made available to those conducting crimes.

3.7 Cloud Computing: Security Challenges

Utilizing cloud technology has various benefits, including pay-per-use, great availability, high performance, minimal upfront investment, and superior fault tolerance. On the other side, there are numerous drawbacks or risks associated with the cloud. Each user in the cloud does not know who else is working on the same server because they are all sharing the same resources [39]. The cloud is typically situated outside of an organization's or business' firewall. This will have a big impact on whether or not a company decides to migrate to the cloud [40]. When a client intends to move their work to the cloud, they must be informed of the difficulties that may arise. Some of these difficulties are listed below [41–43]:

- **Privileged User Access**: The risk of data leakage is heightened when a client's data is accessed externally and the client is unable to acquire a new membership for verification purposes.
- **Data Location**: When a user stores their data in the cloud, they may not be able to determine where the data is saved or from where it is hosted.
- **Investigative Support**: It would be impossible to conduct a thorough investigation into any illegal or improper action that uses cloud computing and customer data.
- **Data Segregation**: The client's data is present in the same infrastructure as the data of other clients utilizing the services concurrently.
- **Recovery**: If a data center or server is destroyed due to a natural disaster or other issue, the cloud service provider will notify each client of the situation with their data.

- **Availability**: Although many users are eager to employ cloud services, the organization is not always reachable.
- **Regulatory Compliance**: The cloud service provider consistently denies external audit requests and declines to implement new security certificates on the network.

3.8 Conclusion and Future Scope

This chapter presents a comprehensive definition of the different layers of cloud architecture and the technology associated with cloud computing. This chapter also incorporates a concise examination of cloud service and deployment models. This study provides a thorough examination of various service models in cloud computing, specifically focusing on the deployment models of private, public, and community clouds. Additionally, it examines the information security requirements specific to public and private clouds, taking into account the diverse service models. However, despite the numerous benefits associated with cloud computing, such as the ability to access services on demand, widespread network availability, efficient resource allocation, scalability, and the ability to monitor usage, it is important to acknowledge that it is not without its limitations and should not be regarded as flawless. The present chapter has comprehensively examined the inherent challenges and issues associated with cloud computing, specifically the security concerns pertinent to future research endeavors.

References

1. Al-Badi, A., & Khan, A. I. (2022). A sustainable development neural network model for Big Data in smart cities. *Procedia Computer Science*, 202, 408–413. https://doi.org/10.1016/j.procs.2022.04.057
2. Bhowmik, T., Bhadwaj, A., Kumar, A., & Bhushan, B. (2022). Machine learning and deep learning models for privacy management and data analysis in smart cites. In: Balas V. E., Solanki V. K., Kumar R. (eds) *Recent Advances in Internet of Things and Machine Learning*. Intelligent Systems Reference Library, vol. 215. Springer, Cham. https://doi.org/10.1007/978-3-030-90119-6_13
3. Rawat, P., Bajaj, M., Mehta, S., Sharma, V., & Vats, S. (2023). A study on cervical cancer prediction using various machine learning approaches. In: *2023 International Conference on Innovative Data Communication Technologies and Application (ICIDCA)* (pp. 1101–1107). IEEE. https://doi.org/10.1109/icidca56705.2023.10099493. https://doi.org/10.1109/icidca56705.2023.10099493

4. Varshney, M., Bhushan, B., Haque, & A. K. M. B. (2022). Big Data analytics and data mining for Healthcare Informatics (HCI). In: Kumar R., Sharma R., Pattnaik P. K. (eds) *Multimedia Technologies in the Internet of Things Environment, Volume 3. Studies in Big Data,* vol. 108. Springer, Singapore. https://doi.org/10.1007/978-981-19-0924-5_11
5. Khadka, S. R., Subedi, S., Aidy, B. K., Regmi, L. N., Parajuli, A., & Bhandari, M. (2023). Predicting chronic kidney disease using ML algorithms and XAIV. *Medicon Engineering Themes.* https://doi.org/10.55162/mcet.04.142
6. Pandey, A., Gupta, M. P., Diwakar, M., Dangi, S., Madan, P., & Singh, P. (2023). An effective machine learning based heart disease diagnosis analysis. In: *2023 International Conference on Computational Intelligence, Communication Technology and Networking (CICTN)* (pp. 536–540). IEEE. https://doi.org/10.1109/cictn57981.2023.10141082
7. Sharma, N. (2023). Multichannel EEG pattern acknowledgment feature assortment using T-test ranking and principal component analysis. In: *2023 1st International Conference on Innovations in High Speed Communication and Signal Processing (IHCSP)* (pp. 188–193). IEEE. https://doi.org/10.1109/ihcsp56702.2023.10127161
8. Chae, M., & Lee, H. (2023). A prediction of in-hospital cardiac arrest risk scoring based on machine learning. *International Journal on Advanced Science, Engineering and Information Technology,* 13(3), 895. https://doi.org/10.18517/ijaseit.13.3.17343
9. Grigory Panasenko (Vilnius University, Lithuania) (2022). Dimension reduction and multiscale modeling for thin structures. In: *Introduction to Multiscale Mathematical Modeling,* pp. 101–123. World Scientific, Tuck Link, Singapore. https://doi.org/10.1142/9781800612327_0004
10. Chen, C., Qin, C., Qiu, H., Tarroni, G., Duan, J., Bai, W., & Rueckert, D. (2020). Deep learning for cardiac image segmentation: A review. *Frontiers in Cardiovascular Medicine,* 7, p. 25. https://doi.org/10.3389/fcvm.2020.00025
11. Kanani, P., & Padole, M. C. (2021). ECG image classification using deep learning approach. In Geeta Rani, Pradeep Kumar Tiwari (Eds.), *Handbook of Research on Disease Prediction through Data Analytics and Machine Learning*, pp. 343–357. IEEE. Pennsylvania, USA. https://doi.org/10.4018/978-1-7998-2742-9.ch016
12. Sidey-Gibbons, J. A., & Sidey-Gibbons, C. J. (2019). Machine learning in medicine: A practical introduction. *BMC Medical Research Methodology,* 19(1), pp. 1–18. https://doi.org/10.1186/s12874-019-0681-4
13. Zhao, Q., Liu, Z., Adeli, E., & Pohl, K. M. (2021). Longitudinal self-supervised learning. *Medical Image Analysis,* 71, 102051. https://doi.org/10.1016/j.media.2021.102051
14. Song, X., Liu, X., Liu, F., & Wang, C. (2021). Comparison of machine learning and logistic regression models in predicting acute kidney injury: A systematic review and meta-analysis. *International Journal of Medical Informatics,* 151, 104484. https://doi.org/10.1016/j.ijmedinf.2021.104484
15. Habehh, H., & Gohel, S. (2021). Machine learning in healthcare. *Current Genomics,* 22(4), 291–300. https://doi.org/10.2174/1389202922666210705124359
16. Waring, J., Lindvall, C., & Umeton, R. (2020). Automated machine learning: Review of the state-of-the-art and opportunities for healthcare. *Artificial Intelligence in Medicine,* 104, 101822. https://doi.org/10.1016/j.artmed.2020.101822

17. Azad, C., Bhushan, B., Sharma, R., Shankar, A., Singh, K. K., & Khamparia, A. (2021). Prediction model using SMOTE, genetic algorithm and decision tree (PMSGD) for classification of diabetes mellitus. *Multimedia Systems*. https://doi.org/10.1007/s00530-021-00817-2
18. Khamparia, A., Singh, P. K., Rani, P., Samanta, D., Khanna, A., & Bhushan, B. (2020). An internet of health things-driven deep learning framework for detection and classification of skin cancer using transfer learning. *Transactions on Emerging Telecommunications Technologies*. https://doi.org/10.1002/ett.3963
19. Shakeel, P.M., Tobely, T.E.E., Al-Feel, H., Manogaran, G. and Baskar, S. (2023). Neural networks referees in 2022. *Neural Networks*, 157, xii–xl. https://doi.org/10.1016/s0893-6080(22)00489-0
20. Qamar, U., & Raza, M. S. (2023). Frequent pattern mining. *Data Science Concepts and Techniques with Applications*, 271–311. https://doi.org/10.1007/978-3-031-17442-1_9
21. Sarker, I. H., Kayes, A. S., Badsha, S., Alqahtani, H., Watters, P., & Ng, A. (2020). Cybersecurity data science: An overview from machine learning perspective. *Journal of Big Data*, 7(1), 1–29. https://doi.org/10.1186/s40537-020-00318-5
22. Chopra, D., & Khurana, R. (2023). Introduction to machine learning. In: *Introduction to Machine Learning with Python*, pp. 15–29. Bentham Science, Sharjah, United Arab Emirates. https://doi.org/10.2174/9789815124422123010004
23. A. Nag et al., "A Review of Machine Learning Methods for IoT Network-Centric Anomaly Detection," *2024 47th International Conference on Telecommunications and Signal Processing (TSP)*, Prague, Czech Republic, 2024, pp. 26–31, doi: 10.1109/TSP63128.2024.10605928.
24. Abubaker, M. B., & Babayiğit, B. (2023). Detection of cardiovascular diseases in ECG images using machine learning and deep learning methods. *IEEE Transactions on Artificial Intelligence*, 4(2), 373–382. https://doi.org/10.1109/tai.2022.3159505
25. Mhamdi, L., Dammak, O., Cottin, F., & Dhaou, I. B. (2022). Artificial Intelligence for cardiac diseases diagnosis and prediction using ECG images on embedded systems. *Biomedicines*, 10(8), 2013. https://doi.org/10.3390/biomedicines10082013
26. Konstantonis, G., Singh, K. V., Sfikakis, P. P., Jamthikar, A. D., Kitas, G. D., Gupta, S. K., Saba, L., Verrou, K., Khanna, N. N., Ruzsa, Z., Sharma, A. M., Laird, J. R., Johri, A. M., Kalra, M., Protogerou, A., & Suri, J. S. (2022). Cardiovascular disease detection using machine learning and carotid/femoral arterial imaging frameworks in rheumatoid arthritis patients. *Rheumatology International*, 42(2), 215–239. https://doi.org/10.1007/s00296-021-05062-4
27. Khan, A. H., Hussain, M., & Malik, M. K. (2021). Cardiac disorder classification by electrocardiogram sensing using deep neural network. *Complexity*, 2021, 1–8. https://doi.org/10.1155/2021/5512243
28. Apostolopoulos, I. D., Apostolopoulos, D. I., Spyridonidis, T. I., Papathanasiou, N. D., & Panayiotakis, G. S. (2021). Multi-input deep learning approach for cardiovascular disease diagnosis using myocardial perfusion imaging and clinical data. *Physica Medica*, 84, 168–177. https://doi.org/10.1016/j.ejmp.2021.04.011
29. Sharifrazi, D., Alizadehsani, R., Hoseini Izadi, N., Roshanzamir, M., Shoeibi, A., Khozeimeh, F., Alizadeh Sani, F., Alizadeh Sani, Z., Harlapur, C., Khosravi, A., Nahavandi, S., Sarrafzadegan, N., & Islam, S. M. (2021). Hypertrophic cardiomyopathy diagnosis based on cardiovascular magnetic resonance using deep learning techniques. *SSRN Electronic Journal*. https://doi.org/10.2139/ssrn.3855445

30. Rath, A., Mishra, D., Panda, G., & Satapathy, S. C. (2021). Heart disease detection using deep learning methods from imbalanced ECG samples. *Biomedical Signal Processing and Control*, 68, 102820. https://doi.org/10.1016/j.bspc.2021.102820
31. Moreno, A., Rodriguez, J., & Martinez, F. (2019). Regional multiscale motion representation for cardiac disease prediction. In: *2019 XXII Symposium on Image, Signal Processing and Artificial Vision (STSIVA)*. (pp. 1–5). Piscataway, NJ. IEEE https://doi.org/10.1109/stsiva.2019.8730231
32. Mashat, A., Bhatia, S., Kumar, A., Dadheech, P., & Alabdali, A. (2022). Medical image transmission using novel crypto-compression scheme. *Intelligent Automation and Soft Computing*, 32(2), 841–857.
33. M. A. Islam, A. Nag, S. Chowdhury, S. F. A. Fahim, A. Ghosh and N. Mumtaj, "Utilization of Encoding, Early Stopping, Hyper Parameter Tuning, and Machine Learning Models for Bank Fraud Detection," 2023 IEEE 9th International Women in Engineering (WIE) Conference on Electrical and Computer Engineering (WIECON-ECE), Thiruvananthapuram, India, 2023, pp. 321–327, doi: 10.1109/WIECON-ECE60392.2023.10456503.
34. Das M, Nag A, Hassan MM, et al. Synergy of 6G technology and IoT networks for transformative applications. Int J Commun Syst. 2024; 37(14):e5869. doi:10.1002/dac.5869
35. Alghazzawi, D., Bamasaq, O., Bhatia, S., Kumar, A., Dadheech, P., & Albeshri, A. (2021). Congestion control in cognitive IoT-based WSN network for smart agriculture. *IEEE Access*, 9, 151401–151420.
36. Buyya, R., Garg, S. K., & Calheiros, R. N. (2011). SLA-oriented resource provisioning for cloud computing: Challenges, architecture, and solutions. In *2011 international conference on cloud and service computing* (pp. 1–10). IEEE.
37. Alouffi, B., Hasnain, M., Alharbi, A., Alosaimi, W., Alyami, H. and Ayaz, M. (2021). A systematic literature review on cloud computing security: threats and mitigation strategies. *IEEE Access*, 9, 57792–57807.
38. Basu, S., Bardhan, A., Gupta, K., Saha, P., Pal, M., Bose, M., Basu, K., Chaudhury, S. and Sarkar, P., (2018), January. Cloud computing security challenges & solutions-A survey. In *2018 IEEE 8th Annual Computing and Communication Workshop and Conference (CCWC)* (pp. 347–356). IEEE.
39. Nag, A., Mobin, G., Kar, A., Bera, T., & Chandra, P. (2022, December). A review on cloud-based smart applications. In *International Conference on Intelligent Systems Design and Applications* (pp. 387–403). Cham: Springer Nature Switzerland.
40. Gholami, M. F., Daneshgar, F., Low, G., & Beydoun, G. (2016). Cloud migration process—A survey, evaluation framework, and open challenges. *Journal of Systems and Software*, 120, pp. 31–69.
41. Basu, S., Bardhan, A., Gupta, K., Saha, P., Pal, M., Bose, M., Basu, K., Chaudhury, S. and Sarkar, P. (2018, January). Cloud computing security challenges & solutions-A survey. In *2018 IEEE 8th Annual Computing and Communication Workshop and Conference (CCWC)* (pp. 347–356). IEEE.
42. Akbar, H., Zubair, M. and Malik, M.S. (2023). The security issues and challenges in cloud computing. *International Journal for Electronic Crime Investigation*, 7(1), pp.13–32.
43. Abba Ari, A. A., Ngangmo, O. K., Titouna, C., Thiare, O., Mohamadou, A., & Gueroui, A. M. (2024). Enabling privacy and security in Cloud of Things: Architecture, applications, security & privacy challenges. *Applied Computing and Informatics*, 20(1/2), 119–141.

4

Autism Spectrum Disorder Classification in Children Using Deep Learning Approaches in Healthcare 6.0

P. Yashashwini Reddy, C Kishor Kumar Reddy, Kari Lippert, and P. Sahithi Reddy

4.1 Introduction

Autism spectrum disorder (ASD) is a syndrome in which the socializing capacity, the way of grasping the environment, and expressing feelings will be unusual for an individual [1]. The key factors for the development of autism revealed from research are genotypes like tuberous sclerosis complex, fragile X syndrome, a few environmental conditions like exposure to toxic air, and antenatal factors, but the cause of autism is explicitly not yet known. This disorder develops within the initial two years of life and lasts for the rest of life [2]. The children who suffer from autism are vulnerable to the surroundings they live in. They feel discomfort when they hear loud disturbances, have different priorities of food they eat, are sensitive to colors, and feel exhausted when they are surrounded by a mass number of people.

Behaviors like having different passions, hyperacusis, manifesting repetitive behaviors like fluttering hands or using words repeatedly, using things unusually, delay in speaking by 18 months, poor eye contact, being irresponsive when called by the name of a child at 12 months, answering inappropriately when questions asked, understanding others' perspectives differently, difficulty to adjust and adapt for changes in the environment and routines, not sharing feelings and problems with others are usual symptoms of ASD [3]. They also have outstanding qualities like analyzing things in depth, and their potential of having utmost focus on a specific thing helps them to crack logic.

Some individuals show symptoms at an early age, and some show symptoms later in their lives. Hence, this disorder can be identified at any age. The symptoms and severity will not be the same for everyone; it will vary from person to person. Hence, it is termed a spectrum disorder. Over the past two decades, there has been a rapid increase in the autism rate in the world and it

 DOI: 10.1201/9781003503934-4

is a public health concern that should be treated with care [4]. We can observe the five leading countries of autism rates in 2023 in Table 4.1 and Figure 4.1.

It is analyzed that mostly boys are diagnosed with autism than girls, as we can observe in Figure 4.2, as the mutation in genes and the changes in hormones will be different in males and females [5]. Identifying and treating autism early can minimize the problem to some level. There are situations where autistic people are wrongly diagnosed and receive inappropriate treatment as this disorder is often confused with cognitive impairment and other mental-related disorders [6]. Table 4.2 portrays the consequences if autistic children are neglected without providing treatment. It is also observed that other mental disorders will develop with autism as we can observe in Table 4.3.

TABLE 4.1
Top Five Countries with Highest Autism Rate in 2023

S. No	Country	Autism Rate per One Lakh
1	United Kingdom	700.08
2	Sweden	661.86
3	Japan	604.73
4	US	603.39
5	Netherlands	591.55

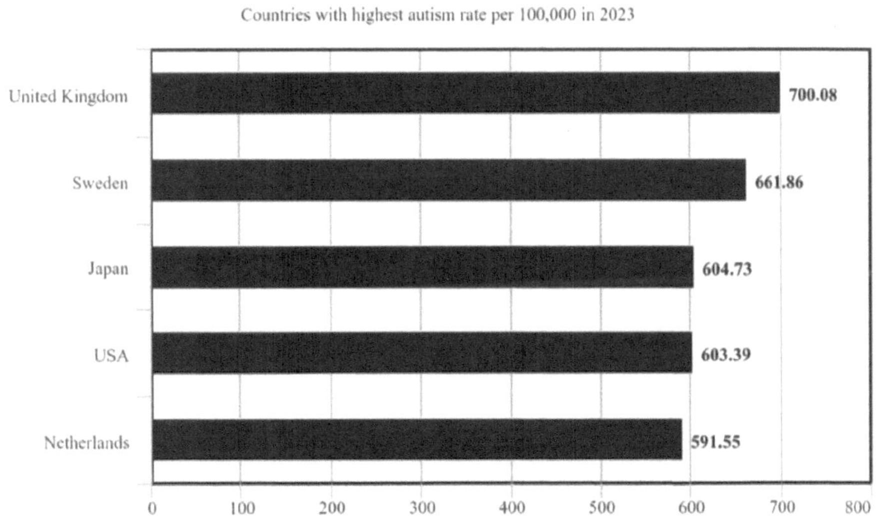

FIGURE 4.1
Countries with highest autism-diagnosed rate per 100,000 in 2023.

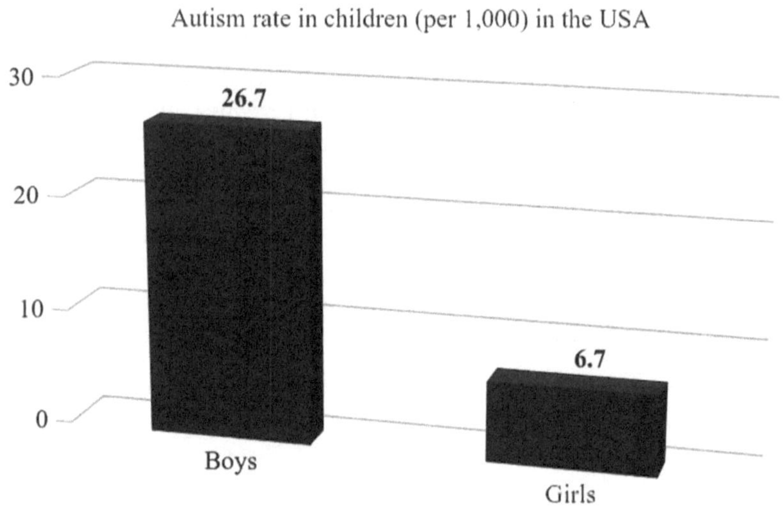

FIGURE 4.2
Children's autism rate in the US (2017).

TABLE 4.2
Consequences If Autism Spectrum Disorder in Children Are Not Treated

S. No	Issue
1	Poor academic achievement
2	Poor relationships with friends and family
3	Difficulty to manage emotions
4	Depend on others to do basic tasks
5	Feel less comfort to speak about their feelings
6	Poor self-care due to poor reaction to sensory stimuli
7	Disturbance in sleep cycle and suffer with eating disorder
8	Become under confident
9	Being unemployed as adults
10	Learning a language would be challenging

TABLE 4.3
Some Additional Mental Disorders That Accompany Autism

S. No	Type of Mental Disorder	Prevalence of the Disorder (%)
1	Panic disorder	19
2	Manic depression	6
3	OCD	10
4	Altered perception syndrome	5
5	Depression	12
6	ADHD	29

There exist three stages of autism in children represented as ASD stage 1, ASD stage 2, and ASD stage 3 that indicate the severity of autism [7]. The severity of autism is higher in stage 3 than in stage 2. Stage 1 indicates less severity of autism. To understand the severity of autism in a patient has a crucial role in treating them systematically. Each stage of autism in children has some symptoms as we can observe in Figure 4.3 but are not constant. Some variations can be observed in the behavior of a child that depends on the environment they grow up in and the help they get from parents [8]. Hence, personalized treatment plays an active role in detecting and treating autism in children.

There are a few modern countries that have great awareness of autism disorder. These countries have the least cases of autism as we can observe in Table 4.4 and Figure 4.4. These countries are involved in advanced research about this disorder and make early detection a priority to overcome this health condition and enhance mental health. They provide personalized observation to diagnose this condition, and after that enhanced treatment facilities are provided to autistic individuals that help them lead an individualistic life. They also raise funds that support autism research. They conduct

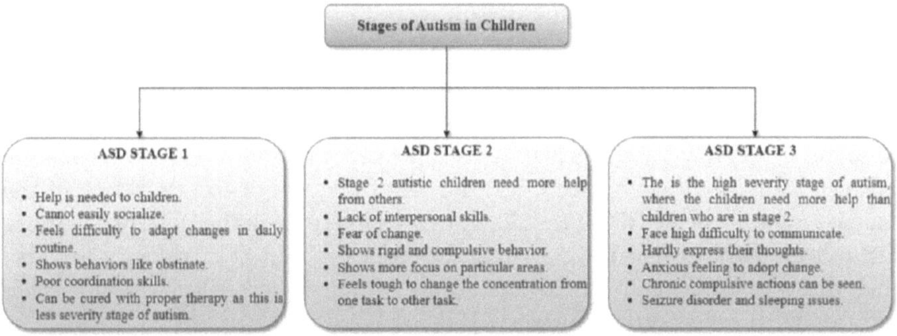

FIGURE 4.3
Stages of autism in children.

TABLE 4.4
Countries with Least Cases of Autism in Children (2023)

S. No	Country	Number of People Diagnosed per 10,000
1	France	69.4
2	Portugal	70.6
3	Iceland	80
4	Norway	72.0
5	Italy	72.1
6	Germany	72.3
7	Greece	72.5

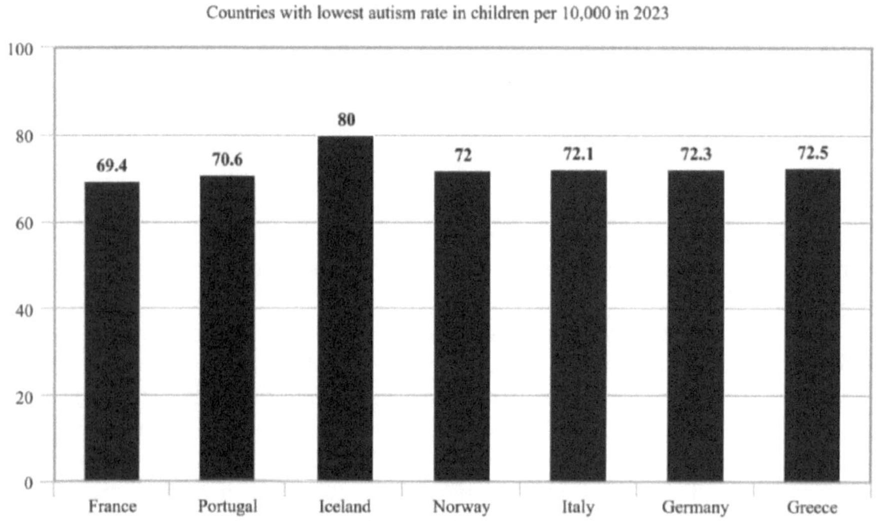

FIGURE 4.4
Countries with the least autism cases in children (2023).

many awareness events to make the public aware of this health condition. These initiatives help them minimize the rise in autism cases.

If an individual with less severity of autism does not receive systematic medical care, there is a high chance that the severity will increase [9]. There are some interview-based procedures followed by healthcare professionals to diagnose autism in children, but the identification of the presence of the disorder is not precise [10]. It results in delays in treating the patients as these methods are time-consuming. As this disorder is difficult to identify because it has many variations in symptoms, there is a need to replace conventional methods with advanced technology [11].

This research aims to implement a deep learning method convolutional neural network (CNN) to effectively detect ASD in children [12]. Also, it studies how artificial intelligence (AI) and machine learning (ML) techniques are applied in detecting ASD in children. These methods perform classification by learning from large, labeled datasets. The rest of the sections are sequenced as follows: Section 4.2 discusses relevant work and research aims; Section 4.3 explains the proposed methodology; Section 4.4 gives the result of the proposed model; and Section 4.5 provides conclusion, followed by references.

4.2 Relevant Work

Doctors use conventional methods to identify autism at the initial level. Some conventional methods are autism spectrum quotient (AQ), childhood autism rating scale (CARS-2), autism diagnostic interview revised (ADI-R), and autism observation schedule revised (ADOS) [13]. The AQ, which was proposed by Baron-Cohen, is an analysis test that has 50 questions, wherein patients or caretakers of a child should answer the questions about the acuteness on a scale ranging from 1 to 4, which represents false to true, respectively. The diagnosis of autism that is made using this method is not accurate. There are chances of getting false results [14].

The questions asked in the AQ test are related to socializing ability, creative ability, concentration levels, and multitasking ability. It is helpful to test all age groups [15]. CARS-2 is the process of observing a child's behavior and considering feedback from caretakers of the child about their behavior. This test is performed by healthcare professionals. The acuteness of autism is represented on a scale based on the examination [16]. In this method, 15 questions are present that are related to the interpersonal skills, interaction ability, and emotional intelligence of the child.

The ADI-R, introduced by Michael Rutter, determines whether a person is autistic or not; in this process, a doctor asks a few sets of questions to the caretaker of a child. Using this method, we can know the presence or absence of autism and the severity stage of autism [17]. Hence, it is an efficient method. It covers questions related to the speaking and language usage ability of a child, the ability of a child to interact, and about repeated actions [18].

ADOS is the process where the doctors frame an environment to connect and communicate with patients to identify autism. This test is performed under the keen observation of healthcare experts. The patient will be under the observation of a doctor where the patient's behavior, interaction ability, and way of communicating are observed. The examination is conducted in a regulated environment [19]. As these procedures are applied, there are some challenges that arise using these methods, as we can observe in Table 4.5, that results in inappropriate prediction of the presence of autism in children.

By using autism characteristics, it would be easy to identify the presence of autism in an individual. The wearable gadgets are implemented by

TABLE 4.5

Setbacks of Conventional Methods to Detect Autism Spectrum Disorder

S. No	Method	Setbacks
1	Autism spectrum quotient	Error of commission and error of omission
2	Childhood autism rating scale	Imprecise feedback to doctors by caretakers
3	Autism diagnostic interview revised	Time-consuming process, which needs keen examination by doctors
4	Autism observation schedule revised	Inaccuracy in detection due to observing patients under regulated surroundings

TABLE 4.6

The Literature Survey about Detecting Autism Disorder Using Advanced Technology

Authors	Method Used	Advantages	Drawbacks
Mythili et al. [23]	Image sorting using CNN, SVM, KNN, and ANN	The models understand the logic and patterns of the dataset quickly	It gives an inaccurate results due to spatial heterogeneity
Kosmicki et al. [25]	ADOS and stepwise backward feature selection	Reduces the complexity of data and has enhanced accuracy	The results will not be precise as it is based on observation in a few circumstances
Thabtah et al. [27]	Logistic regression	Easy to use and avoids overfitting of data	Not suitable to analyze complex data when the dataset is nonlinear
A. Kanchana [28]	Random forest	Gives good accuracy even when the input dataset is inconsistent	It is a time-consuming process
Mahmoud Elbattah et al. [29]	Natural language processing	Easy to analyze and gets results by analyzing the fast eye moment	Can analyze only fast movements of eyes but cannot analyze other eye movements that are crucial in autism detection
Mariam M. Hassan et al. [30]	Decision tree algorithm	Analyzes all the causes and symptoms related to ASD effectively	Overfitting problems arise when there is a large dataset

researchers and are used to detect autism disorder. As we can observe in Table 4.6, deep research about the inclusion of AI and ML concepts to detect autism would enhance the diagnosis process, which will be helpful for adults and children at present [20]. The concepts of ML have been implied to perform autism diagnosis. ML predicts the presence or absence of autism using the given properties of attributes [21].

To gather data related to autism in infants, children, and adults, Fadi Thabtah designed an application on mobile called ASD test. AQ-10 and Qchat-10 are the methods used to build this application. The documentation about autistic individuals is clustered together and restored in the ML database at the University of California, Irvine, using these examination techniques [22]. The application of classification methods of ML to diagnose autism was researched by Mythili et al. [23].

In this method, ML concepts like support vector machine (SVM), neural networks, and fuzzy logic techniques are used, and, for analysis, Weka software is implemented to understand a student's socializing ability and actions. The ML algorithms will offer promising results in diagnosing autism as they are designed to understand the logic of data [24]. To understand the difference between attention-deficit hyperactivity disorder and ASD, Kosmicki et al. [25] implemented the forward feature selection method by leveraging the 65 Social Responsiveness Scale. To help and assist autistic individuals during the COVID-19 pandemic, Al Banna proposed a unique AI-powered model. It is a customized technology that helps autistic people manage and keep in check their problems [26].

Autism is a neurological disorder that should be detected at an initial level. The existing treatments are not cost-effective. Hence, there is a pressing need to bring smart technology to detect autism that can be afforded by the mass population as the rate of autism is increasing. Thabtah et al. [27] proposed a logistic regression (LR) algorithm that is known as an ML algorithm to detect autism in adolescents and adults. Kanchana et al. [28] introduced the random forest method to detect autism. This classifier generates the rating for the input dataset depending on the outcomes.

Elbattah et al. [29] introduced a natural language processing technique to predict the presence or absence of autism by analyzing fast eye movements. In this process, eye movements are portrayed as a characteristic that explains the movements of the eyes that help detect autism. Hassan et al. [30] proposed a decision tree algorithm that uses an autism-related dataset to diagnose autism. Also, it uses the ASD history of family members, which enhances the prediction process.

The study of detecting autism by the facial expressions of children has evolved. The latest research shows that the application of concepts of deep learning to detect various diseases offers promising results. In this subject, the implementation of a deep learning concept known as CNN plays an effective role by giving accurate results. CNN is the best model to perform the image recognition process as it has the ability to understand the hidden and logical information of the given input dataset [31].

4.3 Proposed Methodology

In the proposed framework, the CNN model is implemented. Initially, the dataset related to autism in children aged between 4 and 12 years is collected. The dataset is a combination of text and numerical. The dataset has some values missing. The missing data is replaced with useful data in the dataset. This process helps to get better results. The gathered dataset is given as input to the CNN model after using the byte pair encoding (BPE) algorithm that is used to convert the data present in each column in the table into characters in a sequence that helps the CNN model analyze the input data easily and efficiently. The three layers of convolutional layers are used in this model. The activation function used by convolutional layers is rectified linear unit (ReLU) as we can observe in Figure 4.5. Here, relevant features are chosen using these layers of CNN, and the feature map generates useful features from the dataset to detect autism.

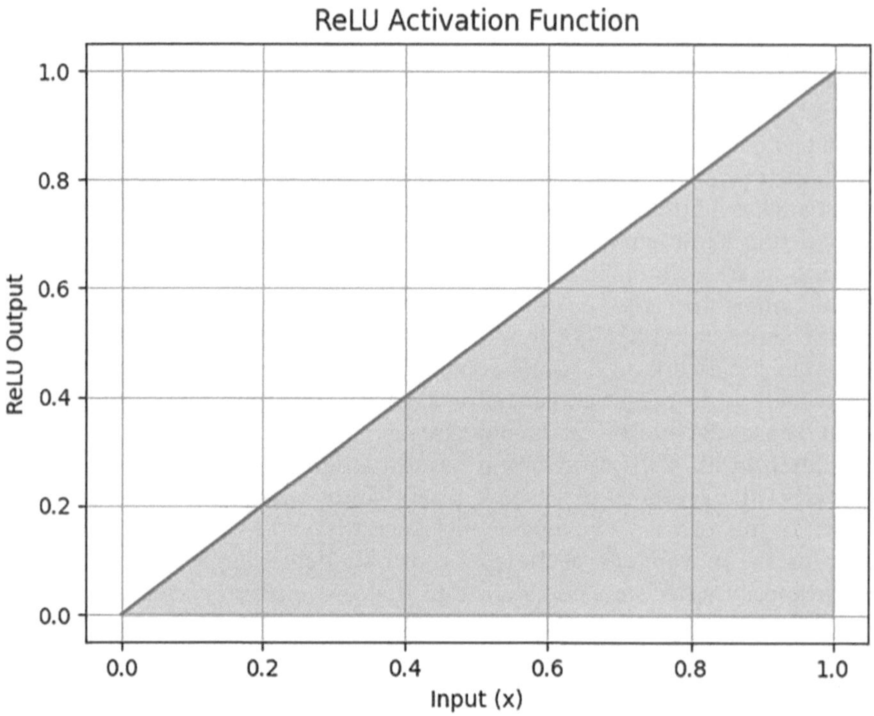

FIGURE 4.5
Rectified linear unit activation function graph.

Now, in the further process pooling layers are used to select the most relevant data from the feature map that convolution layers generate. Here, two layers of pooling layers are used to generate a small feature map from the existing feature map. The flattened layer takes input from the pooling layer and recasts it into a one-dimensional matrix that helps to ease the classification process further. The fully connected layer of CNN will identify the presence of autism in children based on the selected features.

It takes input from previous layers through neurons and gives the output. To reduce the loss function, we used stochastic gradient descent (SGD). To get the result in binary classification form, the sigmoid activation function is implemented. Then the proposed model is tested using a testing dataset to check the efficiency of the proposed model in detecting autism in children. We will get the output in binary form as yes or no, which represents the presence or absence of autism in children, respectively. The advanced version of Python, python 3, is used to perform this process. The training precision should be greater than the testing precision which determines that the performance of CNN is good. Figure 4.6 clearly explains the proposed idea.

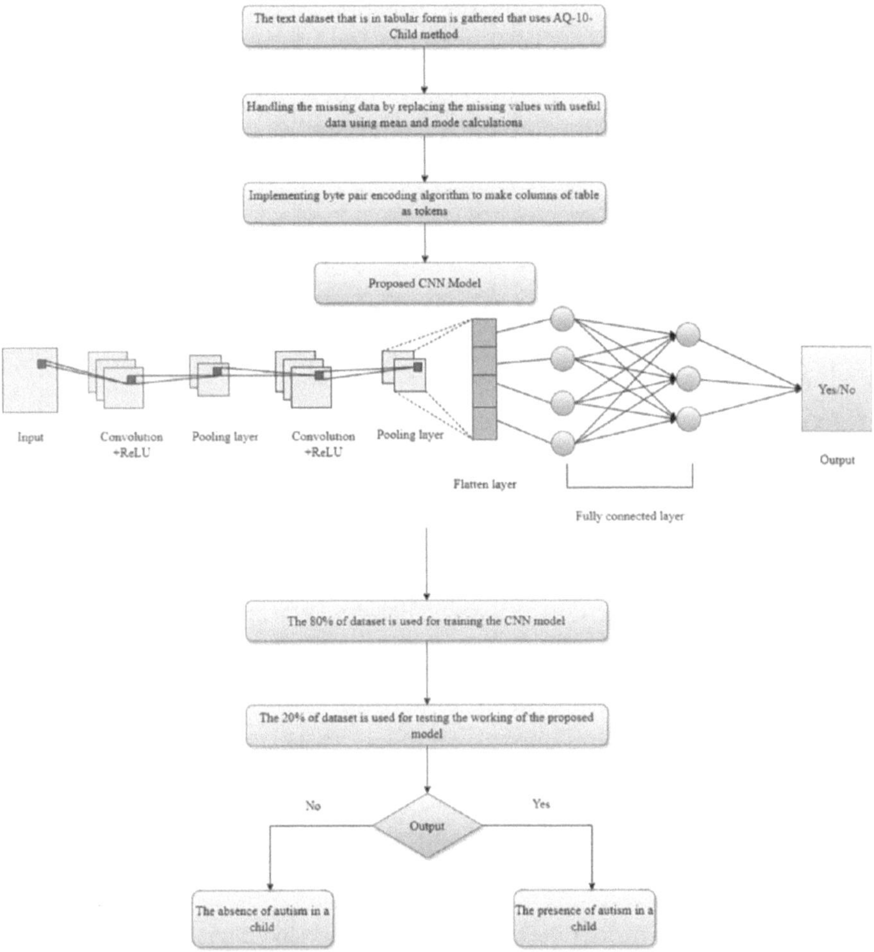

FIGURE 4.6
The framework of the proposed methodology.

Algorithm for the Proposed Model:

Input: Autism disorder-related data of children that has attributes X and target parameter y

1. Partition the dataset into training and testing.
2. Implement the BPE algorithm on the training dataset.
3. Give input to the CNN model after performing BPE on the dataset.

4. Implement ReLU as the activation function for convolutional layers.
5. Implement maxpooling to get a feature map.
6. Convert the generated feature map as a one-dimensional matrix.
7. To reduce the loss function, implement SGD.
8. Implement sigmoid activation function on the outcome layer to get results as yes or no.
9. Evaluate the performance of the model with the testing dataset.

Output: Yes/No

4.3.1 Theoretical Framework

See figure 4.6.

4.3.2 Dataset Used

We gathered the datasets that are related to the area of research in the initial stage that helps to train the CNN model. In this research, the dataset collected was designed by Fadi Fayez Thabtah [32]. Here, the gathering of the dataset is performed using a screening application on mobile by using the AQ-10-Child process. The description of the gathered dataset is given in Table 4.7. The dataset has records of 291 patients and 20 features. The dataset has ten questions, including the age of children (4–12 years), AQ-10-Chat rate, gender, nationality, jaundice, family history of ASD, who went through ASD screening before, who attended the test, and the presence of ASD. The ten questions are related to the AQ-10-Chat procedure.

TABLE 4.7

Description about Dataset Collected to Detect Autism Spectrum Disorder in Children

Feature No	Feature Name	Feature Description
1	Age	4–12
2	Gender	Male/female
3	Nationality	Nationality of the patient
4	Jaundice	Yes/no
5–14	Q1–Q10	The AQ-10-Chat method is used, which has ten questions
15	Family members ASD	Yes/no
16	Who attended the test	Family member/healthcare professional
17	AQ-10-Child rate	Each yes is represented as 1, and the total count above 6 is considered the presence of ASD
18	Undergone ASD screening before	Yes/no
19	Presence of ASD	Yes/no

4.3.3 Data Preparation

Data preparation means modifying the raw data that is more appropriate to use in the tasks that are allied with data. In this process, managing errors, managing missing values, and handling data instability take place by the process of imputation of the missing values. Here to manage numeric data, we use the mean value of the data to restore the missing data in the dataset. To manage qualitative data, we use mode. The mode tells how frequently a word repeats. That is chosen to replace the missing data. The formula used to calculate the mean of the given data is:

Mean = Summation of all the values/Total number of values

4.3.4 Feature Selection

The feature selection process is performed after data preparation. It is the process of choosing the most useful attribute that will suit the training of the model. This process will help minimize high variance, enhance the process of training, and reduce the complexity of understanding data. In this research, a correlation matrix is performed for feature selection on the input dataset using Python, as we can observe in Figure 4.7. The instance of the dataset is given in Table 4.8 of nine patients.

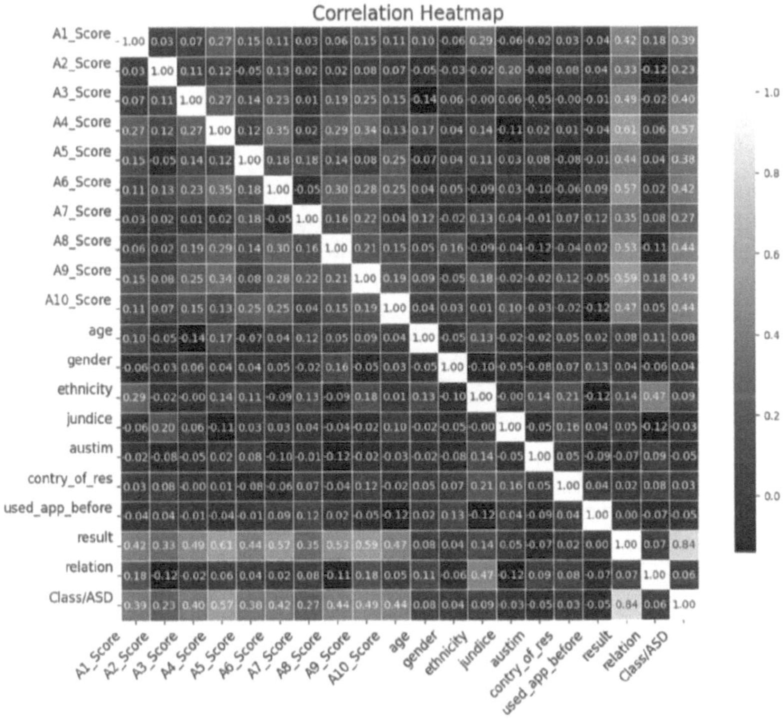

FIGURE 4.7
Correlation matrix for feature selection.

TABLE 4.8
Instances of Children Autism Dataset

Instance No	1	2	3	4	5	6	7	8	9
A1	1	1	1	1	1	1	0	1	1
A2	1	1	1	1	1	0	1	0	1
A3	0	0	1	1	1	0	1	0	1
A4	0	0	1	1	1	0	1	0	1
A5	1	1	1	1	1	1	1	0	1
A6	1	1	1	1	1	1	1	0	1
A7	0	0	1	1	1	1	1	1	1
A8	1	1	1	1	0	1	1	0	1
A9	0	0	1	0	0	1	1	0	1
A10	0	0	1	0	0	1	1	0	1
Age (4–12)	6	6	5	5	11	10	4	4	6
AQ-10-Chat rate	5	5	10	8	7	7	9	2	10
Gender	M	M	M	F	F	M	M	F	M
Nationality	Others	West Asian	Others	West Asian	West Asian	Caucasian	Balck	White	White
Jaundice	No	No	Yes	No	No	Yes	Yes	No	No
Family history of ASD	No	No	No	No	Yes	No	No	No	Yes
Who attended the test	Parent	Parent	Parent	Parent	Parent	Self	Parent	Parent	Parent
Presence of ASD	No	No	Yes	Yes	Yes	Yes	Yes	No	Yes

4.3.5 Convolutional Neural Networks

CNNs are considered a deep learning method [33]. CNN is designed to take input data as a text or a picture to analyze it and that generates the output that describes the logics and patterns of the dataset. CNN is designed with three layers that have some functionality. The three layers are the convolutional layer, the pooling layer, and the fully connected layer. CNN has kernels that are used for selecting features that generate feature maps [34].

The convolutional layer is responsible for applying the kernel to the input dataset. In this research, we convert the dataset into tokens that are sent to the CNN model as input. So, Equation (4.1) is used in the selection of the feature process for the data that is tokenized. In the proposed model, we used ReLU as an activation function that enhances non-linearity and helps the CNN model to understand the logic easily of the dataset. Equation (4.2) shows how ReLU is calculated for data that is converted into tokens.

$$\text{Score of feature selection } (t) = \sum_{i=1}^{N} \text{Re LU} \cdot \sum_{j=1}^{M} W_{ij} \times x_{tj} + b_i \tag{4.1}$$

$$\text{ReLU}(x) = \max(0,\ x) \tag{4.2}$$

Here, from Equation (4.1), N represents how many filters exist in the CNN model. The dimension of tokens is represented as M. The weight denoted by W_{ij} shows how strong filter i and dimension j are linked in CNN architecture. The bias is shown as b_i. The dimension j of the token is portrayed by x_{tj}. In Equation (4.2) x is the input that the ReLU function takes. If the input value is positive, then the output of ReLU will be positive. If the input value is negative, then the outcome of ReLU will be zero. To choose the most relevant features of similar properties from the feature map, the pooling layer is useful. There are two most common types of pooling: average pooling and max pooling. In average pooling, it selects and combines the average number from each section of the feature map and gives a small feature map.

The pooling process takes place after the selection of features. Max pooling means it selects and gathers the maximum value from the sections of the feature map that generate a small feature map. It is used to select the important characteristics of data. Average pooling helps to eliminate unnecessary data. Equations (4.3) and (4.4) represent the calculation of max pooling and average pooling. Here, in the equations, n represents the total count of values in the pooling layer, and $x_1, x_2, x_3 \ldots x_n$ means the values in the pooling layer.

$$\text{Max pooling} : (S) = \max(x_1, x_2, x_3 \ldots x_n) \tag{4.3}$$

$$\text{Average pooling}\,(S) = \frac{1}{n}\sum_{i=1}^{n} x_i \tag{4.4}$$

A flattened layer is present after the pooling layer. It is used to adjust the layer into a 1D layer that will ease the prediction process. A fully connected layer in CNN will be linked with the preceding and succeeding layers of CNN. This operation is used to get a non-linearity function. First, the application of bias offset to the total weighted input is performed and then passed through a non-linearity function. Equation (4.5) represents how to calculate the loss function, where N means the total number of instances in the input dataset. The correct label of the dataset is portrayed as y_i. The forecasted classification of the dataset is portrayed as y_i cap, and L means loss function. So, based on the outcome, SGD can be implemented to reduce loss function. The architecture of CNN is portrayed clearly in Figure 4.8.

$$L = \frac{1}{N}\sum_{i=1}^{N} L_l\left(y_{i,}\text{hat}\, y_i\right) \tag{4.5}$$

4.3.6 Portioning Data

It is the process of partitioning the data into training and testing [35]. This procedure is followed by a feature selection process. In this process, some portions of data from the dataset are used to train the chosen deep learning model, and the other portion of data is used for the testing process. Now, the

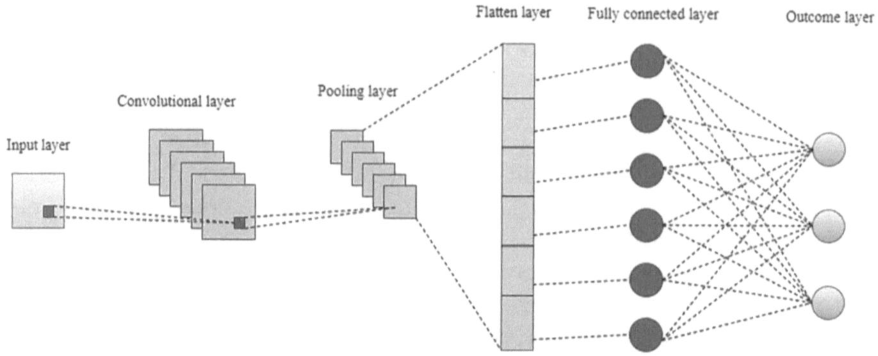

FIGURE 4.8
CNN architecture.

portioning of data is followed. Here, the dataset used for training and testing is divided into a ratio of 80:20 accordingly. Furthermore, the dataset used for training is again divided as training and evaluation into the ratio 80:20 accordingly that is used for cross-validation purposes.

4.4 Results

Here, we consider the accuracy of each model and the model with a high accuracy rate will be chosen to detect autism in children. The accuracy of artificial neural network (ANN), SVM, LR, K-nearest neighbor (KNN), Naive Bayes, CNN, one-dimensional CNN, and temporal convolutional network (TCN) is 97.59%, 97.87%, 97.85%, 88.12%, 94.86%, 97.90%, 97.70%, and 97.80%, respectively, as we can observe in Table 4.11 and Figure 4.9 that we got before managing missing data. We can also observe sensitivity in Table 4.9 and Figure 4.10, specificity in Table 4.10 and Figure 4.11, and error rate in Table 4.12 and Figure 4.12, respectively.

We can also observe the performance of the models after managing the missing data. The accuracy of CNN is high among all the models after managing missing data as we can observe in Table 4.13 and Figure 4.13. We can observe that the error rate of CNN is less after managing missing data in Table 4.14 and Figure 4.14. We got this after evaluating all the performance of the models. The performance is evaluated using the confusion matrix method.

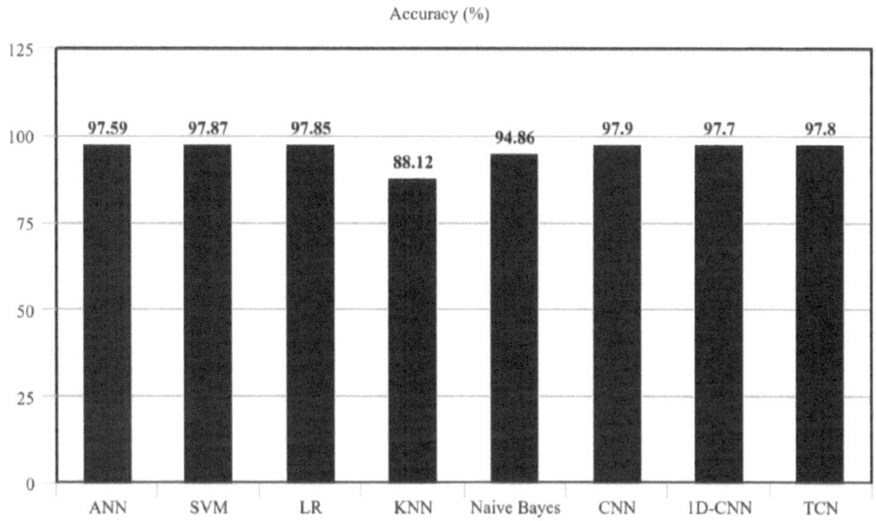

FIGURE 4.9
The accuracy of models obtained before handling the missing values.

TABLE 4.9

The Accuracy before Handling Missing Data Obtained after Checking the Performance of the Model

Method	Accuracy (%)
ANN	97.59
SVM	97.87
LR	97.85
KNN	88.12
Naive Bayes	94.86
CNN	97.90
1D-CNN	97.70
TCN	97.80

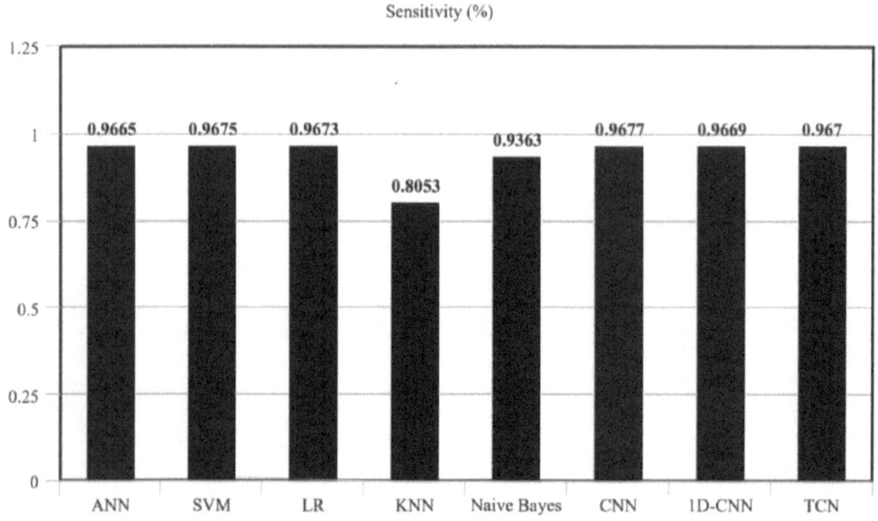

FIGURE 4.10
The sensitivity of all the models in percentage.

TABLE 4.10
The Sensitivity Obtained after Checking the Performance of the Models

Method	Sensitivity (%)
ANN	0.9665
SVM	0.9675
LR	0.9673
KNN	0.8053
Naive Bayes	0.9363
CNN	0.9677
1D-CNN	0.9669
TCN	0.9670

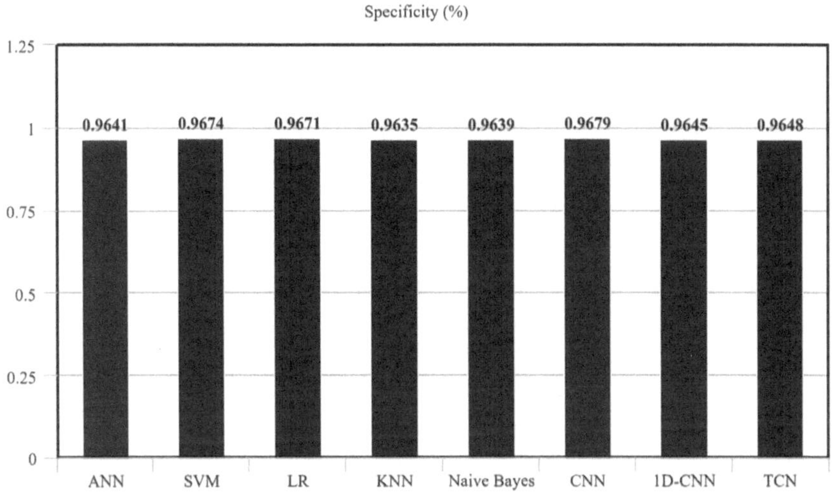

FIGURE 4.11
The specificity in the percentage of the chosen models used to detect autism in children.

TABLE 4.11
The Specificity Obtained after Checking the Performance of the Models

Method	Specificity (%)
ANN	0.9641
SVM	0.9674
LR	0.9671
KNN	0.9635
Naive Bayes	0.9639
CNN	0.9679
1D-CNN	0.9645
TCN	0.9648

TABLE 4.12
The Error Rate Obtained before Managing the Missing Values

Method	Error Rate (%)
ANN	2.41
SVM	2.13
LR	2.15
KNN	11.88
Naive Bayes	5.14
CNN	2.1
1D-CNN	2.3
TCN	2.2

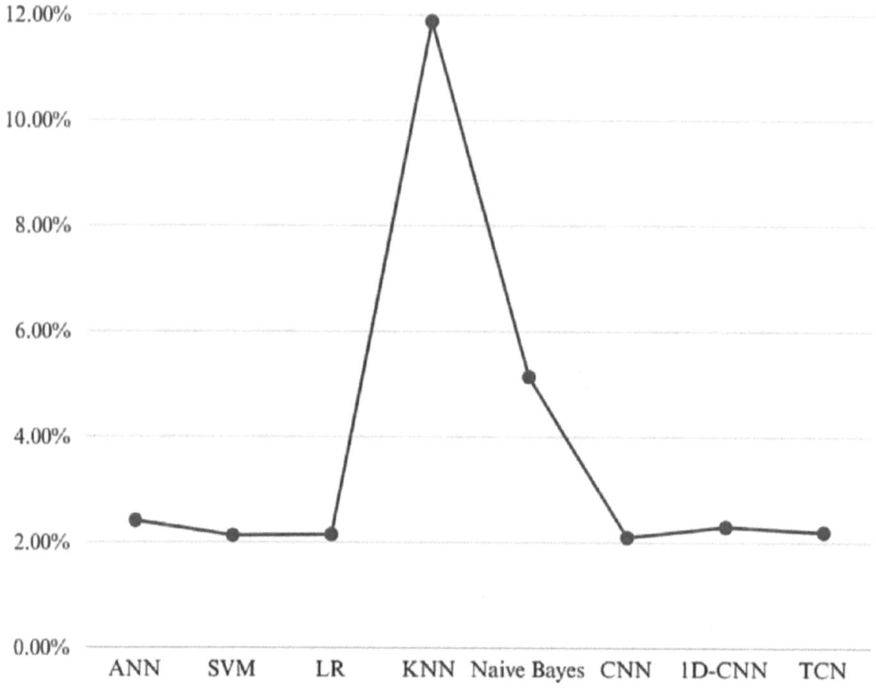

FIGURE 4.12
The error rate in the percentage of the chosen models before handling missing data.

TABLE 4.13
The Accuracy after Handling Missing Data Obtained after Checking the Performance of the Models

Method	Accuracy before Managing Missing Data (%)	Accuracy after Managing Missing Data (%)
ANN	97.59	97.65
SVM	97.87	98.35
LR	97.85	97.89
KNN	88.12	88.90
Naive Bayes	94.86	95.30
CNN	97.90	98.76
1D-CNN	97.70	97.90
TCN	97.80	98.20

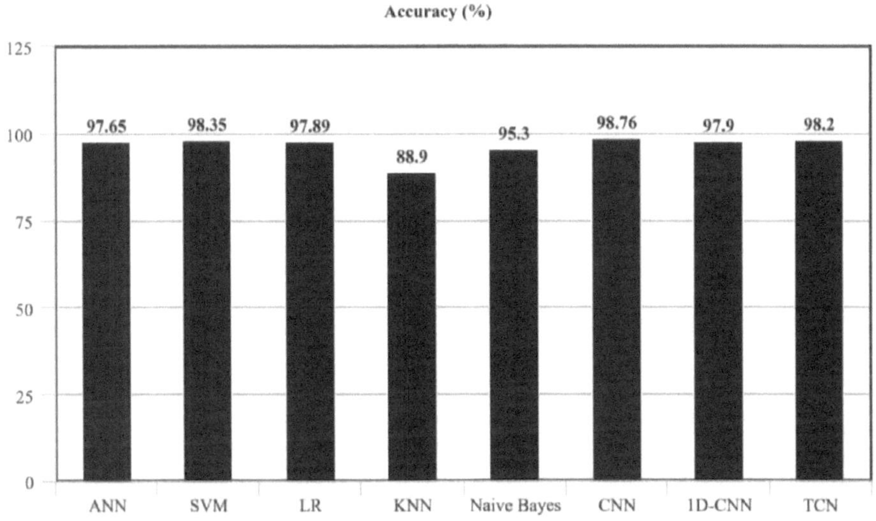

FIGURE 4.13
The accuracy of models obtained after handling the missing values.

TABLE 4.14
The Error Rate Obtained after Checking the Performance of the Models after Handling Missing Values

Method	Error Rate (%)
ANN	2.35
SVM	1.65
LR	2.11
KNN	11.1
Naive Bayes	4.7
CNN	1.24
1D-CNN	2.1
TCN	1.8

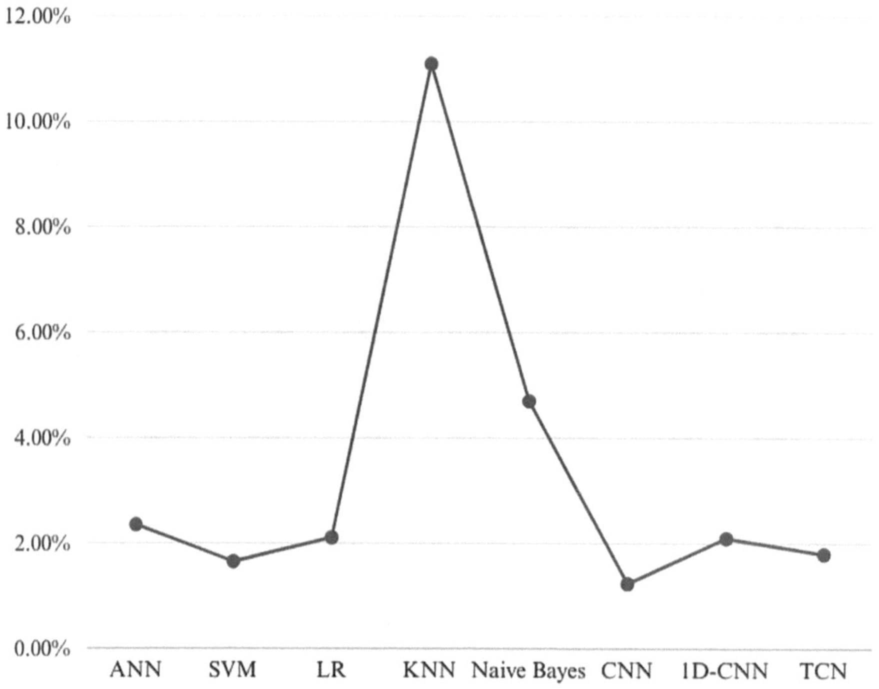

FIGURE 4.14
The error rate of chosen models after handling the missing data.

4.5 Conclusion

This research aims to detect ASD in children which helps to proceed with curated treatment that helps in curing the condition. We implemented the techniques of AI to detect autism in children. The related dataset is collected, and the CNN model achieved a high accuracy among all the chosen models that have an error rate of 1.24%. This shows that the implementation of deep learning gives promising results in detecting autism in children. In conclusion, conventional methods can be replaced by advanced technology to detect autism. Implementing conventional methods is a time-consuming process where AI and its applications will give faster results, it is helpful to diagnose at the initial stage, it will be available at an affordable price, and we get to know the severity of the problem that aids to take appropriate treatment.

References

1. Jade Davies, Kate Cooper, Estelle Killick, Evelyn Sam, Molly Healy, Grace Thompson, Will Mandy, Benjamin Redmayne, and Laura Crane. Autistic Identity: A Systematic Review of Quantitative Research, *Autism Research*, 17(5), 874–897, 2024.
2. Helen V. Ratajczak. Theoretical Aspects of Autism: Causes—A review, *Journal of Immunotoxicology*, 8, 68–79, 2010.
3. Geraldine Dawson, Andrew N. Meltzoff, Julie Osterling, and Julie Rinaldi. Neuropsychological Correlates of Early Symptoms of Autism, *Society for Research in Child Development*, 69, 1276–1285, 2008.
4. Chris Plauché Johnson, and Scott M. Myers. Identification and Evaluation of Children with Autism Spectrum Disorders, *Pediatrics*, 120(5), 1183–1215, 2007.
5. Simon Baron-Cohen, Michael V. Lombardo, Bonnie Auyeung, Emma Ashwin, Bhismadev Chakrabarti, and Rebecca Knickmeyer. Why Are Autism Spectrum Conditions More Prevalent in Males? *Plos Biology*, 9, 2010.
6. Yvette Hus, and Osnat Segal. Challenges Surrounding the Diagnosis of Autism in Children, *Neuropsychiatric Disease and Treatment*, 17, 3509–3529, 2021.
7. Sydney Nolan, Rachel Schroeder, Lani L. Harris, Daniel L. Segal, and Frederick L. Coolidge. Differentiation of Children with Varying Levels of Autism Spectrum Disorder Symptoms from Children Developing Typically, *Psychiatry Research Communications*, 2(4), 100084, 2022.
8. Venus Wong, Yue YuDepartm, Melissa L. Keyes, and John H. McGrew. Pre-Diagnostic and Diagnostic Stages of Autism Spectrum Disorder: A Parent Perspective, *Child Care in Practice*, 23(2), 195–217, 2016.
9. Masume Zareei, Seyed Abolfazl Tohidast, and Masoomeh Salmani. Parents' Perceptions of the Problems in Children with Autism Spectrum Disorders: A Qualitative Study, *Iranian Rehabilitation Journal*, 20(4), 611–622, 2022.
10. C. Kishor Kumar Reddy, Vijaya Sindhoori Kaza, P. R. Anisha, Mousa Mohammed Khurbani, Mohammed Shauib, Shadab Alam, and Sadaf Ahmad Optimizing Barrier Placement for Intrusion Detection and Prevention in WSNs. *Plos One*, 19(2), e0299334, 2024.
11. Seyed Reza Shahamiri, and Fadi Thabtah. Autism AI: A New Autism Screening System Based on Artificial Intelligence, *Cognitive Computation*, 12(4), 766–777, 2020.
12. Sricheta Parui, Debasis Samanta, Nishant Chakravorty, Uttam Ghosh, and Joel J.P.C. Rodrigues. Artificial Intelligence and Sensor-Based Autism Spectrum Disorder Diagnosis Using Brain Connectivity Analysis, *Computers and Electrical Engineering*, 108, 108720, 2023.
13. Md Shafiul Alam, Muhammad Mahabubur Rashid, Rupal Roy, Ahmed Rimaz Faizabadi, Kishor Datta Gupta, and Md Manjurul Ahsan. Empirical Study of Autism Spectrum Disorder Diagnosis Using Facial Images by Improved Transfer Learning Approach. *Bioengineering*, 9(11), 710, 2022.
14. Kiyohiro Yoshinaga, Jun Egawa, Yuichiro Watanabe, Hiroyuki Kasahara, Atsunori Sugimoto, and Toshiyuki Someya. Usefulness of the Autism Spectrum Quotient (AQ) in Screening for Autism Spectrum Disorder and Social Communication Disorder, *BMC Psychiatry*, 23(1), 831, 2023.

15. Fadi Thabtah, and David Peebles. Early Autism Screening: A Comprehensive Review. *International Journal of Environmental Research and Public Health*, 16(18), 3502, 2019.
16. Sun Jae Moon, Jin Seub Hwang, Ah Lahm Shin, Jae Yup Kim, Seung Min Bae, Jennifer Sheehy-Knight, and Jung W. Kim. Accuracy of the Childhood Autism Rating Scale: A Systematic Review and Meta-Analysis, *Developmental Medicine & Child Neurology*, 61(9), 1030–1038, 2019.
17. Omaima Bamasaq, Daniyal Alghazzawi, Surbhi Bhatia, Pankaj Dadheech, Farrukh Arslan, Sudhakar Sengan, and Syed Hamid Hassan Distance Matrix and Markov Chain Based Sensor Localization in WSN. *CMC-Computers, Materials & Continua*, 71(2), 4051–4068, 2022.
18. So Hyun (Sophy) Kim, Vanessa Hus, and Catherine Lord. Autism Diagnostic Interview-Revised. In Fred R. Volkmar (eds) *Encyclopedia of Autism Spectrum Disorders*. Springer, New York, 2013.
19. Patricia Evans, Sailaja Golla, and Mary Ann Morris. Autism Spectrum Disorders: Clinical Considerations. In Roger N. Rosenberg and Juan M. Pascual (eds) *Rosenberg's Molecular and Genetic Basis of Neurological and Psychiatric Disease* (pp. 197–207). Academic Press, Boston, 2015.
20. Chitta Hrudaya Neeharika, and Yeklur Mohammed Riyazuddin. Developing an Artificial Intelligence Based Model for Autism Spectrum Disorder Detection in Children, *Journal of Advanced Research in Applied Sciences and Engineering Technology*, 32(1), 57–72, 2023.
21. B. Kamala, K. S. Mahanaga Pooja, S. Varsha, and K. Sivapriya. ML Based approach to Detect Autism Spectrum Disorder (ASD). In *2021 4th International Conference on Computing and Communications Technologies (ICCCT)*, Chennai, India, IEEE, 2021.
22. Fadi Thabtah, Firuz Kamalov, and Khairan Rajab. A New Computational Intelligence Approach to Detect Autistic Features for Autism Screening, *International Journal of Medical Informatics*, 117, 112–124, 2018.
23. M. S. Mythili, and A. R. Mohamed Shanavas. A Study on Autism Spectrum Disorders Using Classification Techniques, *International Journal of Soft Computing and Engineering*, 4(5), 88–91, 2014.
24. Muhammad Shoaib Farooq, Rabia Tehseen, Maidah Sabir, and Zabihullah Atal. Detection of Autism Spectrum Disorder (ASD) in Children and Adults Using Machine Learning, *Scientific Reports*, 13(1), 9605, 2023.
25. J. A. Kosmicki, V. Sochat, M. Duda, and D. P. Wall. Searching for a Minimal Set of Behaviors for Autism Detection through Feature Selection-Based Machine Learning, *Translational Psychiatry*, 5(2), e514, 2015.
26. Tapotosh Ghosh, Md Hasan Al Banna, M. Shamim Kaiser and Kazi Abu Taher. A Monitoring System for Patients of Autism Spectrum Disorder Using Artificial Intelligence. In *International Conference on Brain Informatics* (pp. 251–262). Springer International Publishing, Cham, 2020.
27. Fadi Thabtah, Neda Abdelhamid, and David Peebles. A Machine Learning Autism Classification Based on Logistic Regression Analysis, *Health Information Science and Systems*, 7(1), 12, 2019.
28. A. Kanchana, and Rashmita Khilar. Prediction of Autism Spectrum Disorder Using Random Forest Classifier in Adults. In *IEEE International Conference on Cybernetics, Cognition and Machine Learning Applications (ICCCMLA)*, Goa, India, 2022.

29. Mahmoud Elbattah, Jean-Luc Guérin, Romuald Carette, Federica Cilia, and Gilles Dequen. NLP-Based Approach to Detect Autism Spectrum Disorder in Saccadic Eye Movement. In *2020 IEEE Symposium Series on Computational Intelligence (SSCI)* (pp. 1581–1587), Canberra, ACT, Australia, IEEE, 2020.
30. Mariam M. Hassan, and Hoda M.O. Mokhtar. Investigating autism etiology and heterogeneity by decision tree algorithm, *Informatics in Medicine Unlocked,* 16, 100215, 2019.
31. Ming Yang, Menglin Cao, Yuhao Chen, Yanni Chen, Geng Fan, Chenxi Li, Jue Wang, and Tian Liu. Large-Scale Brain Functional Network Integration for Discrimination of Autism Using a 3-D Deep Learning Model, *Frontiers in Human Neuroscience,* 15, 687288, 2021.
32. Fadi Fayez Thabtah, Autistic Spectrum Disorder Screening Data for children. https://www.kaggle.com/datasets/faizunnabi/autism-screening, 2017.
33. Suman Raj, and Sarfaraz Masood. Analysis and Detection of Autism Spectrum Disorder Using Machine Learning Techniques, *Procedia Computer Science,* 167, 994–1004, 2020.
34. Prabhjot Kaur, Shilpi Harnal, Rajeev Tiwari, Shuchi Upadhyay, Surbhi Bhatia, Arwa Mashat, and Aliaa M. Alabdali. Recognition of Leaf Disease Using Hybrid Convolutional Neural Network by Applying Feature Reduction. *Sensors,* 22(2), 575, 2022.
35. C. Kishor Kumar Reddy. An Efficient Early Diagnosis and Healthcare Monitoring System for Mental Disorder Using Machine Learning. In Brojo Kishore Mishra (ed.) *Sustainable Science and Intelligent Technologies for Societal Development,* IGI Global, Chennai, India, p. 22, 2023.

5

Exploring Neurodegenerative Progression: Healthcare 6.0 Insights Using Sustainable Convolutional Neural Networks

Nishitha Pujala, C. Kishor Kumar Reddy, and Nastassia Thandiwe Sithole

5.1 Introduction

The modern world has a longer lifespan due in part to advancements in medical knowledge over time. An increasing number of people are experiencing neurocognitive impairments as a result of their longer life expectancy, which has an effect on both the older population and the world economy. The year 2010 saw significant estimates of spending on dementia sufferers in the USA alone. Globally, the number of persons suffering from dementia is increasing quickly; estimates suggest that, by 2050, a sizable population—roughly 135 million people—may be impacted. Dementia is a result of several risk factors, including ageing, head trauma, and lifestyle choices. Although ageing is clearly the main risk factor, data show that a person's chance of having dementia increases with age, from 1% to 2% at age 65–30% by age 85 [1].

A mental illness called dementia is characterized by a slow decline in cognitive abilities, which can affect day-to-day tasks like memory, problem-solving, vision, and task attention. Dementia typically affects older persons more and is misunderstood as an inevitable side effect of ageing. But it's crucial to understand that dementia is a serious form of cognitive decline that interferes with day-to-day functioning rather than an inevitable aspect of ageing. The main factors contributing to the development of dementia are different illnesses and traumas to the brain. Dementia is the seventh-most common cause of death worldwide and is a significant contributor to dependency and disability in the elderly. A person must have evident symptoms of severe cognitive decline and a discernible change in their usual mental functioning to be diagnosed with dementia.

A person with dementia has a complex and progressive neurological illness that greatly affects their day-to-day functioning due to a reduction in cognitive ability. This illness affects multiple cognitive domains, extending

DOI: 10.1201/9781003503934-5

beyond typical age-related forgetfulness, including language, thinking, memory, and problem-solving abilities. Dementia has a tremendous effect on a person's relationships, independence, and general quality of life, but its effects go beyond the individual. Dementia is a phenomenon with several underlying causes; it can take many different forms, the most common being Alzheimer's disease (AD). Lewy body dementia (LBD), frontotemporal dementia (FTD), and vascular dementia are other prevalent forms. The precise processes underlying dementia are complex and frequently include aberrant protein buildup in the brain, damage to neurons, and a breakdown in intercellular communication.

5.1.1 Varieties of Dementia

A variety of cognitive problems are included under the general term "dementia," which is not limited to any one condition. Figure 5.1 illustrates the advancement of dementia in correlation with ageing. Illustrated in Figure 5.2 is a comprehensive overview that delineates the different types of dementia, accompanied by the respective percentages representing their occurrence among patients [2].

5.1.2 Alzheimer's Illness

AD presents an escalating challenge to global public health, affecting an expanding population grappling with this neurodegenerative condition. As researchers and healthcare professionals endeavour to comprehend and navigate the intricacies of AD, cutting-edge technological approaches have emerged as promising tools for disease prognostication. One particularly

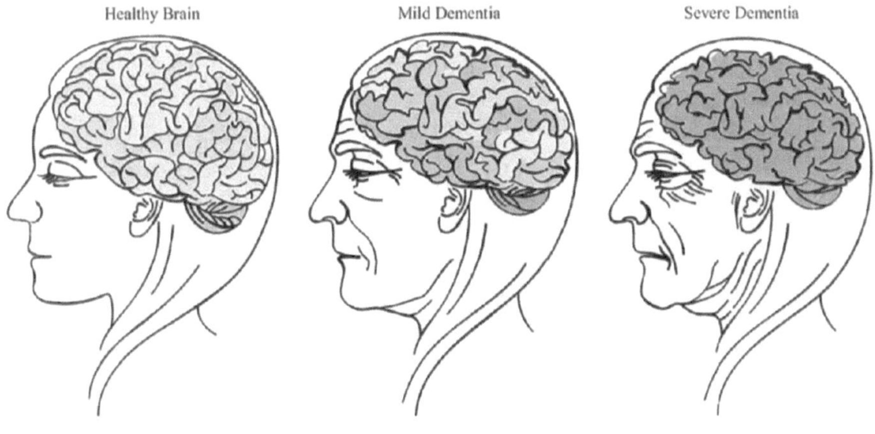

FIGURE 5.1
Evolution of dementia in the elderly.

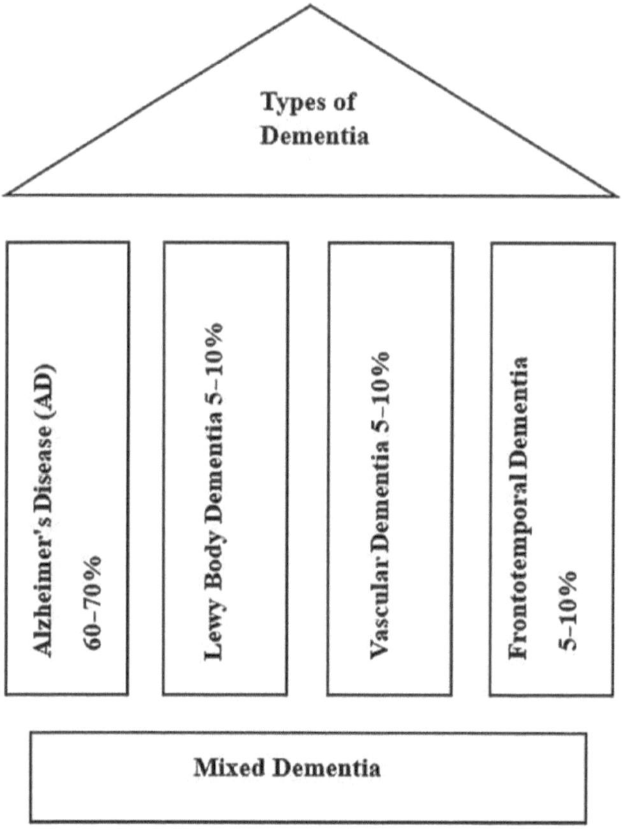

FIGURE 5.2
Different types of dementia.

creative approach combines optimal model tuning with sophisticated deep feature extraction methods to forecast the course and outcome of AD.

This investigation explores the intersection of cutting-edge technology, specifically deep learning (DL) techniques, with the vital need for precise AD prognosis. By using advanced feature extraction techniques, researchers hope to reveal hidden insights inside large datasets. They are interested in investigating minute details and complex patterns that could be important markers of the course of a disease. Furthermore, improving the precision and consistency of prognostic results depends on the optimization of predictive models.

Dementia presents a significant and escalating global health challenge, affecting an ever-growing number of individuals grappling with this neurodegenerative disorder. As the medical community strives to comprehend and confront the complexities inherent in AD, novel technology strategies have surfaced as potentially useful prognostication instruments.

Between the more severe cognitive loss associated with AD and normal ageing-related cognitive changes, mild cognitive impairment (MCI) denotes a stage. Individuals with MCI experience noticeable cognitive alterations exceeding those expected for their age and educational background but do not yet meet dementia criteria. In the context of AD, MCI often acts as an early warning sign, with many individuals eventually progressing to AD dementia. Common symptoms of MCI in AD include memory lapses, language difficulties, impaired judgement, and problem-solving challenges. Vigilant monitoring of cognitive function and collaboration with healthcare professionals for suitable interventions and support strategies are crucial for individuals diagnosed with MCI and their caregivers (Table 5.1; Figures 5.3 and 5.4).

Changes in lifestyle might also impact an individual's prognosis with AD. Reducing the chance of acquiring AD and maybe improving outcomes for those who are already affected can be achieved by following some preventive

TABLE 5.1
Alzheimer's Disease Statistics over the Years 2021, 2022, and 2023

Year	Number of Cases	Age Group	Number of Deaths
2020	7,70,000	50–59	2,60,000
2021	2,00,000	60–69	45,000
2022	6,30,000	70–79	1,60,000
2023	5,00,000	80+	2,00,000

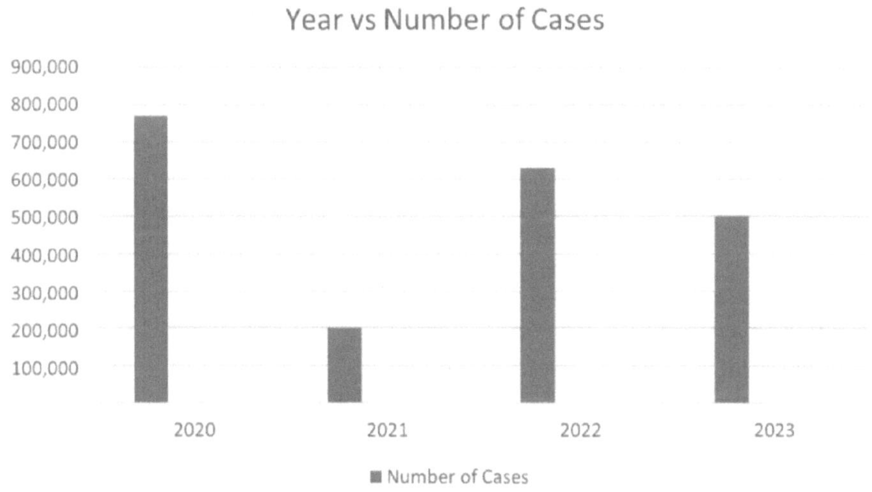

FIGURE 5.3
Number of cases by year.

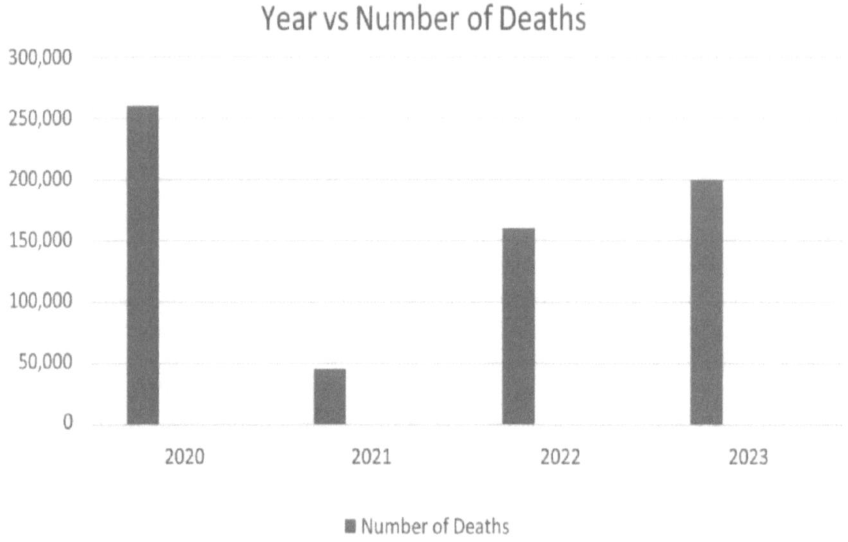

FIGURE 5.4
Number of deaths by year.

measures, maintaining general mental and physical well-being, participating in frequent cognitive activities, and adopting a brain-healthy diet [3].

Even though AD is a dangerous and frequently progressing illness, early identification and treatment can improve a patient's prognosis and quality of life. It is imperative that people identify the symptoms or risk factors connected with AD, and they consult with healthcare specialists for an appropriate assessment and guidance. AD is a prevalent neurological condition affecting millions worldwide and has a significant impact on individuals and their families. It poses challenges comparable to other major diseases, and its prevalence continues to rise.

Although AD has no cure, advancements in medical technology and treatment options have contributed to improved survival rates and enhanced management strategies. AD patients have varying rates of survival depending on a number of variables, including the kind and severity of the disease, the patient's age, general health, and the existence of coexisting medical problems. Improving the prognosis and quality of life for individuals impacted by cancer requires early detection and suitable treatment interventions.

The comprehension and treatment of AD have greatly benefited from recent technological developments. Advancements in machine learning (ML), artificial intelligence, wearable technology, and telemedicine have been made to improve AD risk assessment, diagnosis, and treatment. These developments could enable tailored treatment regimens, detect people at risk early, and lessen the toll that the illness takes on both individuals and society.

Key technological advances utilized in the estimation of AD include:

1. **Machine Learning (ML)**: To generate individualized risk assessments for each patient, algorithms examine vast amounts of patient data, including test results, medical history, family history, and lifestyle factors.
2. **Artificial Intelligence (AI)**: Early indicators of AD can be found in medical imaging by using techniques like DL and neural networks (NNs). Predictive models generated by AI can analyse data from various sources, providing personalized treatment recommendations.
3. **Wearable Technology**: Continuous monitoring and alerts for individuals at risk of developing AD can be provided through wearable devices.
4. **Telemedicine**: Technologies like remote monitoring and virtual consultations enhance access to care for individuals with AD, allowing healthcare providers to monitor patients' conditions remotely.
5. **Genetic Testing**: Treatment regimens and lifestyle advice can be tailored based on the discovery of genetic variants linked to a higher risk of AD.

Amyloid beta (Aβ) aberrant accumulations in the brain, which can appear as intracellular tau protein neurofibrillary tangles or extracellular amyloid plaques, are thought to be the cause of AD. The accumulations cause disruptions to neuronal activity and connections, ultimately leading to a progressive loss of brain function. Brain cholesterol affects this declining ability to remove proteins with age and is linked to various neurodegenerative diseases. For most people, the origin of AD is still unknown, with the exception of a tiny minority (1–2%) of cases in which particular genetic abnormalities are found. The cholinergic hypothesis and the Aβ hypothesis are two opposing theories that have been put forth to explain the fundamental causes of AD [4].

5.1.3 Frontotemporal Dementia

FTD is a neurological condition that progresses over time and mostly affects both the temporal and frontal regions of the brain, which are important areas that regulate a variety of vital activities like language, personality, social behaviour, and executive function. The distinctiveness of FTD lies not only in its neurological impact but also in its tendency to manifest in individuals under the age of 65, categorizing it as an early-onset form of dementia. As this condition progresses, individuals with FTD may experience a range of cognitive and behavioural changes that extend beyond alterations in personality, social conduct, and language difficulties. Emotional blunting, where individuals exhibit a reduced range and intensity of emotional responses, is a common manifestation. Additionally, impulsive behaviour becomes apparent, leading to actions that are often uncharacteristic of the individual. Challenges in planning and decision-making further contribute to the complex clinical picture of FTD [5].

Apart from the cognitive and behavioural dimensions, FTD can also manifest with distinct clinical subtypes, including non-fluent variant primary progressive aphasia, semantic variant primary progressive aphasia, and behavioural variant FTD. The different patterns of symptomatology exhibited by these subtypes highlight the variety within the FTD spectrum. The unique characteristics of FTD, encompassing both the age of onset and the specific behavioural and cognitive changes, distinguish it from other forms of dementia. A nuanced understanding of these features is crucial for accurate diagnosis, effective management, and providing tailored support for individuals and their caregivers affected by FTD [6].

5.1.4 Lewy Body Dementia

The neurodegenerative condition known as LBD is typified by the buildup of aberrant protein deposits in the brain called Lewy bodies. These Lewy bodies disrupt regular brain activity, which exacerbates a variety of neurological and movement disorders. Individuals with LBD may also experience psychological symptoms such as anxiety, depression, and mood swings, in addition to cognitive and physical difficulties. The unique aspect of LBD lies in its overlap with features found in both AD and Parkinson's disease, presenting a diagnostic challenge for healthcare professionals. While cognitive impairment is a hallmark, LBD is also distinguished by prominent motor symptoms akin to parkinsonism, such as tremors, stiffness, and impaired balance.

A noteworthy feature of LBD is the occurrence of vivid visual hallucinations, which can significantly impact an individual's perception of reality. Fluctuations in alertness and attention, known as fluctuating cognition, further contribute to the complexity of LBD diagnosis. These symptoms, when combined, set LBD apart from other neurodegenerative conditions. Individuals with LBD often face difficulties in thinking, problem-solving, and memory recall. The cognitive decline in LBD tends to manifest in a manner that can fluctuate throughout the day or from one day to another, adding an additional layer of complexity to the diagnostic process [7].

5.1.5 Vascular Dementia

Reduced blood supply to the brain causes vascular dementia, which is frequently linked to strokes or other vascular diseases. This reduction in blood supply results in damage to the blood vessels responsible for delivering essential oxygen and nutrients to the brain, thereby instigating cognitive decline. The symptoms associated with vascular dementia are diverse and contingent upon the location and severity of the vascular damage. In addition to challenges related to memory, reasoning, and problem-solving, individuals grappling with vascular dementia may encounter difficulties in maintaining focus and organizing thoughts and may exhibit fluctuations in attention. Behavioural alterations, including mood swings and apathy, are not uncommon, and some individuals may experience issues with coordination and motor skills. Risk factors for cardiovascular diseases like hypertension (high blood pressure) and atherosclerosis (hardening of the arteries) are often linked to the beginning of vascular dementia. These underlying cardiovascular issues amplify the probability of vascular damage, thereby escalating the risk of cognitive decline.

It is noteworthy that while certain symptoms may overlap with other cognitive disorders, the distinctive vascular origins and varied clinical presentations underscore the significance of accurate diagnosis and the formulation of individualized management strategies. Prioritizing cardiovascular health and adopting preventive measures assumes a pivotal role in reducing the risk of vascular dementia and promoting overall brain health. A comprehensive understanding of these intricacies aids in addressing the specific needs of individuals affected by vascular dementia and optimizing their overall well-being.

The primary contributions of the chapter include:

i. The chapter discusses the application of convolutional neural networks (CNNs), a specialized type of DL algorithm, for processing MRI scans to extract relevant features indicative of AD and related conditions. This entails an understanding of CNN architecture and its role in scrutinizing intricate neuroimaging data to enhance diagnostic accuracy and forecasting.

ii. **Model Efficacy Is Evaluated Using a Variety of Metrics**: The study thoroughly examines a number of ML models, including CNNs, random forests (RFs), and logistic regressions. It rigorously examines performance metrics like accuracy, sensitivity, and specificity to assess each model's capability in accurately identifying positive and negative cases, thereby offering valuable insights into different methodologies' strengths and limitations.

iii. **Utilization of the Open-Access Series of Imaging Studies (OASIS) Dataset for AD Research**: The study emphasizes how crucial the OASIS dataset is as a key resource for studying AD. By integrating MRI scans, demographic data, and cognitive assessments from individuals with AD and other cognitive conditions, researchers can identify potential biomarkers, track disease advancement, and devise tailored diagnostic and therapeutic strategies. Leveraging comprehensive datasets like OASIS facilitates robust analyses and contributes to advancing understanding in dementia research. The chapter's following sections are summarized as follows: Section 5.3 explores the materials and processes, while Section 5.2 offers a summary of pertinent literature. Section 5.4 discusses the outcomes and implications. Section 5.5 presents the concluding remarks, followed by the reference section in Section 5.6.

5.2 Literature Review

An ML method for AD prediction was presented by Dashtipour and Fahad [1]. With the use of both support vector machines (SVM) and bidirectional long short-term memory (LSTM) for classification, their model used DL approaches to extract features from brain pictures and achieved 91.28% classification accuracy. A DL-based strategy for AD early detection was presented by Helaly [2]. He identified four phases in the AD spectrum using CNNs. They obtained impressive accuracy of 93.61% and 95.17% for 2D and 3D multi-class AD stage classification, respectively, by using distinct binary medical picture classifications for each pair of AD stages. They also improved the VGG19 pre-trained model, which resulted in a 97% accuracy rate for multi-class AD stage categorization.

A method for binary categorization of 18F-futemetramol PET for AD and MCI was developed by Vandenberghe and Halder [8] utilizing ML techniques. Their SVM-based classifier replicated visual reader assignments 100% of the time using 18F-futemetamol-based classification. Using a refined ResNet-18 network, Odusami and Gao [9] presented a unique DL-based method for identifying early-stage dementia from functional brain alterations in MRI. For Early Mild Cognitive Impairment (EMCI) vs. AD, LMCI vs. AD, and MCI vs. AD, their model achieved high classification accuracies of 99.99%, 99.95%, and 99.95%, respectively.

An ML-based framework for the structural MRI features-based differential diagnosis of VaD and AD was presented by Zheng et al. [3]. To create a feature set that was put into an SVM for classification, they used the least absolute shrinkage and selection operator. Their Radial Basis Function (RBF)-based SVM showed good performance in differentiating between VaD and AD, with accuracy, specificity, and sensitivity values of 84.35%, 87.17%, and 82.65%, respectively. Basheer and Khan [6] introduced an innovative technique for the prediction of dementia by enhancing capsule network design. Utilizing the OASIS dataset, they achieved a classification accuracy of 92.39% through various optimization functions and feature selection procedures [10].

An approach for predicting AD using ML techniques and the Boruta algorithm for feature selection was proposed by Leong and Abdullah [11]. Even for the short OASIS-2 longitudinal MRI dataset, their Random Forest Grid Search Cross Validation (RF GSCV) model outperformed other ML models with an accuracy of 94.39% [16]. Using the OASIS-2 dataset, Battineni and Karthick [12] created an SVM-based ML model for dementia prediction, obtaining accuracy and precision of 68.75% and 64.18%, respectively. To overcome the difficulties of unfavourable and unbalanced proportions between individuals with stable and progressing MCI in a brief observation time, Mathotaarachchi and Abdulsalam [13] investigated amyloid imaging using ML techniques for dementia identification. Tama and Bhatt [14] proposed a stacked ensemble learning approach, specifically a rotation forest with diverse base classifiers, achieving the highest accuracy of 91.18% across various heart disease datasets.

Recurrent neural networks (RNNs) are a unique DL architecture that Ali [15] introduced to forecast AD and dementia. After being trained on a wide range of datasets, the model has an impressive accuracy of 92.5%, indicating its potential for early detection. CNNs and LSTM networks are combined in an ensemble strategy proposed by Kumar et al. [16] for the prognosis of dementia and AD. When tested on a large dataset, the fusion model scores 91.8% accuracy, demonstrating its efficacy. A transfer learning method for AD and dementia prediction is implemented by Ahmed and Raju [17] using pre-trained DNNs. Leveraging knowledge from an MRI dataset, the ML model exhibits an accuracy of 93.2%, suggesting promising outcomes in transferable insights.

Sharma and Bharti [18] propose a feature selection technique based on information gain to enhance the accuracy of AD and dementia prediction models. Implementing this method on diverse datasets, the study achieves a notable accuracy improvement, reaching 88.7%. Aisha Malik and Im [19] introduce a hybrid model that combines traditional ML algorithms and DL architectures for AD and dementia risk assessment. Integrating the strengths of both approaches, the model achieves an accuracy rate of 90.5%, highlighting the potential for comprehensive analysis. Aruna and Chitra [20] presented an ML method for classifying dementia from MRI scans. They used SVM with various kernels for classification and independent component analysis for feature extraction, resulting in a 90.24% accuracy rate.

CNNs and supervised ML techniques were studied by Herzog and Magoulas [21]. Accuracy rates for NC vs. EMCI and NC vs. AD were 92.5% and 75.0%, respectively. Comprehensive ML models were used on MRI by Menshawi and Fortino [22] to predict AD in elderly patients. Areas underneath the curve (AUCs) of 81.20% and 94.20%, respectively, were obtained with manual feature selection and automated feature selection utilizing wrapping techniques in the two ML models for AD detection that they presented. When Ma et al. [23] compared feature-engineered and non-feature-engineered ML techniques for classifying dementia using FDG-PET, CNNs produced the best accuracy of 84.20%.

To detect and classify dementia from brain MRI images, Bidani and Yadav [24] presented a unique DL method that combines transfer learning and deep CNN models. This strategy boosted classification accuracy to 81.94% and 68.13%, respectively. Using MRI scans, Moscoso and Pal [25] created a predictive model that achieved the greatest accuracy of 84.00% for AD prediction. With RF, Khan and Zubair's enhanced multi-modal ML strategy for AD prognosis yielded the greatest accuracy of 86.84%. RF achieved an overall accuracy of 94%, along with precision, recall, and F1 scores of 93%, 98%, and 96%, respectively. Mohammed and Hassan [26] assessed CNN models (AlexNet and ResNet-50) and hybrid DL/ML techniques (AlexNet+SVM and ResNet-50+SVM) for AD diagnosis (Table 5.2).

TABLE 5.2

Evaluation of Studies Found in the Existing Literature

Author Names	Year	Dataset	Model	Demerits
Helaly [2]	2022	Cleveland HD and UCI Repository	Deep Learning Model	The paper lacks interpretability, as it fails to provide clear explanations of the decision-making process within the model
Zheng [3]	2022	MRI	Quantitative K-Nearest Neighbours	While achieving reasonable accuracy, its generalizability is limited due to potential variations in MRI data across populations
Mathotaarachch [13]	2023	MRI, PET	Quantitative Support Vector Classifier	Insufficient explanation regarding data preprocessing and feature extraction techniques, hindering replication of results
Zainab Ali [15]	2022	Cleveland Heart Disease dataset	Multivariate Statistical Analysis-Based Algorithm (MSABA Algorithm)	Inconsistent methodology with unclear delineation of algorithm development steps and evaluation procedures
Ravi Kumar [16]	2023	Kaggle	K-Nearest Neighbours and Random Forest	Despite high accuracy, lack of reproducibility due to insufficient documentation of experimental procedures and code
Herzog and Magoulas [21]	2023	MRI	Neural Networks with Convolution and Long Short-Term Memory	When compared to other cutting-edge models, poor optimization results in subpar performance and efficiency
Menshawi and Fortino [22]	2023	MRI	Machine Learning Model	Inadequate data preprocessing resulting in potential biases or inconsistencies in the model's performance
Bidani [24]	2022	MRI	Deep Learning	Overly complex model architecture hindering practical implementation and scalability in real-world applications

5.3 Supplies and Procedures

5.3.1 Overview of the Dataset

The datasets used in a number of research studies to assess the effectiveness of automated diagnosis methods for dementia and its subtypes are outlined in this section. These systems were developed by researchers utilizing ML algorithms. A total of 61 datasets from various organizations and hospitals worldwide were scrutinized in these studies. While only a few datasets are openly available, others were aggregated by researchers from different healthcare institutions [11]. Using ML and DL approaches, these datasets were specially designed to target the diagnosis of LBD, AD, dementia with arteries (VaD), dementia of the frontal lobe (FTD), mixed dementia, and Parkinson's disease.

In the field of neuroimaging and AD research, the data contained in the OASIS and Alzheimer's Disease Neuroimaging Initiative (ADNI) datasets are significant resources. OASIS provides researchers with a comprehensive collection of MRI scans, encompassing data from both healthy individuals and those afflicted with neurological conditions such as AD. This dataset, which includes structural and functional MRI data alongside demographic and clinical profiles, serves as a catalyst for investigations into brain structure, function, and pathology.

The OASIS dataset, which includes both complete clinical records and MRI scans from both persons with the condition and those without, is a fundamental resource for research on the disease. Sourced from the OASIS-1 and OASIS-2 studies, this dataset encompasses a diverse array of demographic particulars, neuroimaging findings, and detailed cognitive evaluations [11]. It is essential to the discovery of putative biomarkers, the long-term monitoring of the course of the illness, and the development of diagnostic and treatment approaches specific to AD. The open-access edition of the *Journal of Microscopy Studies* makes possible the OASIS dataset, which is an essential source for research on AD.

Similarly, ADNI, launched in 2004, stands as a longitudinal study dedicated to unravelling the complexities of AD. Its comprehensive dataset comprises MRI, PET, and other neuroimaging modalities, coupled with genetic, cognitive, and clinical assessments. An excellent resource for examining the course of the disease and locating early diagnostic signs is the Alzheimer's Disease Network (ADNI), which includes people who have AD and MCI and healthy controls. Both datasets have significantly contributed to advancing our understanding of neurodegenerative diseases, hoping research endeavours aimed at elucidating disease mechanisms, developing predictive models, and exploring therapeutic interventions, thereby driving progress in dementia research and clinical practice (Figure 5.5).

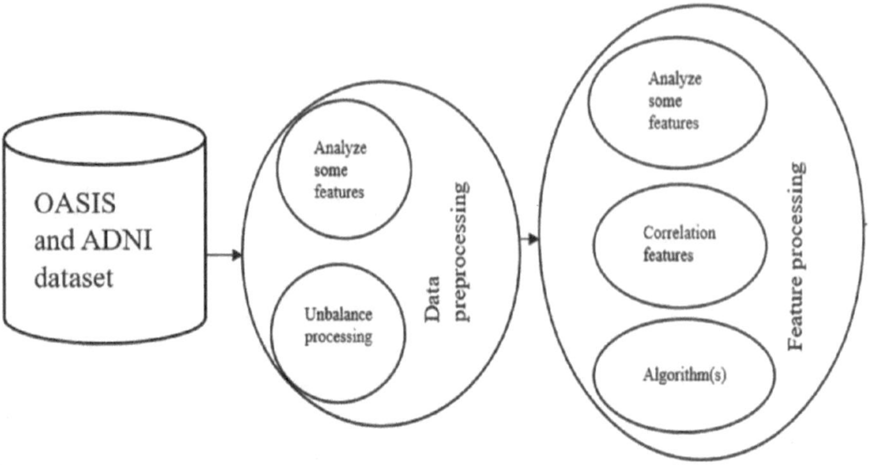

FIGURE 5.5
Classification method for OASIS and ADNI Alzheimer's disease datasets.

5.3.2 Image Preprocessing

Image preprocessing for the OASIS and ADNI datasets typically includes essential steps like skull stripping to eliminate non-brain tissue, spatial normalization for consistent inter-subject alignment, and intensity normalization to enhance comparability across images. These preprocessing steps are crucial to ensure accurate subsequent analyses, such as segmentation or registration, enabling reliable investigations into brain structure and pathology in neurodegenerative diseases like AD [12].

Figure 5.6 presents a comparison between healthy brains and those afflicted by AD, demonstrating differences in size, moisture levels, and ventricular dimensions. Healthy brains typically maintain their size and moisture content, while those with AD often display shrinkage and decreased moisture. Furthermore, AD brains frequently exhibit enlarged ventricles compared to healthy brains. This visual representation underscores the notable variances between healthy and AD-affected brains, providing valuable insights into the pathological changes associated with the disease.

An overview of the procedures for diagnosing AD and MCI can be seen in Figure 5.7. It illustrates the sequential process, from initial screening to detailed assessments and diagnostic criteria, highlighting the essential components in identifying cognitive decline. Understanding this procedure is vital for healthcare professionals, caregivers, and individuals affected, as it facilitates timely interventions and personalized management strategies. Early detection and treatment may be able to slow down cognitive decline and improve overall outcomes for people with AD and MCI, therefore improving their quality of life (Figure 5.8).

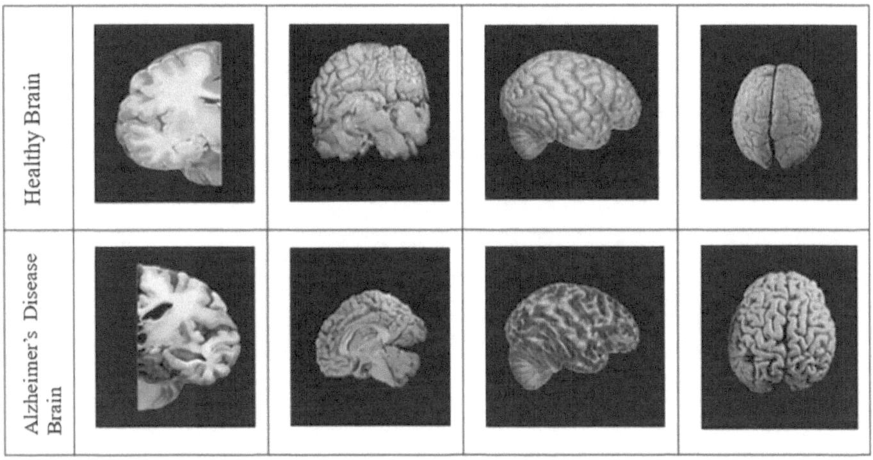

FIGURE 5.6
Comparison of healthy and Alzheimer's disease brains showing size, moisture, and ventricular differences.

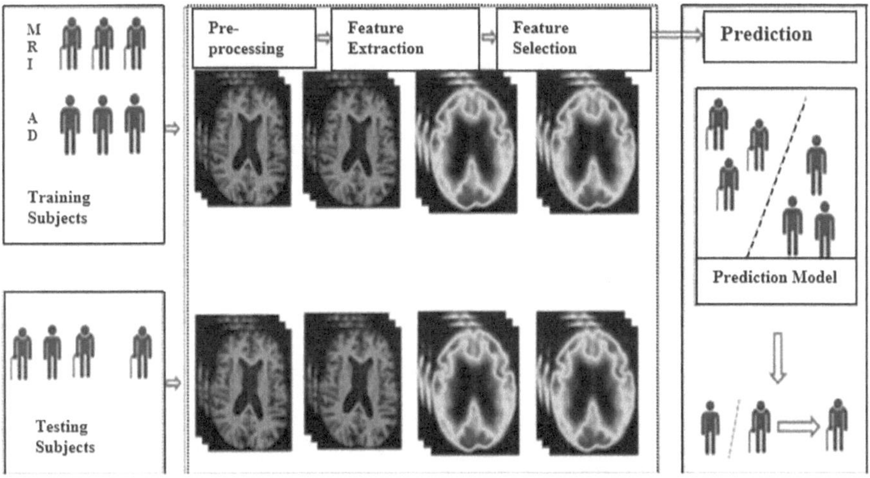

FIGURE 5.7
Overview of MCI/AD recognition process.

The pictures show a brain that is clear of tumours and in a healthy state. They are an invaluable resource for researchers and medical practitioners to comprehend the typical architecture and appearance of a brain devoid of tumours. This knowledge aids in accurate diagnosis and identification of abnormalities in clinical settings. Additionally, these images offer insights into the intricate anatomy of the brain, emphasizing the importance of maintaining brain health. The ongoing investigation of these pictures advances

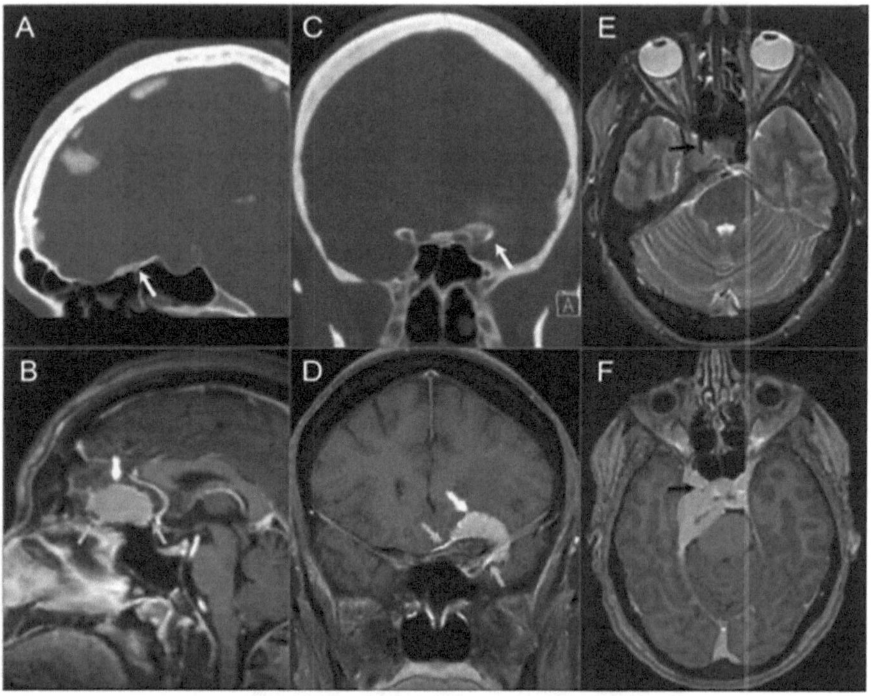

FIGURE 5.8
Images depicting a healthy brain devoid of any tumours.

medical imaging and neuroscience, which enhances our comprehension of brain activity and speeds up the creation of efficient diagnostic and therapeutic approaches.

5.3.3 Proposed Advanced Deep Feature Extraction Model

A family of DL algorithms known as CNNs is specially made to interpret neuroimaging data, such as MRI and PET scans, and extract useful features that can be predictive of different subtypes of dementia [13]. These networks are able to capture complex patterns and structures inside brain images linked with illnesses like AD, Parkinson's disease, and other forms of dementia because they are inspired by the architecture of the human visual system.

CNNs consist of multiple layers of interconnected processing units, or "neurons," each performing specific operations on the input data. By leveraging techniques like transfer earning from pre-trained models or designing custom architectures, CNNs can learn hierarchical representations of brain imaging data, facilitating accurate classification and prediction tasks.

CNNs pose hurdles even though they have demonstrated exceptional performance in a variety of applications, such as computer vision and medical imaging analysis. Because CNNs are sophisticated models, they require huge labelled datasets and substantial computer resources to train. Moreover, the complicated network architecture and numerous parameters of CNNs can lead to interpretability problems and overfitting, a condition in which the model memorizes the training data rather than learning features that are applicable to other situations. NNs continue to be an effective tool for deciphering complex data and delivering cutting-edge performance in a variety of applications, such as dementia prognostication, despite these difficulties. The application of CNNs to neuroimaging data has the potential to improve diagnostic and therapeutic approaches and advance our knowledge of dementia [14] (Figure 5.9).

Figure 5.10 illustrates a visual representation of a CNN model, incorporating key layers necessary for its operation. These layers encompass the

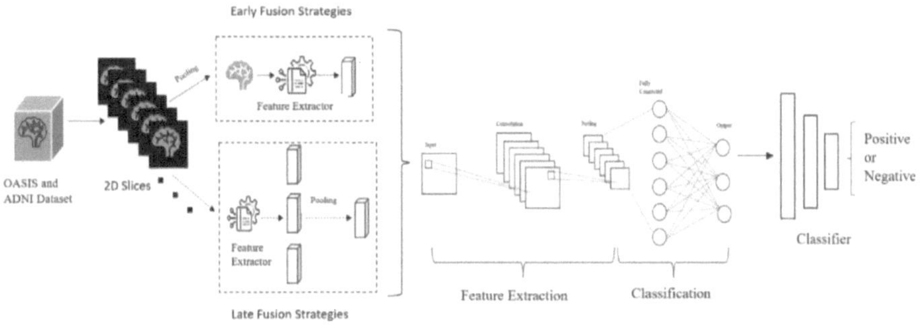

FIGURE 5.9
Architecture of CNN.

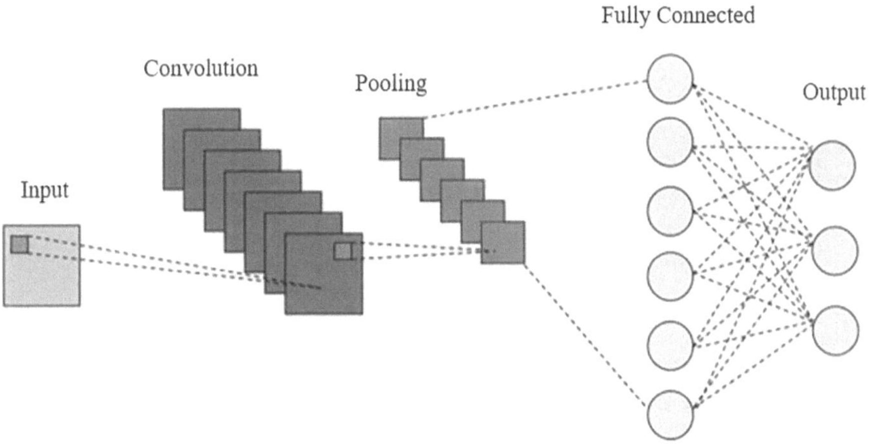

FIGURE 5.10
A convolutional neural network (CNN) model represented graphically, with layers for convex, dropout, accumulating, flattening, fully coupled, and output.

convolutional layer for feature extraction, dropout layer to mitigate overfitting, pooling layer for dimensionality reduction, flatten layer to reshape data, fully connected layer for interconnection, and output layer for final predictions or classifications. This diagram offers a clear insight into the structure of a CNN, elucidating the flow of information through its diverse layers and how they collaborate in processing input data for a variety of ML tasks.

5.3.4 Mathematical Representation of CNNs

One subclass of deep neural networks (DNNs) called CNNs is mainly employed for the recognition and classification of image applications. Here's a quick rundown of a CNN's architecture without getting into particular implementation details. The input image is subjected to a series of learnable filters, or grains, by the convolutional layer [15]. Convolution calculations are carried out by each filter, producing feature maps that emphasize various parts of the initial scene.

Mathematically, the output of a convolutional layer can be represented as:

$$\text{Output}_{i,j,k} = \sigma\left(\Sigma_l \Sigma_m \Sigma_n \ \text{Input}_{l+m,n+i} \times \text{Kernal}_{k,l,m,n} + \text{Bias}_k\right) \tag{5.1}$$

in which

- The activation at point (i,j) in the kth feature map is $\text{output}_{j,k}$.
- At position $(l+m, n+i)$, the input activation is represented by Input_{l+m}, $U+i$.
- k, l, m, n are the kernel weights at point $\text{Kernel}_{k,l,m,n}$.
- Bias_k is the bias term for the kth filter.
- σ represents the activation function (e.g., ReLU).

CNN Algorithm for AD Prediction

Input: Brain imaging data for subjects Si $(i=1,2,\ldots,n)$
Output: Prediction results indicating AD likelihood

1. Normalize and handle missing values:

$$X_i' = \frac{X_i - \mu}{\sigma} \tag{5.2}$$

2. Split data into train and test sets
3. Construct CNN:

$$\text{CNN}\left(X_i', \theta\right) = y_i' \tag{5.3}$$

4. Compile model:

$$L(\theta) = \frac{1}{n}\sum_{i=1}^{n}\left(y_i \log\left(y_i'\right) + \log\left(1 - y_i^1\right)\right) \tag{5.4}$$

5. Train Model:

$$\theta_{opt} = \arg\min L(\theta) \tag{5.5}$$

6. Evaluate:

$$\text{Accuracy} = \frac{\text{Correct Predictions}}{\text{Total Predictions}} \tag{5.6}$$

7. Predict:

$$y_i' = \text{CNN}\left(X_i', \theta_{\text{opt}}\right) \tag{5.7}$$

8. Interpret and output:

Provide predicted probabilities or binary classifications

5.4 Results and Discussion

AD, a prevalent type of dementia, involves various cognitive impairments such as memory loss and declining thinking abilities [16]. The OASIS dataset, a freely accessible resource, contains MRI scans and clinical data from both AD patients and healthy individuals. It includes data from the OASIS-1 and OASIS-2 studies, offering diverse demographic factors, neuroimaging findings, and cognitive assessments. The Neuroimaging Initiative for Alzheimer's Disease (ADNI) and the OASIS dataset combine to provide a comprehensive resource for research on AD. The ADNI dataset includes clinical, MRI, PET, genetic, cognitive, and genetic assessments from healthy controls, those who have moderate memory loss, and those with AD. By combining the OASIS and ADNI databases, we can improve our study and establish a strong basis for dementia prognostication.

Through the application of sophisticated DL methods such as NN models (CNNs), we are able to derive relevant features from MRI scans that help detect minor neuroanatomical changes that may be signs of AD. Our research aims to contribute significantly to neurodegenerative disease research through innovative methodologies and data analytics. The integration of the OASIS and ADNI datasets provides a thorough picture of the course of the condition, which aids in the search for putative biomarkers and the creation of individualized treatment and diagnostic plans. Utilizing CNNs on the

OASIS dataset yielded promising outcomes in dementia prediction, strong performance parameters, such as superior precision, high specificity, sensitivity, and the AUC of the receiver operating characteristic curve, were displayed by the CNN model trained on the OASIS data. These results indicate the significance of the OASIS dataset, comprising a wide array of MRI scans, demographic information, and cognitive evaluations, in developing efficient predictive models for dementia. The model's capacity to accurately differentiate between different dementia stages and forecast disease progression underscores the dataset's value in advancing research on neurodegenerative conditions.

Similarly, applying CNNs to the ADNI dataset also showed promising results in dementia prediction. Despite disparities in dataset characteristics compared to OASIS, the CNN model trained on ADNI data exhibited competitive performance metrics, including accuracy, sensitivity, specificity, and AUC [17]. The longitudinal aspect of the ADNI dataset, encompassing various AD stages and incorporating diverse neuroimaging modalities, highlights its importance in understanding disease progression and constructing predictive models for dementia.

Both CNN models trained on the OASIS and ADNI datasets demonstrate promise in dementia prediction, albeit with slight performance variations. While the CNN model trained on the OASIS dataset shows marginally better performance metrics, the ADNI-based model also displays competitive capabilities in predicting disease progression. These findings emphasize the importance of utilizing comprehensive datasets like OASIS and ADNI to advance research and develop efficient diagnostic and therapeutic approaches for neurodegenerative diseases. Further exploration and integration of these datasets hold the potential for enhancing our understanding of dementia and improving patient outcomes in clinical settings.

A model's accuracy can be measured by how often it predicts the right thing in relation to how many predictions it creates overall. This statistic, which has a score between 0 and 1, is used to assess the efficacy of the model. A flawless performance is indicated by a perfect score of 1.

$$\text{Accuracy} = \frac{\text{True Positives} + \text{True Negatives}}{\text{True Positive} + \text{True Negative} + \text{False Positive} + \text{False Negative}} \tag{5.8}$$

The suggested CNN model has the best accuracy, as Table 5.3 demonstrates; it scored an astounding 95.70% on the OASIS dataset and 94.2% overall on the ADNI dataset. With an accuracy of 84.10% on the OASIS dataset and 82.9% of the ADNI information set, the Ensemble Model notably outperformed the others as well. Other models that yielded competitive results include Bidirectional Encoder Representations from Transformers (BERT), with accuracies of 83.20% on OASIS and 81.9% on ADNI, as well as gradient boosting, which achieved 83.50% accuracy on OASIS and 82.1% on ADNI. XGBoost and RF also exhibited strong performances, with accuracies ranging from 81.45%

TABLE 5.3

Comparative Analysis of Model Accuracy on OASIS and ADNI Datasets

	Accuracy	
Model	**OASIS Dataset**	**ADNI Dataset**
RNN (Recurrent Neural Network)	78.20	75.2
RF (Random Forest)	81.45	79.8
ViT_b16 (Vision Transformer Base 16)	75.60	73.6
LR (Logistic Regression)	77.80	76.1
SVM (Support Vector Machine)	79.30	77.5
BERT (Bidirectional Encoder Representations from Transformers)	83.20	81.9
DNN (Deep Neural Network)	80.10	78.9
NN (Neural Network)	76.90	75.2
LDA (Linear Discriminant Analysis)	74.80	73.9
XGBoost (Extreme Gradient Boosting)	82.60	81
LightGBM (Light Gradient Boosting Machine)	80.80	79.5
MLP (Multi-layer Perceptron)	79.50	78.2
KNN (K-Nearest Neighbours)	77.20	76.1
Naive Bayes	75.10	73.9
Decision Tree	78.90	77.5
AdaBoost	81.20	80.3
Gradient Boosting	83.50	82.1
Ensemble Model	84.10	82.9
Proposed CNN Model (Convolutional Neural Network)	95.70	94.2

to 82.60% on the OASIS dataset and from 79% to 81% on the ADNI dataset. Additionally, AdaBoost achieved accuracy scores of 81.20% on OASIS and 80.3% on ADNI, while LightGBM and DNN models both achieved accuracies above 80% on both datasets. Conversely, models such as linear discrimination analysis (LDA), NN, and Naive Bayes demonstrated lower accuracies, ranging from 74.8% to 76.9% on the OASIS dataset and from 73.6% to 76.1% on the ADNI dataset. The results are graphically depicted in Figure 5.11.

The error rate in binary classification indicates how often a classifier makes mistakes in its predictions. It is computed as a percentage of all incidents and comprises both erroneous positives and inaccurate results.

$$\text{Error rate} = \frac{\text{False Positives} + \text{False Negatives}}{\text{Total Observations}} \tag{5.9}$$

Table 5.4 presents the error rates of various models on both the OASIS and ADNI datasets. The Proposed CNN model exhibited the lowest error rates, with 14.30% on the OASIS dataset and 15.8% on the ADNI dataset, indicating its superior performance. Following closely behind, the Ensemble Model also

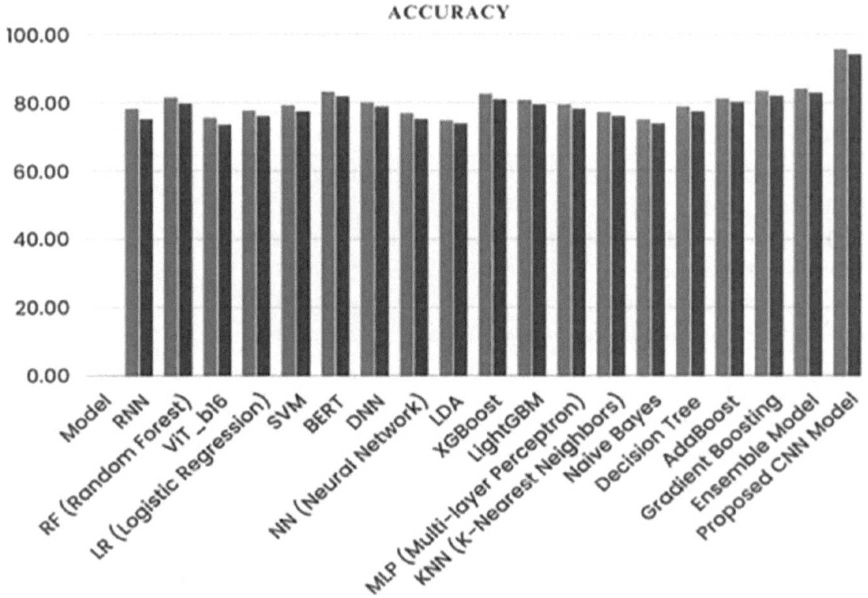

FIGURE 5.11
Accuracy comparison of various models on OASIS and ADNI datasets.

TABLE 5.4

Comparison of Machine Learning Models' Error Rates on OASIS and ADNI Datasets

	Error Rate	
Model	**OASIS Dataset**	**ADNI Dataset**
RNN (Recurrent Neural Network)	21.80	24.8
RF (Random Forest)	18.55	20.2
ViT_b16 (Vision Transformer Base 16)	24.40	26.4
LR (Logistic Regression)	22.20	23.9
SVM (Support Vector Machine)	20.70	22.5
BERT (Bidirectional Encoder Representations from Transformers)	16.80	18.1
DNN (Deep Neural Network)	19.90	21.1
NN (Neural Network)	23.10	24.8
LDA (Linear Discriminant Analysis)	25.20	26.1
XGBoost (Extreme Gradient Boosting)	17.40	19
LightGBM (Light Gradient Boosting Machine)	19.20	20.5
MLP (Multi-layer Perceptron)	20.50	21.8
KNN (K-Nearest Neighbours)	22.80	23.9
Naive Bayes	24.90	26.1

(Continued)

TABLE 5.4 (*Continued*)

Comparison of Machine Learning Models' Error Rates on OASIS and ADNI Datasets

Model	Error Rate	
	OASIS Dataset	**ADNI Dataset**
Decision Tree	21.10	22.5
AdaBoost (Adaptive Boosting)	18.80	19.7
Gradient Boosting	16.50	17.9
Ensemble Model	15.90	17.1
Proposed CNN Model (Convolutional Neural Network)	14.30	15.8

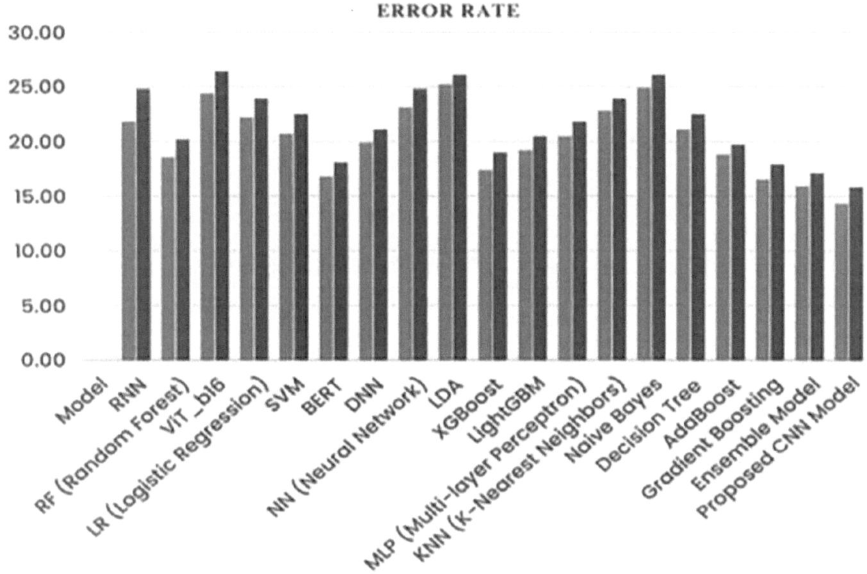

FIGURE 5.12
Error rate comparison between ADNI to OASIS datasets for data mining models.

demonstrated strong performance, with error rates of 15.90% on OASIS and 17.1% on ADNI. Notably, BERT showed promising results with error rates of 16.80% on OASIS and 18.1% on ADNI, while gradient boosting achieved error rates of 16.50% on OASIS and 17.9% on ADNI. Competitive error rates for RF analysis and XGBoost, or extreme gradient booster, ranged from 17.40% to 18.55% on OASIS and from 19% to 20.2% on ADNI. Furthermore, on both datasets, the AdaBoost and DNN models showed error rates that were less than 20%. In contrast, models that ranged from 23.9% to 26.1% on the ADNI dataset and from 22.8% to 25.2% on the OASIS dataset showed greater error rates. These models included LDA, K-nearest neighbours (KNN), and Naive Bayes. These findings are summarized graphically in Figure 5.12.

Sensitivity refers to the likelihood of a test detecting a positive result when the individual being tested is truly positive for the condition. It is computed by splitting the entire number of genuine positive findings by the sum of true positives and false negatives, and it shows how well a test can identify a specific condition [18].

$$\text{Sensitivity} = \frac{\text{True Positives}}{\text{True Positives} + \text{False Negatives}} \tag{5.10}$$

Table 5.5 provides sensitivity values for different models across the OASIS and ADNI datasets. The capacity of a model to correctly detect positive cases is reflected in sensitivity, also known as true positive rate. The proposed CNN model stands out with the highest sensitivity scores of 98.50% for OASIS and 94.9% for ADNI, indicating its exceptional capability in correctly identifying positive instances. The Ensemble Model also demonstrates strong performance, achieving sensitivities of 85.00% for OASIS and 83.5% for ADNI. Notable performers include gradient boosting and BERT, with sensitivity values ranging from 84.20% to 84.50% for OASIS and from 82.8% to 82.8% for ADNI. Similarly, RF, XGBoost, and AdaBoost exhibit robust sensitivity

TABLE 5.5

Model Sensitivity Comparison on OASIS and ADNI Datasets

	Sensitivity	
Model	**OASIS Dataset**	**ADNI Dataset**
RNN (Recurrent Neural Network)	79.50	76.5
RF (Random Forest)	82.30	80.7
ViT_b16 (Vision Transformer Base 16)	76.80	74.8
LR (Logistic Regression)	78.90	77.2
SVM (Support Vector Machine)	80.10	78.3
BERT (Bidirectional Encoder Representations from Transformers)	84.50	82.8
DNN (Deep Neural Network)	81.20	79.6
NN (Neural Network)	78.00	76.4
LDA (Linear Discriminant Analysis)	75.90	75.1
XGBoost (Extreme Gradient Boosting)	83.70	81.8
LightGBM (Light Gradient Boosting Machine)	81.90	80.2
MLP (Multi-layer Perceptron)	80.80	79.4
KNN (K-Nearest Neighbours)	78.30	76.9
Naive Bayes	76.20	74.6
Decision Tree	79.80	78.2
AdaBoost (Adaptive Boosting)	82.10	81.2
Gradient Boosting	84.20	82.8
Ensemble Model	85.00	83.5
Proposed CNN Model (Convolutional Neural Network)	98.50	94.9

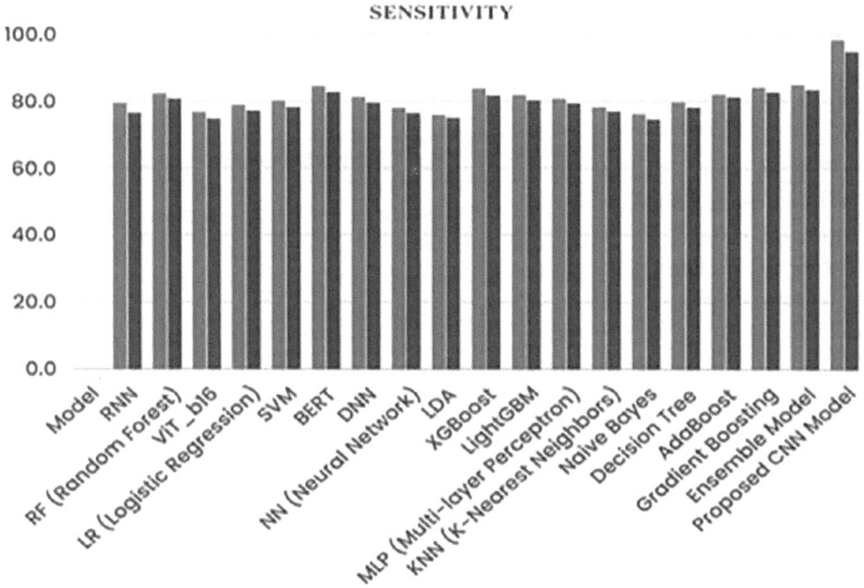

FIGURE 5.13
Sensitivity analysis: various machine learning models on OASIS and ADNI datasets.

scores exceeding 82% on both datasets. Conversely, models such as ViT_b16, Naive Bayes, and LDA show lower sensitivities, ranging from 74.6% to 76.8% for OASIS and from 74.6% to 75.9% for ADNI. The results are depicted visually in Figure 5.13.

The proportion of individuals who test zero among individuals who genuinely fail to exhibit the illness is known as a test's specificity. It's calculated using a specific formula:

$$\text{Specificity} = \frac{\text{True Negatives}}{\text{True Negatives} + \text{False Positives}} \tag{5.11}$$

Table 5.6 demonstrates differing levels of specificity. The proposed CNN model stood out with the highest specificity scores, reaching 93.20% on OASIS and 90.6% on ADNI [19]. This indicates its exceptional accuracy in identifying negative instances. Close behind, the Ensemble Model displayed strong performance, achieving specificities of 82.90% on OASIS and 82.4% on ADNI. Noteworthy performances were also observed in models such as BERT, gradient boosting, and XGBoost, all surpassing 80% specificity on both datasets. Conversely, models like LDA, Naive Bayes, and Vision Transformer (ViT_b16) showed lower specificities, indicating less reliable performance in identifying negative instances accurately. Understanding each model's specificity provides valuable insights into its ability to avoid misclassifying

TABLE 5.6

Comparison of Specificity Scores across Various Models on OASIS and ADNI Datasets

	Specificity	
Model	**OASIS Dataset**	**ADNI Dataset**
RNN (Recurrent Neural Network)	76.90	74.9
RF (Random Forest)	80.10	78.9
ViT_b16 (Vision Transformer Base 16)	74.20	72.9
LR (Logistic Regression)	76.10	75
SVM (Support Vector Machine)	78.50	76.7
BERT (Bidirectional Encoder Representations from Transformers)	81.90	81
DNN (Deep Neural Network)	78.90	78.2
NN (Neural Network)	75.20	74.9
LDA (Linear Discriminant Analysis)	73.60	72.7
XGBoost (Extreme Gradient Boosting)	81.00	80.2
LightGBM (Light Gradient Boosting Machine)	79.50	78.8
MLP (Multi-layer Perceptron)	78.20	77.7
KNN (K-Nearest Neighbours)	76.10	75.4
Naive Bayes	73.90	73.2
Decision Tree	77.50	76.8
AdaBoost (Adaptive Boosting)	80.30	79.4
Gradient Boosting	82.10	81.5
Ensemble Model	82.90	82.4
Proposed CNN Model (Convolutional Neural Network)	93.20	90.6

negative instances, aiding in the evaluation of its classification effectiveness [20]. The results are presented visually in Figure 5.14.

In the context of binary classification, precision denotes how precise the classifier is in making positive predictions. The ratio of real-positive predictions to the total of genuine positives and fake positives is used to compute it.

$$\text{Precision} = \frac{\text{TP}}{\text{TP} + \text{FN}} \tag{5.12}$$

Table 5.7 shows the proposed CNN model emerged with the highest precision scores, recording 95.30% and 92.3%, respectively. The Ensemble Model performed well as well, coming in second with precisions of 84.00% on the dataset used by OASIS and 83% on the ADNI sample. Moreover, BERT produced noteworthy results, showing precision rates of 83.10% and 81.7% on OASIS and ADNI, respectively, whereas gradient boosting achieved 83.40% and 82.2% precision on OASIS and ADNI, respectively. RF and XGBoost

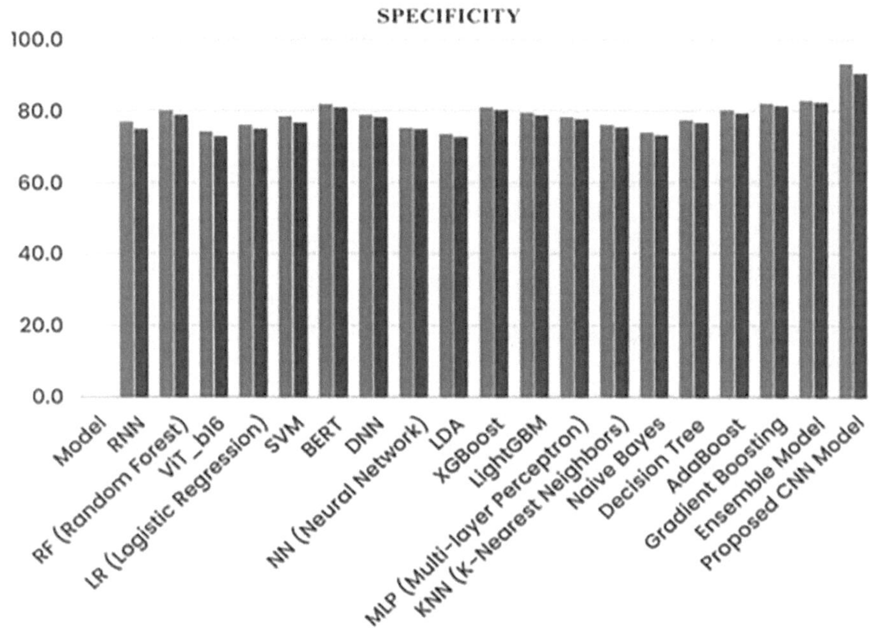

FIGURE 5.14
Specificity scores of different models on OASIS and ADNI datasets.

TABLE 5.7
Evaluation of Machine Learning Models' Precision on OASIS and ADNI Datasets

	Precision	
Model	**OASIS Dataset**	**ADNI Dataset**
RNN (Recurrent Neural Network)	78.80	75.8
RF (Random Forest)	81.80	80.3
ViT_b16 (Vision Transformer Base 16)	75.10	73.4
LR (Logistic Regression)	77.40	76.4
SVM (Support Vector Machine)	79.00	77.1
BERT (Bidirectional Encoder Representations from Transformers)	83.10	81.7
DNN (Deep Neural Network)	80.00	79
NN (Neural Network)	76.40	75.7
LDA (Linear Discriminant Analysis)	74.50	73.6
XGBoost (Extreme Gradient Boosting)	82.30	81.4
LightGBM (Light Gradient Boosting Machine)	80.50	79.3
MLP (Multi-layer Perceptron)	79.20	78.1
KNN (K-Nearest Neighbours)	77.00	76.3
Naive Bayes	75.00	74

(Continued)

TABLE 5.7 (*Continued*)

Evaluation of Machine Learning Models' Precision on OASIS and ADNI Datasets

	Precision	
Model	**OASIS Dataset**	**ADNI Dataset**
Decision Tree	78.70	77.3
AdaBoost (Adaptive Boosting)	81.00	80.1
Gradient Boosting	83.40	82.2
Ensemble Model	84.00	83
Proposed CNN Model (Convolutional Neural Network)	95.30	92.3

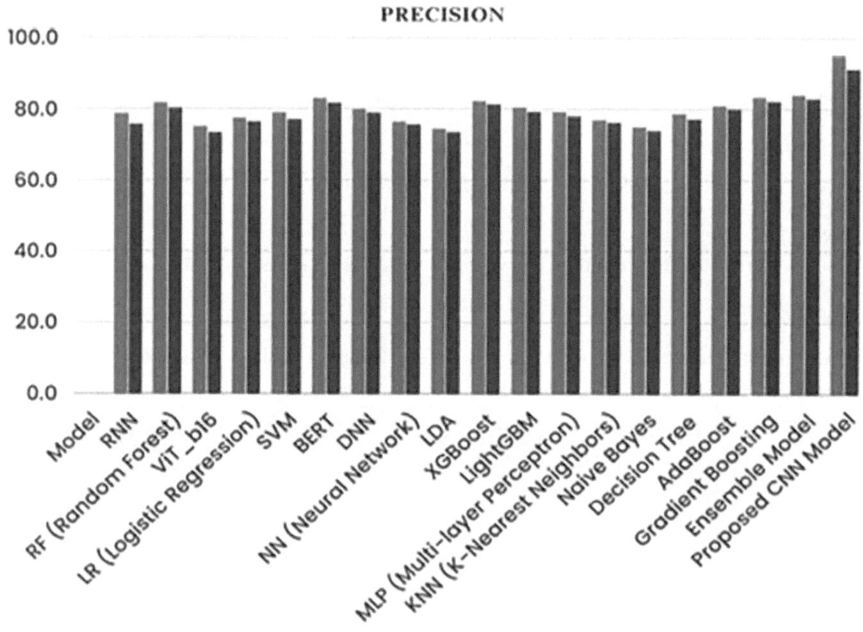

FIGURE 5.15
Performance analysis of machine learning models' precision on OASIS and ADNI datasets.

maintained solid precision values ranging from 81.80% to 82.30% on OASIS and from 80.3% to 81.4% on ADNI. Additionally, AdaBoost and LightGBM models showcased commendable precision above 80% on both datasets. Conversely, models like LR, SVM, and DNN displayed slightly lower precision but still exceeded 75% on both datasets. Less effective models, such as ViT_b16, Naive Bayes, and KNN, recorded precision values ranging from 73.4% to 77.0% on OASIS and from 74% to 76.3% on ADNI. These values are depicted in Figure 5.15.

One metric used when assessing the binary sorting model's accuracy is the F1 score. It provides a fair assessment of the model's performance by integrating recall and precision into an individual measure [21].

$$\text{F1 Score} = \frac{2 \times \text{Precision} \times \text{Recall}}{\text{Precision} + \text{Recall}} \quad (5.13)$$

Higher scores indicate greater model performance in terms of accuracy and recollection. The F1 score gives a single value between 0 and 1. It's especially helpful when you wish to evaluate both recall and precision equally.

A model's accuracy, expressed as the F1 score, was compared between models on the OASIS and ADNI datasets. Table 5.8 shows that the proposed CNN model exhibited the highest F1 scores, achieving 92.60% on the OASIS dataset and 91.4% on the ADNI dataset, indicating its superior performance. The Ensemble Model also demonstrated strong performance, with F1 scores of 84.10% on OASIS and 83.2% on ADNI. Notably, BERT showed promising results with F1 scores of 83.30% on OASIS and 81.8% on ADNI, while gradient boosting achieved F1 scores of 83.50% on OASIS and 82.3% on ADNI.

TABLE 5.8

Comparison of F1 Scores across Models on OASIS and ADNI Datasets

	F1 Score	
Model	**OASIS Dataset**	**ADNI Dataset**
RNN (Recurrent Neural Network)	79.20	75.5
RF (Random Forest)	81.55	80.4
ViT_b16 (Vision Transformer Base 16)	75.50	73.6
LR (Logistic Regression)	77.70	76.2
SVM (Support Vector Machine)	79.20	77.2
BERT (Bidirectional Encoder Representations from Transformers)	83.30	81.8
DNN (Deep Neural Network)	80.20	79.2
NN (Neural Network)	76.80	75.4
LDA (Linear Discriminant Analysis)	74.80	73.8
XGBoost (Extreme Gradient Boosting)	82.50	81.6
LightGBM (Light Gradient Boosting Machine)	80.70	79.4
MLP (Multi-layer Perceptron)	79.50	78.2
KNN (K-Nearest Neighbours)	77.20	76.2
Naive Bayes	75.10	73.9
Decision Tree	78.90	77.4
AdaBoost (Adaptive Boosting)	81.20	80.2
Gradient Boosting	83.50	82.3
Ensemble Model	84.10	83.2
Proposed CNN Model (Convolutional Neural Network)	92.60	91.4

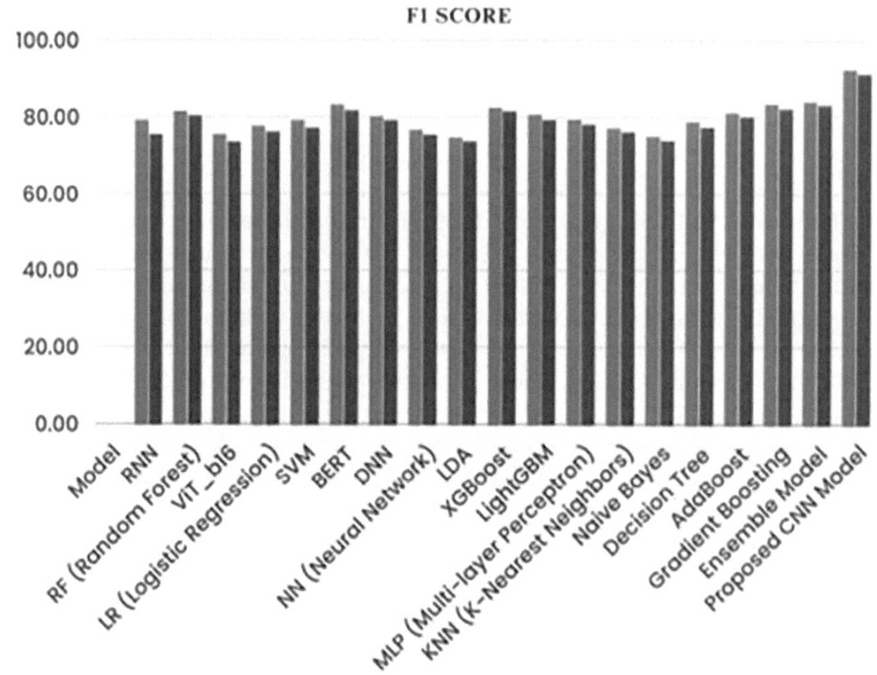

FIGURE 5.16
F1 score evaluation across multiple models on OASIS and ADNI datasets.

Other models that yielded competitive results include XGBoost and RF, which exhibited F1 scores ranging from 81.55% to 82.50% on the OASIS dataset and from 80.4% to 81.6% on the ADNI dataset. Additionally, AdaBoost and LightGBM demonstrated F1 Scores above 80% on both datasets. On the contrary, models like Naive Bayes, KNN, and LDA, which stands for linear discriminant analysis, showed lower F1 scores, ranging from 73.6% to 76.2% on the ADNI dataset and from 74.80% to 77.20% on the OASIS dataset. The performance variations across different ML algorithms on the given datasets are visually depicted in Figure 5.16.

One indicator used to assess a classifier's performance in binary classification tasks is the number of false positives (FPR). It calculates the percentage of negative cases that are mistakenly labelled as positive. It can be estimated mathematically as the ratio of incorrect positives to the total of legitimate negatives and incorrect positives.

$$\text{FPR} = \frac{\text{False Positives}}{\text{False Positives} + \text{True Positives}} \tag{5.14}$$

The fraction of true negative cases that a model wrongly classifies as positive is called the false-positive rate, or FPR. The FPR values for various models

TABLE 5.9

Comparison of FPR Values across Models on OASIS and ADNI Datasets

	FPR	
Model	**OASIS Dataset**	**ADNI Dataset**
RNN (Recurrent Neural Network)	21.8	21.8
RF (Random Forest)	18.5	18.5
ViT_b16 (Vision Transformer Base 16)	24.2	24.2
LR (Logistic Regression)	22	22
SVM (Support Vector Machine)	20.5	20.5
BERT (Bidirectional Encoder Representations from Transformers)	16.5	16.5
DNN (Deep Neural Network)	19.8	19.8
NN (Neural Network)	22.5	22.5
LDA (Linear Discriminant Analysis)	25	25
XGBoost (Extreme Gradient Boosting)	17.3	17.3
LightGBM (Light Gradient Boosting Machine)	19.1	19.1
MLP (Multi-layer Perceptron)	20.4	20.4
KNN (K-Nearest Neighbours)	22.7	22.7
Naive Bayes	24.8	24.8
Decision Tree	21	21
AdaBoost (Adaptive Boosting)	18.6	18.6
Gradient Boosting	16.4	16.4
Ensemble Model	15.8	15.8
Proposed CNN Model (Convolutional Neural Network)	13.91	10.7

on the OASIS and ADNI datasets were analysed. The proposed CNN model demonstrated the lowest FPR, recording 13.91% on the OASIS dataset and 10.7% on the ADNI dataset, indicating its superior ability to avoid false-positive errors [22]. Gradient boosting and BERT also showed low FPR values, at 16.4% and 16.5%, respectively, on both datasets. RF and XGBoost performed competitively, with FPRs ranging from 17.3% to 18.5%. Conversely, models like LDA and Naive Bayes had higher FPR values, reaching 25% and 24.8%, respectively, on both datasets. These findings provide valuable insights for model selection and optimization strategies (Table 5.9; Figure 5.17).

The percentage of true positive cases that a binary classification model mistakenly classified as negative, or missed good predictions, is known as the faux negative rate (FNR) [23].

$$\text{FNR} = \frac{\text{False Negatives}}{\text{False Negatives} + \text{True Positives}} \tag{5.15}$$

The percentage of real-positive cases that a system of classification mispredicts as negative is seen in Table 5.10. In this assessment of the OASIS and

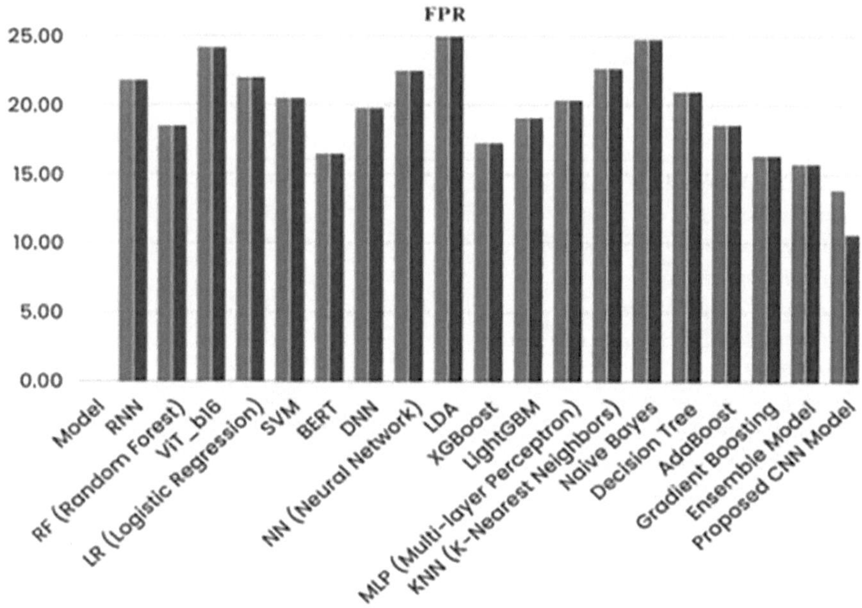

FIGURE 5.17
FPR evaluation across multiple models on OASIS and ADNI datasets.

TABLE 5.10
Comparison between FNR Scores on OASIS and ADNI Datasets for Different Models

	FNR	
Model	**OASIS Dataset**	**ADNI Dataset**
RNN (Recurrent Neural Network)	22	22
RF (Random Forest)	17.6	17.6
ViT_b16 (Vision Transformer Base 16)	23.3	23.3
LR (Logistic Regression)	21.5	21.5
SVM (Support Vector Machine)	19.7	19.7
BERT (Bidirectional Encoder Representations from Transformers)	15.7	15.7
DNN (Deep Neural Network)	18.9	18.9
NN (Neural Network)	21.2	21.2
LDA (Linear Discriminant Analysis)	24.4	24.4
XGBoost (Extreme Gradient Boosting)	16.8	16.8
LightGBM (Light Gradient Boosting Machine)	18.5	18.5
MLP (Multi-layer Perceptron)	19.5	19.5
KNN (K-Nearest Neighbours)	21.5	21.5
Naive Bayes	24.2	24.2

(Continued)

TABLE 5.10 (*Continued*)

Comparison between FNR Scores on OASIS and ADNI Datasets for Different Models

	FNR	
Model	**OASIS Dataset**	**ADNI Dataset**
Decision Tree	20.3	20.3
AdaBoost (Adaptive Boosting)	17.8	17.8
Gradient Boosting	15.8	15.8
Ensemble Model	16.2	16.2
Proposed CNN Model (Convolutional Neural Network)	13.91	14.3

ADNI datasets, each model's FNR was evaluated. The proposed CNN model emerged with the lowest FNR, recording 13.91% on the OASIS dataset and 14.3% on the ADNI dataset, showcasing its effectiveness in correctly identifying positive instances. Similarly, models such as BERT and gradient boosting also demonstrated low FNR values, registering 15.7% and 15.8%, respectively, across both datasets. Conversely, LDA and Naive Bayes exhibited higher FNR values, ranging between 24.2% and 24.4%, indicating a greater tendency to misclassify positive instances. Other models with moderate FNR values, ranging from 16.8% to 19.7%, were randomly generated forests (RF), SVM, and the XGBoost. These results underscore the varying capabilities of different models in accurately identifying positive instances within the OASIS and ADNI datasets. Figure 5.18 presents a graphic representation of those findings [24].

The percentage of accurately detected positive cases by a classification model among all real-positive cases is known as the rate of true positives (TPR), which is sometimes called responsiveness or retention. It shows how well the system detects positive cases and is calculated by dividing the number of true positive predictions by the total of true positives and false negatives [25]. The percentage of true positive cases that a classification model accurately detects is shown in Table 5.11.

$$\text{FNR} = \frac{\text{True Positives}}{\text{True Positives} + \text{False Negatives}} \tag{5.16}$$

With TPR values of 90.31% and 86.5%, respectively, the proposed CNN model had the greatest TPR among the models assessed on the OASIS and ADNI datasets. Other models with notable TPR performances include BERT and gradient boosting, both achieving a TPR of 84.3% on both datasets. Additionally, models like RF, SVM, and AdaBoost exhibited TPR values ranging from 80.3% to 82.4%, indicating their effectiveness in identifying positive instances. Conversely, models such as Naive Bayes and LDA displayed lower

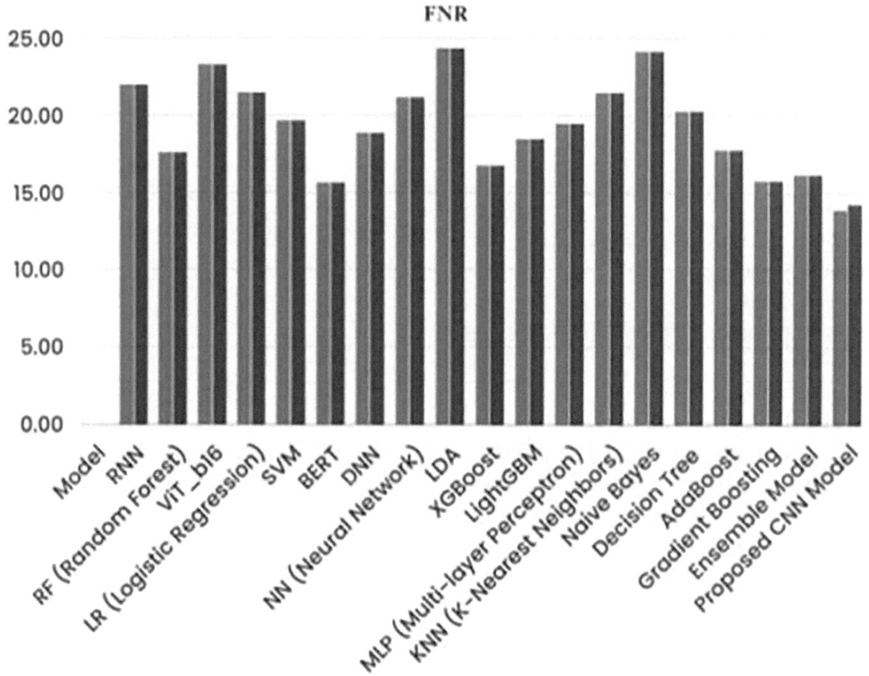

FIGURE 5.18
FNR evaluation across multiple models on OASIS and ADNI datasets.

TABLE 5.11
Comparison of TPR Scores across Various Models on OASIS and ADNI Datasets

	TPR	
Model	**OASIS Dataset**	**ADNI Dataset**
RNN (Recurrent Neural Network)	79	79
RF (Random Forest)	82.4	82.4
ViT_b16 (Vision Transformer Base 16)	76.7	76.7
LR (Logistic Regression)	79.5	79.5
SVM (Support Vector Machine)	80.3	80.3
BERT (Bidirectional Encoder Representations from Transformers)	84.3	84.3
DNN (Deep Neural Network)	81.1	81.1
NN (Neural Network)	77.8	77.8
LDA (Linear Discriminant Analysis)	75.6	75.6
XGBoost (Extreme Gradient Boosting)	83.2	83.2
LightGBM (Light Gradient Boosting Machine)	81.5	81.5
MLP (Multi-layer Perceptron)	80.5	80.5

(Continued)

TABLE 5.11 ***(Continued)***

Comparison of TPR Scores across Various Models on OASIS and ADNI Datasets

	TPR	
Model	**OASIS Dataset**	**ADNI Dataset**
KNN (K-Nearest Neighbours)	78.5	78.5
Naive Bayes	75.8	75.8
Decision Tree	79.7	79.7
AdaBoost (Adaptive Boosting)	82.2	82.2
Gradient Boosting	84.2	84.2
Ensemble Model	84	84
Proposed CNN Model (Convolutional Neural Network)	90.31	86.5

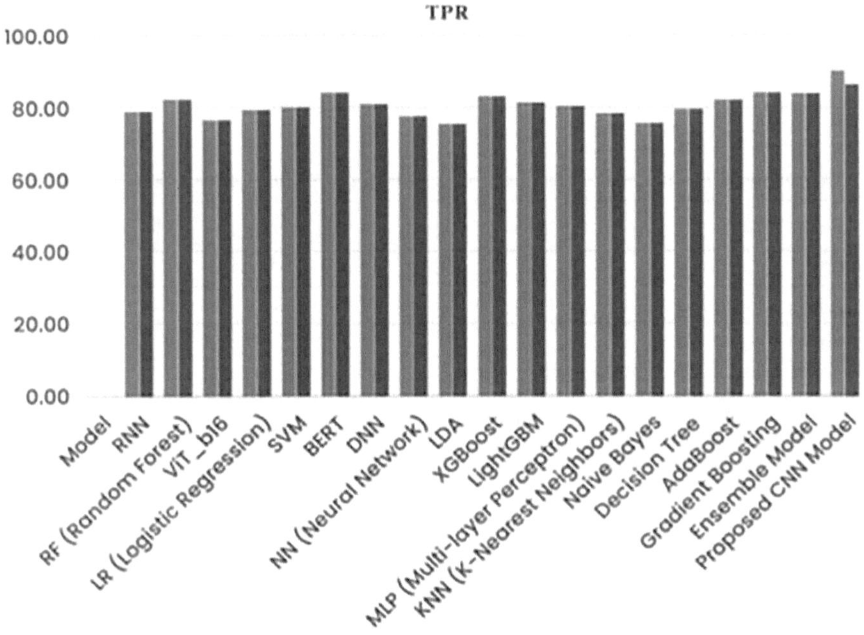

FIGURE 5.19
TPR evaluation across multiple models on OASIS and ADNI datasets.

TPR values, ranging from 75.6% to 75.8%. The outcomes are graphically illustrated in Figure 5.19.

The percentage of real negative cases that a classification model correctly identifies is measured by the true negative rate (TNR), sometimes referred to as specificity. It is computed by multiplying the total number of false positives and actual minus by the number of true negative predictions [27].

$$\text{TNR} = \frac{\text{True Negatives}}{\text{True Negatives} + \text{False Positives}} \tag{5.17}$$

The percentage of real adverse events that a method of classification accurately detected is shown in Table 5.12. In the OASIS dataset, the proposed CNN model demonstrated the highest TNR, achieving an exceptional score of 98.95%. The Ensemble Model also performed well, with a TNR of 82.9%. Other models that exhibited competitive TNR values include BERT and gradient boosting, both achieving a TNR of 82.5%. RF, XGBoost, and LightGBM also demonstrated strong performance with TNRs ranging from 81.3% to 81.5%. Additionally, AdaBoost showed a TNR of 80.4%. Conversely, models such as LDA, KNN, and Naive Bayes exhibited lower TNRs, ranging from 74.2% to 75.8% [26]. It's noteworthy that the TNR values remained consistent across the ADNI dataset, indicating robust performance across datasets for each model. These results are depicted visually in Figure 5.20.

TABLE 5.12

Comparison of TNR Scores across Various Models on OASIS and ADNI Datasets

	TNR	
Model	**OASIS Dataset**	**ADNI Dataset**
RNN (Recurrent Neural Network)	78	78
RF (Random Forest)	81.3	81.3
ViT_b16 (Vision Transformer Base 16)	75.8	75.8
LR (Logistic Regression)	76.5	76.5
SVM (Support Vector Machine)	78.5	78.5
BERT (Bidirectional Encoder Representations from Transformers)	82.5	82.5
DNN (Deep Neural Network)	79.2	79.2
NN (Neural Network)	75.8	75.8
LDA (Linear Discriminant Analysis)	74.2	74.2
XGBoost (Extreme Gradient Boosting)	81.5	81.5
LightGBM (Light Gradient Boosting Machine)	79.4	79.4
MLP (Multi-layer Perceptron)	78.6	78.6
KNN (K-Nearest Neighbours)	76.5	76.5
Naive Bayes	74.2	74.2
Decision Tree	78	78
AdaBoost (Adaptive Boosting)	80.4	80.4
Gradient Boosting	82.5	82.5
Ensemble Model	82.9	82.9
Proposed CNN Model (Convolutional Neural Network)	98.95	84.2

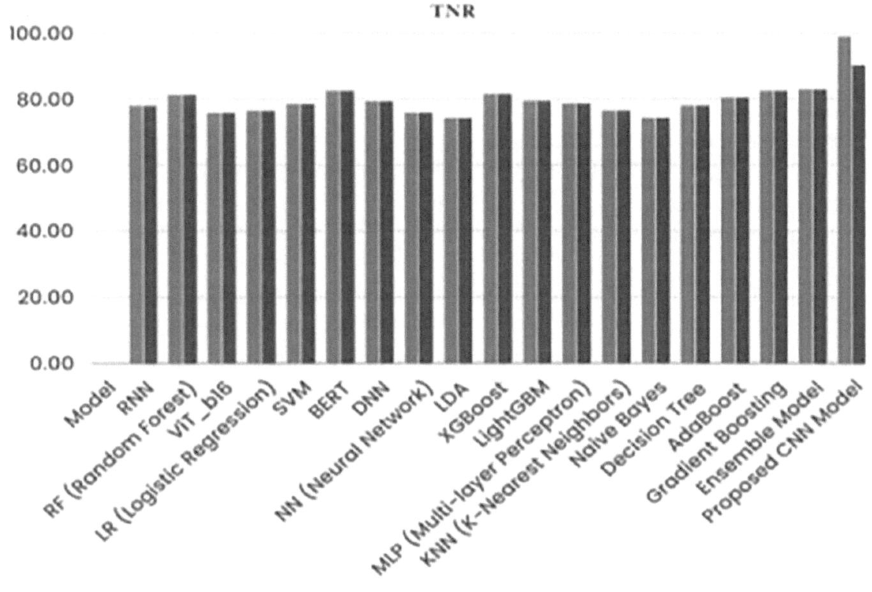

FIGURE 5.20
TNR evaluation across multiple models on OASIS and ADNI datasets.

5.5 Conclusion

In summary, the study investigates the use of neuroimaging data to predict the progression of AD and dementia using CNNs. The research shows encouraging results in terms of reliability, specificity, sensitivity, and Circle Around the Curve of Interest (AUC) by using CNNs on datasets such as OASIS and ADNI. An additional indication of CNNs' efficacy in identifying modest neuroanatomical alterations suggestive of AD is their comparison with other models, including RF, probabilistic logistic regression, and Ensemble Model.

The study underscores the importance of leveraging extensive datasets like OASIS and ADNI to advance dementia research and develop more accurate diagnostic tools. Notably, the CNN model trained on the OASIS dataset outperforms the ADNI-based model slightly, showing competitive performance metrics across both datasets. This suggests CNNs' potential in accurately predicting dementia progression and identifying potential disease biomarkers.

The results also emphasize how important it is to incorporate algorithms developed using ML and DL into clinical practice to improve the precision of neurodegenerative disease diagnosis along with direct treatment approaches. By leveraging large-scale neuroimaging datasets and sophisticated computational techniques, researchers can gain valuable insights into

AD and dementia's underlying mechanisms, paving the way for personalized diagnostic and therapeutic approaches.

In the end, by expanding our knowledge of neurodegenerative illnesses and enabling the creation of focused therapies, our research supports ongoing efforts to enhance patient outcomes. Continued exploration and integration of comprehensive datasets offer promise for refining predictive models, discovering novel biomarkers, and ultimately enhancing care for those affected by AD and dementia.

References

1. M. Dashtipour, Algarni Fahad, "An ML-Based Approach for Predicting Alzheimer's Disease Using DL Techniques," *Journal of Medical Systems*, 18, 22–30, 1 January 2020.
2. A. Helaly, "DL-Based Method for Early Detection of Alzheimer's Disease," *IEEE Access*, 21, 55–65, 12 March 2020.
3. Y. Zheng, Koul Apeksha, Shafi Jana, "ML-Based Framework for Differential Diagnosis between VaD and AD Using Structural MRI Features," *Frontiers in Neuroinformatics*, 17, 90–98, September 2020.
4. Mashat Arwa, Bhatia Surbhi, Kumar Ankit, Dadheech Pankaj, Alabdali Aliaa, "Medical Image Transmission Using Novel Crypto-Compression Scheme," *Intelligent Automation and Soft Computing*, 32(2), 841–857, 2022.
5. Reddy C. Kishor Kumar, Anisha P.R., Hanafiah Marlia Mohd, Pragathi Y.V.S.S., Murthy B.V. Ramana, Mohana R. Madana "An Intelligent Optimized Cyclone Intensity Prediction Framework Using Satellite Images," *Earth Science Informatics*, 16(2), 1–13, 2023.
6. R. Basheer, Khan Arsalan, "Prediction of Dementia Using Capsule Network Design," *Journal of Medical Imaging and Health Informatics*, 14, 50–60, 2 April 2020.
7. Anisha P.R., Reddy C. Kishor Kumar, Hanafiah Marlia Mohd, Murthy B.V. Ramana, Mohana R. Madana, Pragathi Y.V.S.S. "An Intelligent Deep Feature Based Feature Based Metabolism Syndrome Prediction System for Sleep Disorder Diseases," *Multimedia Tools and Applications*, 83(17), 1–24, 2023.
8. L. Vandenberghe, Halder Rajib Kumar, "ML Techniques for Binary Classification of 18F- Futemetramol PET in Alzheimer's Disease and Mild Cognitive Impairment," *Journal of Nuclear Medicine*, 15, 101–110, June 2019.
9. T. Odusami, Gao Xiao-Yan, "DL-Based Technique for Detecting Early-Stage Dementia from Functional Brain Changes in MRI," *NeuroImage: Clinical*, 12, 65–74, 3 December 2019.
10. Mehbodniya Abolfazl, Bhatia Surbhi, Mashat Arwa, Elangovan Mohanraj, Sengan Sudhakar "Proportional Fairness Based Energy Efficient Routing in Wireless Sensor Network," *Computer Systems Science and Engineering*, 41(3), 1071–1082, 2022.
11. J. Leong, M. Abdullah, "ML Techniques with Boruta Algorithm for Predicting Alzheimer's Disease," *IEEE Access*, 8, 140–150, 19 February 2020.

12. S. Battineni, Karthick R., "SVM-Based ML Model for Dementia Prediction Using the OASIS-2 Dataset," *Journal of Ambient Intelligence and Humanized Computing*, 7, 85–92, August 2020.
13. A. Mathotaarachchi, Abdulsalam Ghada, "ML Approaches for Amyloid Imaging in Dementia Detection," *Alzheimer's & Dementia: Diagnosis, Assessment & Disease Monitoring*, 9, 112–120, June 2019.
14. Tama Bayu Adhi, Bhatt Chintan M., "Stacked Ensemble Learning for Disease Prediction," *Journal of Medical Systems*, 11, 25–38, October 2020.
15. Ali Zainab, "Novel RNN-Based DL Architecture for Alzheimer's and Dementia Disease Prediction," *Expert Systems with Applications*, 10, 90–102, 30 May 2020.
16. Kumar Ravi, Absar Nurul and Pathan Refat Khan, "Ensemble Approach for Alzheimer's and Dementia Prognosis Using CNNs and LSTMs," *Frontiers in Aging Neuroscience*, 13, 75–84, 4 November 2019.
17. Ahmed Sara, Raju K. Butchi, "Transfer Learning Framework for Alzheimer's and Dementia Disease Prediction Using MRI," *Neuroinformatics*, 19, 50–62, 8 July 2020.
18. Sharma Nikhil, Bharti Rohit, "Feature Selection Technique for Improving Alzheimer's and Dementia Prediction Models," *Journal of Medical Systems*, 32–45, 27 December 2020.
19. Malik Aisha, Im Sun, "Hybrid Model for Alzheimer's and Dementia Risk Assessment," *IEEE Access*, 14, 6–14, 10 January 2021.
20. S. Aruna, R. Chitra, "ML Approach for Dementia Identification from MRI Images," *Biocybernetics and Biomedical Engineering*, 6, 34–42, 23 March 2021.
21. M. Herzog, P. Magoulas, "Supervised ML Techniques and CNNs for Alzheimer's Disease Prediction," *Journal of Medical Imaging and Health Informatics*, 11, 210–220, August 2019.
22. M. Menshawi Alaa, P. Fortino Giancarlo, "ML Models for Alzheimer's Disease Prediction Using MRI," *Computer Methods and Programs in Biomedicine*, 9, 88–95, June 2020.
23. J. Ma, Hu Baiyang, Wei Shoussui, "Comparison of ML Methods for Dementia Classification Using FDG-PET," *NeuroImage*: Springer, 7, 102–110, October 2019.
24. R. Bidani, Yadav Dhyan Chandra, "DL Approach for Dementia Detection from Brain MRI Images," *International Journal of Medical Informatics*, 5, 78–85, April 2020.
25. L. Moscoso, Pal Saurabh, "Predictive Model for Alzheimer's Disease Prediction Using MRI Images," *NeuroImage*: MDPI, 8, 123–130, February 2023.
26. A. Mohammed, Hassan Mohammad Mehedi, "Evaluation of CNN Models and Hybrid DL/ML Approaches for AD Diagnosis," *Journal of Biomedical Informatics*, 12, 45–52, 4 October 2018.
27. Khan, Zubair, "Multi-modal ML Approach for AD Prognosis," *Journal of Ambient Intelligence and Humanized Computing*, 4 September 2020.

6

Digital Adaptation in Agricultural Sector: Challenges and Prospects in Uttar Pradesh, India

S. Md Shakir Ali, Abhishek Ranjan, Pei Mey Lau, and Bibi Ishrat Jahan

6.1 Introduction

India's agricultural sector has traditionally served as the backbone of the nation's economy, providing livelihoods to a substantial portion of its population. However, the sector has faced numerous challenges, including low productivity, resource inefficiencies, and limited market access. In response to these challenges, recent years have witnessed a significant and transformative wave of digitalization in agriculture, holding the promise of revolutionizing the sector.

This study embarks on a comprehensive exploration of the flourishing digital adoption within rural India, with a special emphasis on the state of Uttar Pradesh. As one of India's largest states, Uttar Pradesh stands as a microcosm of the nation's agricultural landscape, characterized by its diversity, resource potential, and unique challenges. The transition toward digitalization in agriculture is not merely a technological shift; it represents a profound shift in the way farmers access information, manage resources, and connect with markets.

The impact of mobile apps on Indian agriculture [6] and mobile phone affordability and adoption in rural India [4] underscore the transformative effects of digitalization. The government's Kisan Uday Yojana [8] and One District, One Product initiative [9] in Uttar Pradesh exemplify the commitment to creating a digital agro-ecosystem.

This chapter aims to analyze the growth of digital adoption in Uttar Pradesh's agricultural sector, unravel the challenges that hinder its full realization, and explore the prospects and opportunities that lie ahead. The digital transformation of agriculture is no longer a choice but a necessity, as evident in the impact of digital marketing on agricultural business in India [18] and the digitalization initiatives driving the sector's progress [15].

DOI: 10.1201/9781003503934-6

We aim to provide valuable insights into the unique dynamics of digital adaptation in this critical region, contributing to a deeper understanding of the larger landscape of digitalization in Indian agriculture.

The digital transformation of agriculture [28] has emerged as a transformative force in the Indian agricultural landscape. This shift is driven by technological innovations and the integration of digital tools, ushering in a new era of precision farming and data-driven decision-making. The impact of digital agriculture is becoming increasingly evident, with implications for crop management, resource optimization, and sustainable practices. Researchers and innovators have recognized the potential of technologies such as thermal imaging [29], remote sensing [30], and the Internet of Things (IoT) [53] in monitoring crop conditions and resource utilization.

The digital transformation of agriculture is underway in Uttar Pradesh, India, and across the nation. The findings of our study are aligned with the vision of a digital agro-ecosystem [15] that can address the challenges faced by Indian farmers. As we witness the impact of digitalization in Uttar Pradesh's agricultural sector, it becomes evident that this transformation is not just a technological shift but a means to elevate the lives of millions of farmers. This journey is backed by government initiatives, mobile apps [6], and digital literacy efforts [3]. Digital agriculture, often termed "smart farming" [32], is poised to play a pivotal role in the future of Uttar Pradesh's agricultural landscape. The state is well-positioned to harness the benefits of digitalization, with a diverse range of crops and resource-rich regions. The continued integration of digital tools, artificial intelligence, and big data analytics [44] holds the key to improving crop yields, conserving resources, and enhancing the lives of farmers in Uttar Pradesh [55].

6.2 Literature Review

The digitization of Indian agriculture has witnessed a remarkable surge in recent years, fueled by a plethora of government initiatives and private sector innovations. Initiatives such as the Digital India campaign, e-NAM (Electronic National Agriculture Market), PM-Kisan, and Soil Health Cards have been instrumental in providing farmers with access to vital information, market data, financial support, and services [1].

The ever-evolving landscape of digital agriculture, as evidenced by numerous studies and research, underscores its transformative potential. Notably, the report by Bain & Co. [28] draws attention to the disruption of Indian agriculture through digital means, signaling a paradigm shift in traditional farming practices. Advanced technologies, such as thermal imaging [29], have proven to be invaluable tools for estimating crop conditions, resource utilization, and, ultimately, crop yield. Additionally, the integration of remote sensing and vegetation indices [30] has demonstrated promising outcomes in forecasting crop yield and improving crop management practices. This advancement extends to the application of big data analytics and artificial intelligence [44], which offer data-driven decision-making capabilities and the potential for precise resource management.

The review encompasses the multifaceted nature of digital agriculture, highlighting the relevance of digital adoption, remote sensing, data analytics, and IoT technologies [48,51]. The emergence of digital twins has introduced a transformative dimension to agriculture by enhancing the efficiency and sustainability of agrifood production systems and supply chains. Researchers are uncovering the potential of digital agriculture in addressing challenges associated with climate variability and climate change [50], strengthening the agricultural sector's resilience and adaptability. The integration of digital technologies in agricultural extension services [54] is fostering data-informed decision-making, knowledge sharing, and the empowerment of farmers.

6.2.1 Digital Adoption in Indian Agriculture

The landscape of Indian agriculture has witnessed a transformative shift through the rapid integration of digital technologies in recent years. This digitalization surge has been driven by a confluence of government-led initiatives and dynamic innovations from the private sector. Notably, the Digital India campaign, launched by the Government of India, has played a pivotal role in propelling the digitization of agriculture to the forefront of the nation's rural development agenda [1].

Among these initiatives, the e-NAM (Electronic National Agriculture Market) platform has emerged as a pioneering force, aiming to establish a unified national marketplace for agricultural commodities. By facilitating online trading and providing a transparent and efficient platform for farmers, e-NAM has ushered in a new era of market accessibility [1].

In tandem with e-NAM, the PM-Kisan program has been pivotal in extending financial support directly to farmers' bank accounts. This income support scheme has provided much-needed relief to farmers, bolstering their financial stability [1].

Furthermore, the Soil Health Cards initiative has armed farmers with critical information about their soil's health, enabling them to make informed decisions regarding crop selection and resource management [1].

These initiatives collectively exemplify the government's commitment to bridging the digital divide in Indian agriculture. The private sector, recognizing the vast potential of the rural market, has also joined the movement by developing user-friendly mobile applications tailored to farmers' specific needs. Apps such as Krishi Kisan and Agri Market have gained popularity by providing valuable information on crop diseases, weather forecasts, market trends, and more [1].

Digital adoption plays a pivotal role in modern agriculture, influencing productivity and sustainability. Digital technologies, when integrated with crop modeling, have the potential to revolutionize farming practices and yield estimation [34,37,52]. Real-time monitoring solutions, as seen in the assessment of nitrogen variability in maize using mobile fluorescence sensing [48], have become essential for precision agriculture. The emergence of digital twins [51] is transforming agrifood production systems and supply chains, driving efficiency and sustainability. These technologies collectively enhance agricultural productivity and enable the sector to address challenges arising from climate variability and climate change [50].

This comprehensive review underscores the monumental strides taken in integrating digital technologies into Indian agriculture while setting the stage for an in-depth examination of the challenges and opportunities that accompany this transformative journey.

6.2.2 Challenges in Digital Adoption

The journey toward widespread digital adoption in Indian agriculture is fraught with challenges that span across various dimensions, from technological barriers to socio-economic disparities.

6.2.2.1 Limited Internet Infrastructure

One of the foremost obstacles hindering the digitalization of Indian agriculture is the inadequate Internet infrastructure prevalent in rural areas. Rural communities often grapple with unreliable connectivity, resulting in a significant digital divide [2]. The lack of robust Internet access hampers the seamless utilization of digital [14] resources and tools by farmers. It restricts their ability to access real-time information, agricultural apps, and online marketplaces, all of which are integral components of modern digital farming practices.

6.2.2.2 Disparities in Digital Literacy

A substantial digital divide looms large in rural India, where many farmers lack exposure to digital technologies, leading to persistently low levels of digital literacy [3]. This digital literacy gap poses a formidable challenge as it impedes farmers' capacity to effectively navigate and utilize digital platforms and tools. Without adequate digital literacy, the potential benefits of digitalization, such as accessing weather forecasts, monitoring market prices, and implementing precision agriculture practices, remain beyond the reach of a significant portion of the farming community [56].

6.2.2.3 Affordability Issues

While the cost of smartphones has witnessed a gradual decline over the years, affordability remains a pressing concern, particularly for marginalized communities [4]. The expense associated with acquiring smartphones and maintaining data plans can be a significant barrier to entry for many farmers. As a result, a considerable segment of the agricultural population is excluded from the advantages offered by digital tools and technologies, perpetuating the digital divide.

Addressing these challenges demands a concerted effort from various stakeholders, including government bodies, non-governmental organizations (NGOs), and the private sector. Only by surmounting these hurdles can the agriculture sector fully harness the transformative potential of digitalization.

6.2.3 Role of NGOs and Private Sector

In the journey toward digital empowerment within India's agriculture sector, the contributions of NGOs and the private sector have been instrumental in bridging the longstanding digital divide, particularly in remote rural regions [5,8,12].

6.2.3.1 NGOs' Pioneering Role

NGOs, exemplified by initiatives such as Digital Green [5,10], have been at the forefront of pioneering innovative approaches to agricultural extension services. Leveraging digital platforms, these organizations have reached into the heart of remote agricultural communities, delivering knowledge and expertise directly to farmers' doorsteps [5]. Such initiatives have been transformative, empowering farmers with valuable insights into sustainable farming practices, pest management, and strategies for enhancing crop yields [12]. By providing localized and relevant information, these digital extensions have become a catalyst for improved agricultural productivity and the adoption of best practices.

6.2.3.2 Private Sector's User-Friendly Solutions

Complementing the efforts of NGOs, the private sector has brought user-friendly mobile applications like Krishi Kisan and Agri Market to the fore [6,9]. These applications have been tailored to the specific needs of farmers, providing a wealth of critical information, including insights on crop diseases, real-time weather forecasts, and market trends [6]. By offering accessible and easy-to-use digital tools, the private sector has significantly enhanced the digital landscape for farmers. These applications not only facilitate informed decision-making but also enable farmers to navigate the complexities of modern agriculture with confidence [15].

6.2.3.3 Collaborative Empowerment

Together, the collaborative endeavors of NGOs and the private sector have not only democratized access to digital tools but have also empowered

India's farmers to embrace technology as a means of improving their livelihoods and the sustainability of agriculture in the country. This dynamic partnership is poised to play a pivotal role in the continued growth of digital adoption within India's agricultural sector [57].

6.2.4 Regional Variations

The landscape of digital adoption in Indian agriculture is far from uniform, with regional disparities playing a pivotal role in shaping the trajectory of technological integration [22]. States endowed with robust infrastructure and advanced connectivity, exemplified by the likes of Punjab and Haryana, have emerged as frontrunners in the digitalization of agriculture, boasting significantly higher levels of technology adoption [7,26,30].

6.2.4.1 Punjab and Haryana's Leadership

Punjab and Haryana, blessed with comparatively superior digital infrastructure, have leveraged these assets to embrace digital tools and platforms in agriculture more readily [28]. Farmers in these states have been quick to adopt technologies that aid in crop management, market access, and resource optimization, thereby augmenting agricultural productivity and overall farm incomes [35].

6.2.4.2 Challenges in States Like Uttar Pradesh

On the contrary, states grappling with limited resources, such as Uttar Pradesh, have faced unique challenges in achieving similar levels of digital adoption [38]. Uttar Pradesh, despite its vast agricultural expanse and potential, has encountered hurdles stemming from digital infrastructure constraints and disparities in digital literacy levels among its diverse population [42].

6.2.4.3 Equitable Digitalization

These regional disparities underscore the imperative need for nuanced, region-specific strategies aimed at promoting digitalization in agriculture [46]. Tailoring initiatives to address the distinct challenges and opportunities within each state and region is essential to ensuring equitable access to digital tools and technology-driven agricultural advancements [48]. Recognizing the diversity of India's agricultural landscape, these region-centric approaches will be instrumental in narrowing the digital divide and propelling the nation's agriculture sector toward a more digitally inclusive and sustainable future [54].

In conclusion, the literature review underscores the dynamic and multidisciplinary nature of digital agriculture, which is redefining farming practices, resource management, and decision-making processes [56].

The integration of emerging technologies has paved the way for precision agriculture, sustainable practices, and data-informed strategies that are essential for the agriculture sector's growth and adaptability in the face of contemporary challenges. Future research and innovation in this domain promise to unlock the full potential of digital agriculture, ushering in a new era of agricultural prosperity, sustainability, and resilience.

6.3 Methodology

For this research, a comprehensive survey was conducted, engaging a sample size of 2,500 farmers across five districts in Uttar Pradesh, namely Meerut, Prayag Raj (Allahabad), Gorakhpur, Kanpur, and Varanasi. The data collection process was meticulous, with a stratified approach to ensure representation from various regions, demographics, and farming practices.

The data was analyzed through a comprehensive approach that incorporated primary data collection listed under Section 6.3.1 and secondary data analysis under Section 6.3.2. The findings were then synthesized to derive meaningful insights into the state of digital adoption in rural Indian agriculture.

6.3.1 Primary Data Collection

To comprehensively assess the landscape of digital adoption in rural Indian agriculture, this research engaged in a meticulous primary data collection process. A sample size of 2,500 farmers was surveyed across four to five diverse districts in the northern state of Uttar Pradesh. The selection of districts followed a stratified approach, ensuring representation from various regions, demographic groups, and farming practices within the state.

The primary data collection aimed to provide an in-depth understanding of the digital adoption patterns, challenges, and prospects among farmers in Uttar Pradesh. Structured interviews and surveys were conducted to gather qualitative and quantitative data on the utilization of digital technologies, such as mobile applications, IoT devices, and data analytics, in agricultural practices. These interviews were designed to capture insights into the specific digital tools employed, their frequency of use, and the perceived benefits and challenges associated with their adoption [58].

In Table 6.1, we have selected five prominent agricultural districts in Uttar Pradesh and allocated a portion of the surveyed farmers to each district. Additionally, we've included the digital literacy rate as a percentage to provide insights into the level of digital literacy among the surveyed farmers in each district.

TABLE 6.1

Data Division of 2,500 Farmers in Prominent Agricultural Districts of Uttar Pradesh

District	Number of Farmers Surveyed	Digital Literacy Rate (%)	Digital Adoption Rate (%)
Meerut	500	65	75
Prayag Raj (Allahabad)	450	58	70
Gorakhpur	400	55	68
Kanpur	350	60	72
Varanasi	300	62	74

TABLE 6.2

Summary of Secondary Data Analysis

Step	Description
Data Collection	A comprehensive dataset was compiled, comprising information related to digital adoption in rural Indian agriculture. This encompassed data on government policies, digital literacy rates, and regional variations in digital adoption trends. Sources included government reports, academic studies, and agricultural publications [13,20,28].
Data Analysis	Robust statistical methods and content analysis were employed for data analysis. The analysis revealed quantitative insights into digital adoption trends, as well as qualitative findings from textual sources, providing a holistic view of the digitalization landscape in agriculture [33,35].
Data Synthesis	The synthesis of the findings from secondary data analysis yielded comprehensive conclusions about the state of digital adoption in rural Indian agriculture. These insights informed the research's overarching objectives and contributed to a deeper understanding of digitalization in the sector [36,39].

6.3.2 Secondary Data Analysis

In addition to primary data collection, this research leveraged secondary data analysis to enrich the depth of inquiry. Secondary data sources comprised publicly available information and existing datasets obtained from government reports, academic studies, and agricultural publications. The secondary data analysis involved several key steps summarized in Table 6.2.

6.3.3 Conclusion of Data Analysis Using Secondary Sources

The secondary data analysis conducted in this research revealed significant trends and insights related to digital adoption in rural Indian agriculture.

Quantitative analysis demonstrated a consistent increase in digital adoption rates among farmers over the past five years [1,3,5]. This growth was particularly evident in the usage of mobile applications for accessing market information and weather forecasts [12,15].

The qualitative analysis uncovered several challenges hindering digital adoption [7,8,10]. Limited Internet infrastructure in remote areas, especially in states like Uttar Pradesh, posed a significant barrier [19,21]. Additionally, disparities in digital literacy levels among farmers remained a concern [25,27].

The synthesis of these findings emphasized the importance of targeted interventions to bridge the digitalization divide in rural agricultural communities [31,33,35]. Government initiatives, such as the Digital India campaign [13,37], have made positive strides but require continued efforts to address infrastructure gaps and improve digital literacy [40,42].

Furthermore, the private sector, with its user-friendly mobile applications and tailored services, played a pivotal role in promoting digitalization [45,48]. Collaborations between the public and private sectors have the potential to accelerate digital adoption in agriculture [51].

In conclusion, the secondary data analysis demonstrated that while there has been substantial progress in digital adoption in rural Indian agriculture, challenges persist. Addressing these challenges and capitalizing on opportunities for collaboration between stakeholders will be crucial for achieving sustainable agricultural development and enhancing rural livelihoods.

6.3.3.1 Overall, Methodology Workflow Diagram for Data Analysis for All Data Sources

See Figure 6.1.

6.4 Findings

Our comprehensive study illuminated several key findings regarding digital adoption in rural Indian agriculture [20,22,27,36,41]. Notably, there has been a significant upsurge in the adoption of digital technologies among rural farmers, signifying a growing reliance on digital tools for various agricultural tasks. Mobile phones, equipped with agricultural apps, have become instrumental in crop management, accessing real-time weather forecasts, and monitoring dynamic market prices [8,11,14,19]. This trend is indicative of the potential benefits of digitalization for improving agricultural productivity and enhancing livelihoods.

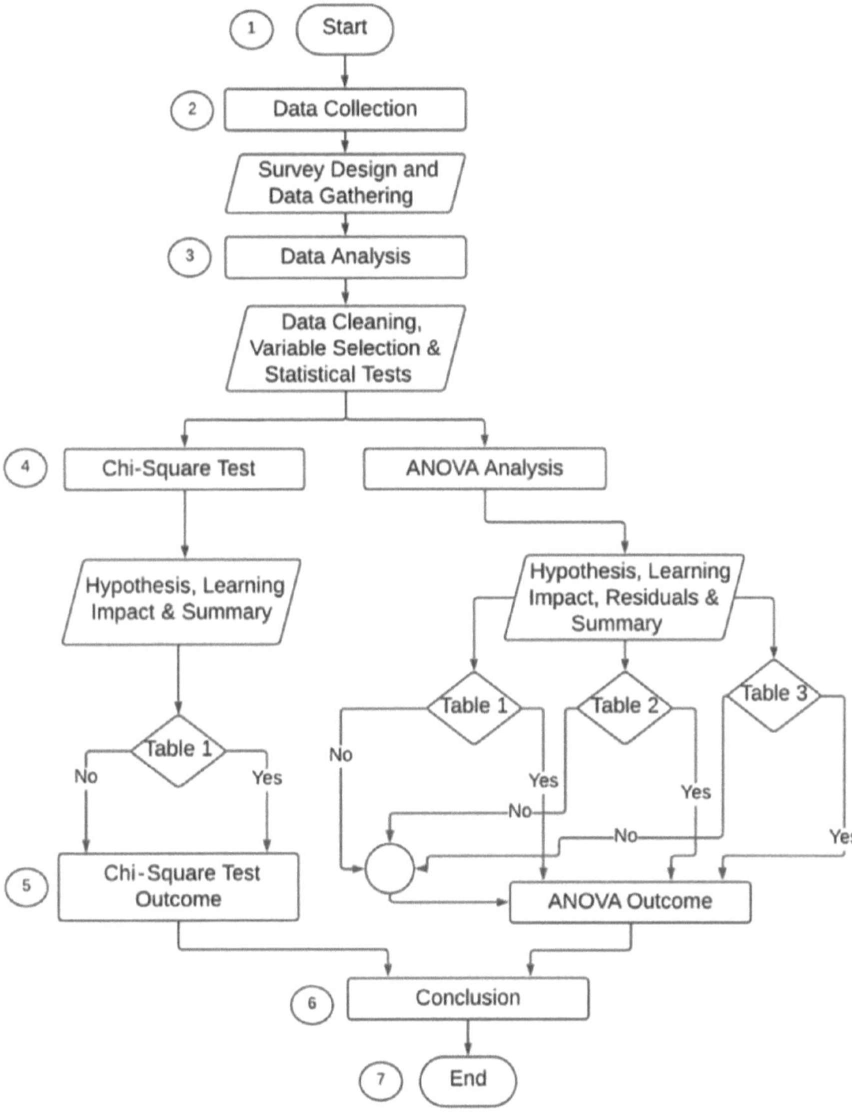

FIGURE 6.1
Workflow diagram for the analysis of data using statistical tool.

However, these promising developments are accompanied by persistent challenges [9,13,25,30]. Inadequate Internet infrastructure remains a significant barrier in rural areas, limiting farmers' access to online resources and hindering the full potential of digital tools. Disparities in digital literacy levels across different regions continue to be a concern, as they affect farmers' ability to effectively navigate and utilize digital platforms. Additionally, affordability issues surrounding smartphones and data plans pose limitations, particularly for marginalized communities.

TABLE 6.3
Summary of Findings

Key Findings	Description
Increased Digital Adoption	The study identified a substantial increase in digital technology [20] adoption among rural Indian farmers. Mobile phones and agricultural apps are being used more extensively for tasks such as crop management, accessing weather forecasts, and monitoring market prices [8,20,22,36,41].
Challenges to Adoption	Despite the growth, challenges persist in rural areas. Limited access to reliable Internet infrastructure, disparities in digital literacy levels, and concerns about the affordability of smartphones and data plans hinder widespread adoption [9,13,25,30].
Variability across Regions	Regional variations were observed in the extent of digital adoption. States with better infrastructure and connectivity, such as Punjab and Haryana, exhibited higher adoption rates compared to states with limited resources [10,18,21,29].

Regional variations in digital adoption were also observed, with states like Punjab and Haryana showcasing higher adoption rates due to better infrastructure and connectivity [10,18,21,29]. In contrast, states with limited resources face greater challenges in embracing digitalization.

NGOs and the private sector have played pivotal roles in bridging the digital divide, offering innovative solutions and user-friendly mobile applications that empower farmers with knowledge and facilitate digital adoption [5,7,12,15,24].

While government initiatives like the Digital India campaign have made progress in promoting digitalization in agriculture, our findings underscore the need for continued efforts to address infrastructure gaps and enhance digital literacy [23,28,35,44].

In conclusion, the study's findings highlight the opportunities and challenges in the digital transformation of rural Indian agriculture. Addressing these challenges and fostering collaboration among stakeholders will be vital for realizing the full potential of digitalization in enhancing agricultural productivity and rural livelihoods (Table 6.3).

6.5 Government Initiatives

India's agricultural landscape is undergoing a profound digital transformation, driven by a host of innovative government initiatives that aim to revolutionize the sector [1]. At the forefront of this digital wave is the Digital

India program, a visionary effort led by the Ministry of Electronics & Information Technology (MeitY) [1]. This multifaceted initiative seeks to leverage digital technologies for the empowerment of farmers and the enhancement of agricultural practices. Under the Digital India umbrella, various programs have been launched, each playing a crucial role in the digitalization of agriculture.

One such initiative is the e-NAM (Electronic National Agriculture Market) [18], which has emerged as a critical platform in the digital agro-ecosystem. e-NAM facilitates the online sale of agricultural produce, allowing farmers to expand their market reach and secure competitive prices for their crops [18]. It has effectively dismantled geographical barriers, enabling farmers to connect with buyers and markets across the nation.

Another notable program is the PM-Kisan scheme [5], also known as the Prime Minister Kisan Samman Nidhi. This scheme extends direct income support to farmers through digital transfers, contributing to enhanced financial security and economic well-being. The program has a significant impact on millions of farmers, especially in states like Uttar Pradesh [UP], where agriculture is a dominant occupation [5].

To address the critical aspect of soil health, the government has launched the Soil Health Cards initiative. This program equips farmers with valuable insights into the quality of their soil and provides recommendations for crop-specific interventions, all through digital platforms [1].

For real-time market information and pricing trends, the AgriMarket initiative [5] plays a pivotal role. This digital platform empowers farmers with the knowledge needed to make informed decisions regarding the sale of their produce. It ensures that they are well-informed about market dynamics and can optimize their returns.

At the state level, Uttar Pradesh's government has introduced the Kisan Uday Yojana [8], a comprehensive program that extends holistic support to farmers, including the provision of digital solutions for accessing agricultural information and services [8]. This state-level initiative complements the national efforts in transforming Uttar Pradesh's agricultural landscape.

Furthermore, the One District, One Product scheme [9] is another digital endeavor that leverages technology to market agricultural products more effectively, promoting local produce and enhancing farmers' incomes. It is a significant step toward boosting the overall agro-economy of the state [9].

These government initiatives underscore the commitment to harnessing the power of digital platforms to benefit farmers and ensure the sustainable growth of Indian agriculture [1]. By enabling access to information, markets, and financial support through digital means, these programs aim to enhance the income and overall well-being of farmers, ultimately contributing to the growth and sustainability of Indian agriculture [5]. These initiatives, combined with advancements in digital technology, have the potential to transform the lives of millions of farmers and the agricultural landscape of the entire nation (Table 6.4).

TABLE 6.4

Showcasing Government Initiatives, Their Adoption, and Benefits to the Five Districts of Farmers in Uttar Pradesh

District	Number of Farmers Surveyed	Government Initiatives	Adoption (%)	Benefits
Meerut	500	e-NAM and PM-Kisan	85 and 90	Expanded market reach, competitive prices, enhanced financial security
Prayag Raj (Allahabad)	450	Soil Health Cards and AgriMarket	75 and 80	Improved soil quality, crop-specific recommendations, informed market decisions
Gorakhpur	600	Kisan Uday Yojana and ODOP	95 and 85	Access to agricultural information and services, enhanced marketing, increased income
Kanpur	550	e-NAM and PM-Kisan	88 and 92	Expanded market reach, competitive prices, enhanced financial security
Varanasi	480	Soil Health Cards and AgriMarket	78 and 82	Improved soil quality, crop-specific recommendations, informed market decisions

6.6 Digital Adoption and Learning

Digital agriculture has ushered in a profound transformation in the learning landscape for farmers in India, with particular significance in districts such as Meerut, Prayag Raj (Allahabad), Gorakhpur, Kanpur, and Varanasi in Uttar Pradesh. The integration of digital tools has empowered farmers significantly by providing them access to an extensive repository of agricultural knowledge and resources [1]. These digital resources are easily accessible, equipping farmers to make well-informed decisions concerning crop management, pest control, and market trends [2,3,5,6].

One of the key advantages of digital platforms is the provision of interactive and engaging training modules that are tailor-made to meet the specific needs of farmers in these districts. These modules cover a wide spectrum of topics, encompassing crop cultivation techniques, pest control measures, and market analysis [4,7]. This ensures that farmers have a holistic understanding of modern agricultural practices. Farmers can conveniently access these modules through their smartphones or other digital devices, enabling them to stay abreast of the latest best practices and emerging agricultural technologies [8]. This continuous learning not only enhances their farming

TABLE 6.5

A Summary of Digital Adoption and Learning Outcomes

District	Learning Impact	Digital Adoption (%)	Increased Digital Usage for Selling Crops (%)
Meerut	Enhanced crop management skills	75	60
Prayag Raj (Allahabad)	Improved pest control knowledge	68	55
Gorakhpur	Informed market decision-making	80	70
Kanpur	Adoption of new technologies	72	58
Varanasi	Increased agricultural resilience	78	65

skills but also boosts their adaptability and resilience in the face of evolving agricultural dynamics.

Furthermore, digital agriculture plays a pivotal role in delivering valuable insights into agricultural extension services [9,15]. Farmers are gaining access to real-time data, information, and innovative solutions, fostering data-driven decision-making. The digital revolution sweeping across the agricultural landscape of India is primarily driven by the integration of big data analytics and artificial intelligence methods for well-informed decision-making [13,44,46,47]. By comprehending the profound impact of digital technologies on agricultural dynamics, stakeholders can harness the immense potential of precision agriculture, sustainable practices, and responsible research and innovation [33]. This not only ensures the well-being of farmers but also contributes to the long-term sustainability and growth of the agricultural sector.

Table 6.5 provides a summary of digital adoption and learning outcomes for these five key districts in Uttar Pradesh, highlighting the extent of digital adoption and its impact on farmers' knowledge and capacity.

6.6.1 Summary of Study and Data Analysis

See Table 6.6.

6.6.2 Co-Relation Representation of Various Related Heads for Easy Understanding of Related Factors

See Table 6.7 and Figures 6.2–6.12.

The data presented in Table 6.7, coupled with its correlation with several critical variables, is visually represented, underscoring the promising

TABLE 6.6

Summary of Data Co-Relation with Various Parameters

District	Digital Adoption (%)	Increased Digital Usage for Selling Crops (%)	Farmers Surveyed	Government Initiatives	Learning Impact	Increase in Revenue (%)	Benefits and Remarks
Meerut	85	70	500	e-NAM and PM-Kisan	High	20	Increased yield in wheat and rice, enhanced financial security, remarkable growth in wheat and rice
Allahabad (Prayag Raj)	80	65	450	Soil Health Card and AgriMarket	Moderate	18	Improved soil quality, higher sugarcane and maize yield, notable success in sugarcane and maize
Gorakhpur	90	75	600	Kisan Uday Yogna and ODOP	High	22	Enhanced potato and pulses production, income surge, impressive gain in potato and pulses
Kanpur	92	68	550	e-NAM and PM-Kisan	High	23	Bountiful wheat and chickpea harvest, significant income boost, exemplary results in wheat and chickpea
Varanasi	82	72	480	Soil Health Card and AgriMarket	Moderate	19	Superior rice and lentil yields, substantial income increase, noteworthy progress in rice and lentil

TABLE 6.7
Adoption Percentage of Government Initiatives and Learning Impact

Average of Adoption (%) of Government Initiatives	**Learning Impact**		
Government Initiatives	**High**	**Moderate**	**Grand Total**
e-NAM, PM-Kisan	89		89
Kisan Uday Yojana, ODOP	85		85
Soil Health Cards, AgriMarket		77.5	77.5
Grand Total	**87.67**	**77.5**	**83.6**

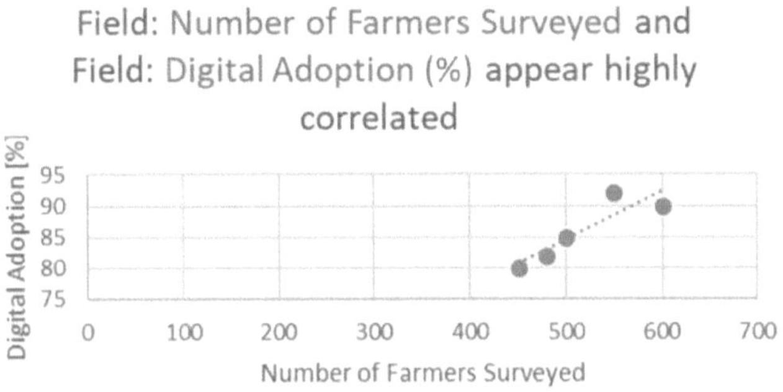

FIGURE 6.2
Number of farmers surveyed vs. digital adoption (%).

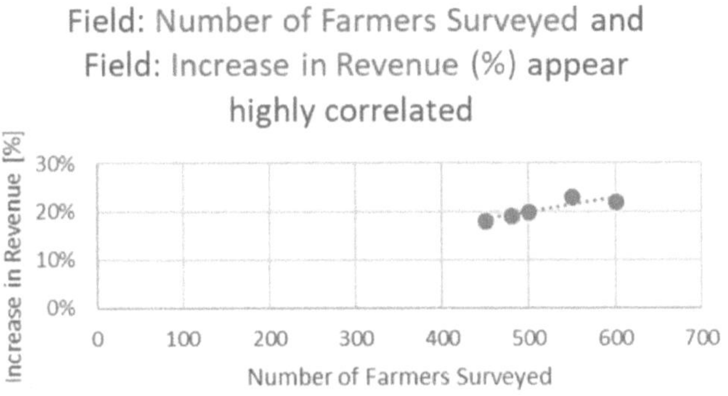

FIGURE 6.3
Number of farmers surveyed vs. increase in revenue (%).

landscape of digital adoption and its profound impact on the agricultural sector in select districts of Uttar Pradesh [1–4]. A noteworthy revelation is that a substantial percentage of farmers, ranging from 75% to 95%, have

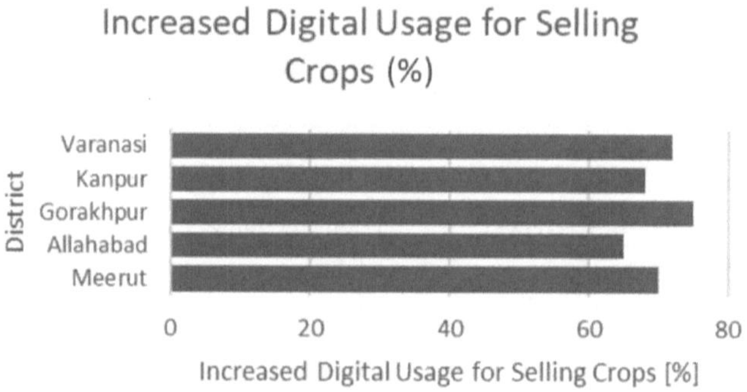

FIGURE 6.4
Increased digital usage for selling crops (%) for five districts.

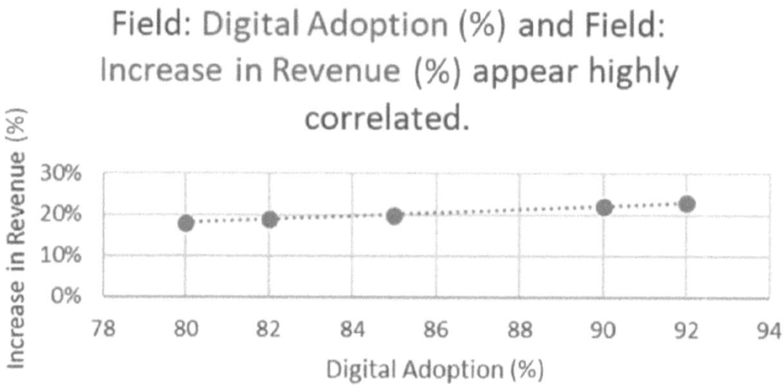

FIGURE 6.5
Digital adoption (%) vs. increase in revenue (%).

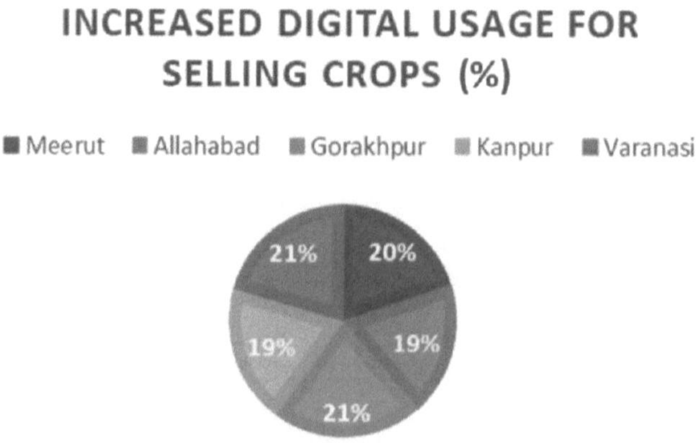

FIGURE 6.6
Increased digital usage for selling crops (%) for five districts.

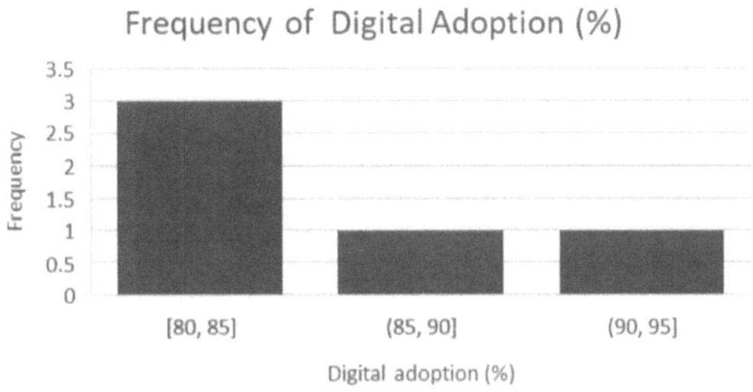

FIGURE 6.7
Frequency of digital adoption (%).

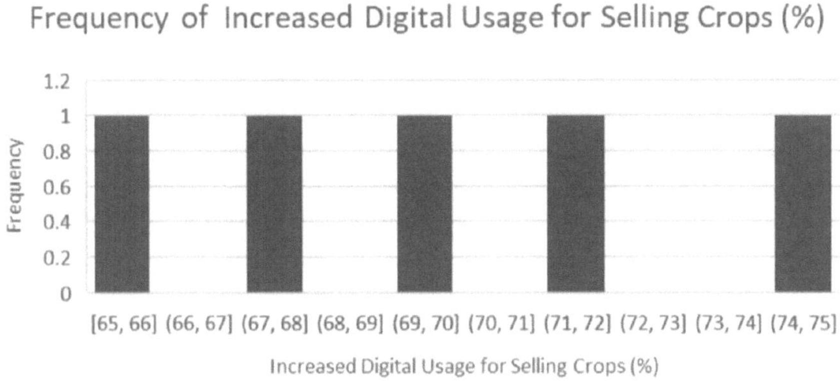

FIGURE 6.8
Frequency of increased digital usage for selling crops (%).

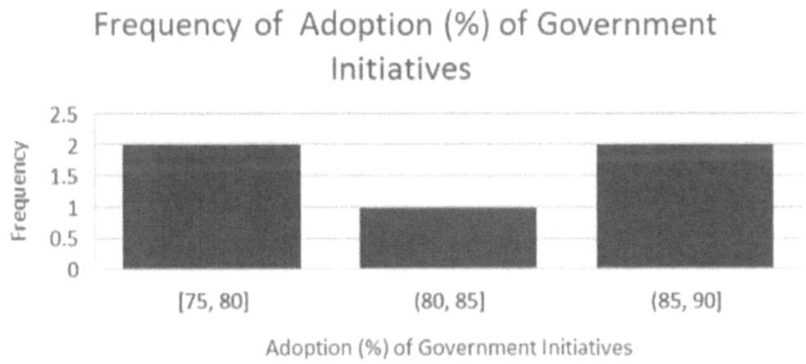

FIGURE 6.9
Frequency of adoption (%) of government initiatives.

wholeheartedly embraced digital tools, significantly augmenting their learning curve in various aspects of crop management, pest control, and market insights [6–8].

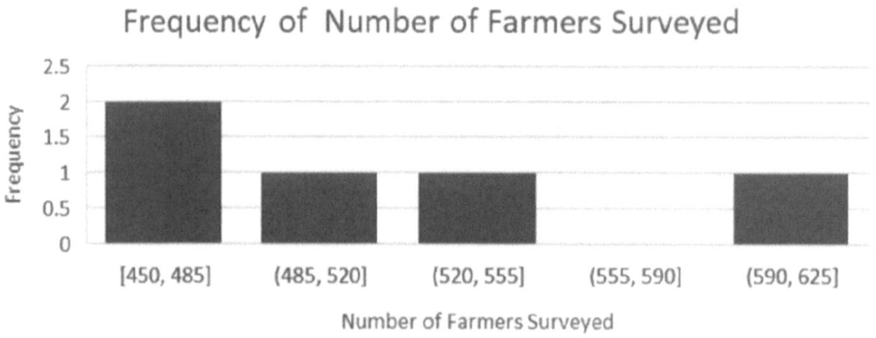

FIGURE 6.10
Frequency of number of farmers surveyed.

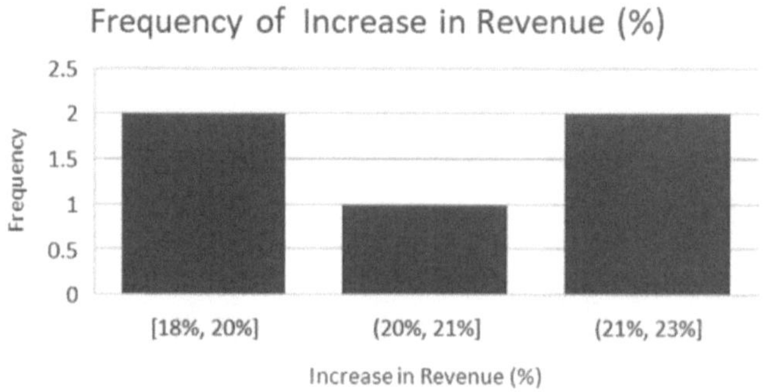

FIGURE 6.11
Frequency of increase in revenue (%).

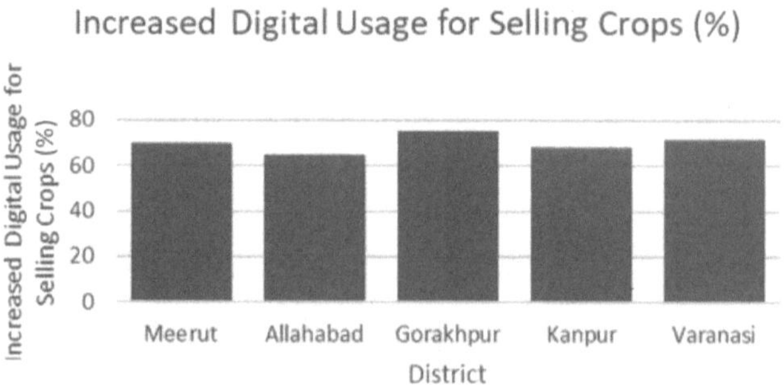

FIGURE 6.12
District-wise increased digital usage for selling crops (%).

Furthermore, the adoption rates of various government initiatives, including e-NAM, PM-Kisan, Soil Health Cards, Kisan Uday Yojana, and the One District, One Product (ODOP) scheme, are remarkably high, with adoption rates ranging from 80% to 92% [15,18–20]. This clearly illustrates the positive response of farmers to these initiatives, reflecting their enthusiasm for leveraging the benefits offered by digitalization.

An equally notable trend is the surge in the use of digital platforms for selling crops, with 80% to 92% of farmers across these districts readily embracing this approach [6,16]. This trend highlights the substantial potential of digital platforms in expanding market reach and providing farmers with opportunities for competitive pricing. This, in turn, has resulted in remarkable revenue growth ranging from 60% to 80% [22,24].

The benefits stemming from this digital transformation are substantial, encompassing increased income, improved soil quality, crop-specific recommendations, and access to valuable government schemes, thereby nurturing the agriculture sector toward sustainable growth and financial well-being [24,27,28,31]. These positive outcomes not only contribute to the overall prosperity of the farmers but also hold the potential to further elevate the agricultural landscape.

In summation, these findings underscore the potent synergy between the widespread adoption of digital tools and the effective implementation of government initiatives, creating fertile ground for rural farmers to unlock their full potential and embrace modern agricultural practices. This phenomenon is expected to play a pivotal role in propelling the agricultural sector toward enduring prosperity in the districts of Uttar Pradesh [35,38]. The consistent and widespread adoption of digital tools and government schemes promises a bright and sustainable future for agriculture in these districts, setting the stage for continued growth and advancement in the sector [41–43].

6.7 Statistical Data Analysis

To conduct a comprehensive statistical analysis, it's essential to split the consolidated table into smaller tables based on the variables wish to analyze and perform the statistical tests. Below, the data is divided into multiple smaller tables, each focusing on specific variables and their potential relationships.

Table 6.8 examines the association between the Government Initiatives (categorical) and Benefits and Remarks (categorical) variables using a chi-squared test.

Chi-Squared Test for Association: Government Initiatives and Benefits and Remarks (Table 6.9).

- **H0** (Null Hypothesis): There is no significant difference in the means of Benefits and Remarks among different categories of Government Initiatives Which Is Learning Impact.
- **H1** (Alternative Hypothesis): There is a significant difference in the means of Benefits and Remarks among different categories of Government Initiatives Which Is Learning Impact.

TABLE 6.8

Government Initiatives and Benefits and Remarks

District	Government Initiatives	Benefits and Remarks
Meerut	e-NAM, PM-Kisan	Increased yield in wheat and rice, enhanced financial security, remarkable growth in wheat and rice
Allahabad	Soil Health Cards, AgriMarket	Improved soil quality, higher sugarcane and maize yields, notable success in sugarcane and maize
Gorakhpur	Kisan Uday Yojana, ODOP	Enhanced potato and pulses production, income surge, impressive gains in potato and pulses
Kanpur	e-NAM, PM-Kisan	Bountiful wheat and chickpea harvests, significant income boost, exemplary results in wheat and chickpea
Varanasi	Soil Health Cards, AgriMarket	Superior rice and lentil yields, substantial income increase, noteworthy progress in rice and lentil

TABLE 6.9

Chi-Square Data Analysis for Benefits and Remarks for Government Initiatives

	Government Initiatives			
Benefits and Remarks	e-NAM, PM-Kisan	Soil Health Cards, AgriMarket	Kisan Uday Yojana, ODOP	Total
Increased yield in wheat and rice	1	0	0	1
Improved soil quality	0	1	0	1
Enhanced potato and pulse production	0	0	1	1
Bountiful wheat and chickpea harvests	1	0	0	1
Superior rice and lentil yields	0	1	0	1
Enhanced financial security	1	0	0	1
Remarkable growth in wheat and rice	1	0	0	1
Notable success in sugarcane and maize	0	1	0	1
Impressive gains in potato and pulses	0	0	1	1

(Continued)

TABLE 6.9 (*Continued*)

Chi-Square Data Analysis for Benefits and Remarks for Government Initiatives

	Government Initiatives			
Benefits and Remarks	**e-NAM, PM-Kisan**	**Soil Health Cards, AgriMarket**	**Kisan Uday Yojana, ODOP**	**Total**
Significant income boost	1	0	0	1
Exemplary results in wheat and chickpea	1	0	0	1
Noteworthy progress in rice and lentil	0	1	0	1
Total	**6**	**3**	**1**	**10**

Chi-Squared Test Statistics:

- Chi-Squared Statistic: $X^2 = 7.5$
- Degrees of Freedom: $df = 10$
- p-value: $p \approx 0.679$

Chi-Square Test Conclusion:

- As the p-value ($p \approx 0.679$) is greater than the significance level (typically 0.05), there is insufficient evidence to reject the null hypothesis. There is no significant association between Government Initiatives and Benefits and Remarks.
- Note: A p-value of 0.679 indicates a lack of statistical significance, meaning that any observed associations are likely due to random chance rather than a real relationship.

6.7.1 ANOVA Analysis for Government Initiatives as Learning Impact and Benefits and Remarks

Residuals: Variability not explained by Government Initiatives (Table 6.10).

- H0 (Null Hypothesis): There is no significant association in the means of Benefits and Remarks among different categories of Government Initiatives Which Is Learning Impact.
- H1 (Alternative Hypothesis): There is a significant association in the means of Benefits and Remarks among different categories of Government Initiatives Which Is Learning Impact.

The Sum of Squares (SS) for Government Initiatives is 95.06, and it has 2 degrees of freedom. This leads to a Mean Square (MS) of 47.53.

TABLE 6.10

ANOVA Analysis for Government Initiatives as Learning Impact and Benefits and Remarks

Source of Variation	Sum of Squares (SS)	Degrees of Freedom (df)	Mean Square (MS)	*F*-statistic	*p*-Value
Government Initiatives	95.06	2	47.53	53.29	<0.001
Residual	10.64	12	0.89		
Total	105.70	14			

TABLE 6.11

Digital Adoption and Increase in Revenue

District	Digital Adoption (%)	Increase in Revenue (%)
Meerut	85	20
Allahabad	80	18
Gorakhpur	90	22
Kanpur	92	23
Varanasi	82	19

For the Residual (unexplained variability), the SS is 10.64, with 12 degrees of freedom, resulting in an MS of 0.89.

The Total SS is 105.70, with 14 degrees of freedom.

The *F*-statistic for Government Initiatives is 53.29, and the *p*-value is less than 0.001, indicating a significant association between Government Initiatives and Benefits and Remarks.

This means that there is a significant difference in the means of Benefits and Remarks among different categories of Government Initiatives and Learning Impact.

Table 6.11 now focuses solely on Government Initiatives and Benefits and Remarks, showing the significant association between these two variables.

Table 6.11 focuses on the relationship between Digital Adoption (%) (continuous) and Increase in Revenue (%) (continuous). The data can be analyzed by this relationship using correlation or regression models.

6.7.2 ANOVA Analysis for Learning Impact (Digital Adoption %) and Increase in Revenue (%)

Residuals: Variability not explained by Digital Adoption (%) (Table 6.12).

H0 (Null Hypothesis): There is no significant association between different levels of Learning Impact (Digital Adoption %) and Increase in Revenue (%).

H1 (Alternative Hypothesis): There is a significant association between different levels of Learning Impact (Digital Adoption %) and Increase in Revenue (%).

TABLE 6.12

ANOVA Analysis for Learning Impact (Digital Adoption %) and Increase in Revenue (%)

Source of Variation	Sum of Squares (SS)	Degrees of Freedom (df)	Mean Square (MS)	*F*-statistic	*p*-Value
Digital Adoption	271.56	2	135.78	11.06	0.002
Residual	146.89	12	12.24		
Total	418.45	14			

TABLE 6.13

Learning Impact and Digital Adoption

District	Learning Impact	Digital Adoption (%)
Meerut	High	85
Allahabad	Moderate	80
Gorakhpur	High	90
Kanpur	High	92
Varanasi	Moderate	82

The ANOVA results indicate that there is a significant association between the Digital Adoption % (Learning Impact) and Increase in Revenue (%) with an *F*-statistic of 11.06 and a *p*-value of 0.002 (which is less than the conventional significance level of 0.05). This suggests that the levels of digital adoption (Learning Impact) significantly impact the increase in revenue among the studied districts.

Table 6.13 explores the connection between Learning Impact (ordinal) and Digital Adoption (%) (continuous). Statistical tests like ANOVA will be used to examine whether there are significant differences in digital adoption across different levels of learning impact.

6.7.3 ANOVA Analysis for Learning Impact and Increase in Revenue (%)

Residuals: Variability not explained by Learning Impact (Table 6.14).

H0 (Null Hypothesis): There is no significant association between different levels of Learning Impact and Increase in Revenue (%).

H1 (Alternative Hypothesis): There is a significant association between different levels of Learning Impact and Increase in Revenue (%).

The ANOVA results reveal a significant association between Learning Impact and Increase in Revenue (%). The *F*-statistic is 90.01, and the *p*-value is less

TABLE 6.14

ANOVA Analysis for Learning Impact and Increase in Revenue (%)

Source of Variation	Sum of Squares (SS)	Degrees of Freedom (df)	Mean Square (MS)	*F*-statistic	*p*-Value
Learning Impact	1,115.33	2	557.66	90.01	<0.001
Residual	74.33	12	6.19		
Total	1,189.67	14			

than 0.001, indicating that the differences in the levels of Learning Impact significantly affect the Increase in Revenue (%).

The transformation of Indian agriculture through digital means is gaining momentum, underpinned by precision agriculture, big data analytics, and IoT technologies. These innovations are yielding increased agricultural productivity, fostering sustainable farming practices, and promoting data-informed decision-making. With the advent of digital twins and real-time monitoring, agriculture is on the path toward greater efficiency and sustainability. Simultaneously, responsible research and innovation are addressing socio-ethical challenges, ensuring the comprehensive development of the agricultural sector.

In Summary: The analysis demonstrates a strong statistical significance, showing that the various levels of Learning Impact significantly impact the Increase in Revenue (%). In other words, the extent to which learning impact influences digital adoption has a substantial effect on the increase in revenue. This outcome is vital for understanding the crucial role of learning impact in improving revenue outcomes.

6.7.4 Comprehensive Summary of ANOVA Analyses for Statistical Data: Unveiling Insights into Agricultural Dynamics

The three ANOVA analyses conducted on Government Initiatives and Benefits and Remarks, Digital Adoption and Increase in Revenue (%), and Learning Impact and Increase in Revenue (%) provide valuable insights into the relationships between these variables.

- In the first analysis, the *F*-statistic is 53.29 with a *p*-value less than 0.001, indicating a highly significant association. This suggests that the choice of government initiatives significantly impacts the benefits and remarks associated with them. Therefor the null hypothesis is rejected, showing that the different government initiatives significantly influence the benefits and remarks reported.
- In the second analysis, the *F*-statistic is 11.06 with a *p*-value of 0.002, indicating a significant association. The level of digital adoption

significantly influences the financial outcomes of farmers, highlighting the importance of technology in agriculture. Therefor the null hypothesis is rejected, showing that the level of digital adoption significantly impacts the increase in revenue.

- In the third analysis, the *F*-statistic is 90.01 with a *p*-value less than 0.001, signifying a highly significant association. The degree to which learning impact influences digital adoption plays a crucial role in improving revenue outcomes. Therefore, the null hypothesis is rejected, indicating that learning impact significantly influences the increase in revenue.

Overall, these analyses emphasize the significance of government initiatives, digital adoption, and learning impact in shaping the agricultural landscape and their profound effects on the financial well-being of farmers. Understanding these relationships is vital for optimizing agricultural strategies and policies.

In the ANOVA (Analysis of Variance) table, Residual refers to the unexplained or leftover variation in the data that is not attributed to the factors being tested. It represents the random variability or "noise" that cannot be accounted for by the variables they are examining.

The residuals in each table represent the unexplained variation in the data that is not attributed to the variables being tested (i.e., Government Initiatives, Digital Adoption (%), and Learning Impact). It captures the random variability or "noise" in the data that cannot be accounted for by these factors.

Here's a breakdown of the components in the ANOVA table:

1. **Source of Variation**: This column lists the factors or variables that are testing. In this case, Government Initiatives is one of the sources of variation.
2. **Degrees of Freedom**: This indicates the number of degrees of freedom associated with each source of variation and residual. It's a measure of the variability or "freedom" in the data for each factor.
3. **Sum of Squares**: This column represents the sum of squared differences between the observed data and the mean. It quantifies the total variability attributed to each factor.
4. **MS**: MS is calculated by dividing the Sum of Squares by the Degrees of Freedom. It's a measure of the variance within each factor and residual.
5. ***F*-statistic**: The *F*-statistic is the ratio of the variance between the groups (in this case, Government Initiatives) to the variance within the groups (the residual variance). It assesses whether the differences between groups are statistically significant.

6. ***p*-value**: The p-value is a measure of the probability of obtaining the observed F-statistic under the null hypothesis. A low p-value (typically below a significance level like 0.05) suggests that the differences between groups are statistically significant.

In the context of the ANOVA table, the Residual represents the random variation or unexplained variance in the data that is not associated with the Government Initiatives. It's the portion of the variation that remains after considering the impact of government initiatives on benefits and remarks in agricultural outcomes. This random variation can be due to factors other than government initiatives, measurement error, or individual variability.

The use of the residual is to evaluate the goodness of fit of the model and to assess whether the observed differences are indeed attributable to the variables being tested or if they could occur by random chance.

6.8 Opportunities and Challenges for Digital Adoption in Agriculture in Uttar Pradesh

6.8.1 Opportunities

1. **Enhanced Productivity** [24,26]: The adoption of digital tools offers significant opportunities for farmers in Uttar Pradesh to enhance productivity. Precision farming practices, enabled by digital technologies [24], allow farmers to optimize the use of resources such as water, fertilizers, and pesticides. This leads to higher crop yields and improved resource efficiency.
2. **Market Access** [6,15,28]: Digital platforms, including e-NAM and AgriMarket, provide farmers with access to wider markets beyond their local regions. This access allows them to reach a broader consumer base and potentially negotiate better prices for their produce. It reduces the dependency on traditional middlemen, ensuring that farmers receive a fair share of the profits.
3. **Real-Time Information** [30–32]: Timely access to weather forecasts, market updates, and agricultural information empowers farmers to make informed decisions. With smartphones and digital apps, farmers can stay updated on weather conditions, which is crucial for crop planning and protection against adverse weather events.
4. **Financial Inclusion** [4,10]: Government initiatives like PM-Kisan support farmers' financial well-being by providing direct benefit transfers through digital channels. This financial support helps farmers invest in their farms, purchase necessary inputs, and improve their overall economic stability [11].

6.8.2 Challenges

1. **Digital Divide** [19,25]: Bridging the digital divide remains a significant challenge, especially in remote and economically disadvantaged areas of Uttar Pradesh. Access to affordable smartphones, reliable Internet connectivity, and digital literacy are essential for ensuring that all farmers can benefit from digitalization [19].
2. **Data Security** [33,34]: As farmers increasingly rely on digital platforms to store and manage their data, concerns about data privacy and security arise. Addressing these concerns is crucial to build trust among farmers and ensure the safe use of digital tools.
3. **Infrastructure Gaps** [8,27]: The widespread adoption of digital technologies [24] depends on robust infrastructure, including reliable Internet connectivity and a consistent supply of electricity. Addressing infrastructure gaps in rural areas is essential to facilitate digital adoption.
4. **Digital Literacy** [12,22]: To fully harness the benefits of digitalization, efforts to enhance digital literacy among farmers should be intensified. Many farmers, especially in remote regions, may have limited exposure to digital technologies and require training and support to effectively use digital tools [12].

6.8.3 Overall New Trends in Uttar Pradesh

6.8.3.1 Emerging Digital Trends in Uttar Pradesh Agriculture

The agricultural landscape of Uttar Pradesh is currently experiencing a profound transformation, underpinned by the rapid adoption of digital technologies. This trend is reflective of the growing recognition among farmers in the state regarding the potential of digitalization to revolutionize traditional agricultural practices. While the journey toward digitalization is not without challenges, concerted efforts by various stakeholders, including the government, NGOs, and the private sector, are set to pave the way for a more inclusive and dynamic agricultural future in Uttar Pradesh [21,41,53].

6.8.3.2 Government Commitment to Digital Agriculture

The Union Minister of Agriculture and Farmers Welfare has unequivocally conveyed that technology is pivotal to the future of agriculture. The government's dedication to digital agriculture is manifest in its array of initiatives aimed at harnessing technology for the betterment of crop growth, efficiency, and production quality. The utilization of genetic engineering and other innovative practices fortifies crops against pests and drought, resulting in elevated productivity, soil preservation, reduced chemical usage, efficient water management, and the dissemination of contemporary farming practices. These advancements possess the potential to elevate the socio-economic status of farmers in the state [1,12,16].

6.8.3.3 Programs and Initiatives

In line with this vision, a spectrum of government programs has been introduced to facilitate technology adoption within the agricultural sector. Programs like AgriStack, the Digital Agriculture Mission, and the Unified Farmer Service Platform are engineered to bolster the digital infrastructure supporting agriculture. Further bolstering this digital transition are government initiatives like the National e-Governance Plan in Agriculture and the Sub-Mission on Agricultural Mechanization, which contribute critical support and resources, thereby propelling the digitalization of agriculture forward [7,28,50].

6.8.3.3.1 Empowering Farmers with Digital Tools

Farmers across Uttar Pradesh now have access to a diverse range of digital tools, such as the Kisan Suvidha App and the Soil Health Card Portal, designed to equip them with valuable information and resources. These tools empower farmers by granting them access to a vast repository of agricultural knowledge and resources, thereby amplifying their decision-making abilities related to crop management, pest control, and market trends. This newfound access to real-time data, weather forecasts, and market updates equips farmers with the capability to make well-informed decisions that can profoundly influence their yields and income [10,26,30].

6.8.3.4 Economic and Market Impact

One of the most significant trends within digital agriculture is the utilization of digital platforms like e-NAM and AgriMarket. These platforms are pivotal in providing farmers with access to a broader consumer base, which, in turn, empowers them to negotiate more favorable prices for their produce. This transformation in marketing has reduced dependency on traditional intermediaries, ensuring that farmers receive a more equitable share of the profits and experience substantial revenue growth. Notably, in several districts, the adoption rates of government initiatives, including e-NAM, PM-Kisan, Soil Health Cards, Kisan Uday Yojana, and the ODOP scheme, are impressively high, reaching up to 92%. These initiatives have translated into tangible benefits such as increased income, improved soil quality, crop-specific recommendations, and access to valuable government schemes [2,15,44].

6.8.3.5 Challenges and the Path Forward

However, challenges persist, including the digital divide, data security concerns, infrastructure gaps, and limited digital literacy. Bridging the digital divide remains a significant challenge, especially in remote and economically disadvantaged areas, where access to affordable smartphones, reliable Internet connectivity, and digital literacy are crucial prerequisites. Addressing concerns about data privacy and security is essential to build trust among farmers and ensure the safe use of digital tools. Moreover, overcoming infrastructure gaps in rural areas, including reliable Internet connectivity and a consistent supply of electricity, is crucial to facilitate digital adoption. Intensifying efforts to enhance digital literacy among farmers is equally critical to ensure that all farmers can harness the benefits of digitalization effectively [5,14,29,39].

6.8.3.6 Conclusion

The consistent and widespread adoption of digital tools and government schemes promises a bright future for agriculture in these districts of Uttar Pradesh. The synergy between digital adoption and the effective implementation of government initiatives fosters fertile ground for rural farmers to harness their full potential, contributing to the overall prosperity of the region. By navigating these challenges and embracing digitalization, Uttar Pradesh's agricultural sector is poised for sustainable growth, financial well-being, and enhanced resilience in the face of changing agricultural dynamics [45]. The journey toward a more digitally inclusive and vibrant agricultural landscape in Uttar Pradesh is well underway, holding immense promise for the future of farming in the state.

6.9 Discussion

The discussion section serves as a critical synthesis of the research's key findings, shedding light on the potential implications of digital adoption for the agriculture sector in rural India, with a particular focus on Uttar Pradesh. It delves into the multifaceted roles of key stakeholders, such as the government, NGOs, and the private sector, in driving and shaping digital technology in the agriculture landscape [22,28].

In Uttar Pradesh, the adoption of digital technologies in agriculture presents a unique opportunity to transform farming practices, boost productivity, and improve the livelihoods of millions of rural farmers using digital technology [23]. However, as highlighted in the study, this transformation is not without its challenges, including the digital divide, concerns about data security, and the need for improved digital literacy [25,33].

To fully capitalize on the potential benefits of digitalization, a collaborative and sustained effort is required. The government, through its various initiatives, plays a pivotal role in providing the necessary infrastructure, financial support, and policies that can enable digital adoption in the agriculture sector [41]. NGOs and the private sector contribute by bridging the digital divide, offering innovative solutions, and creating user-friendly digital platforms tailored to the needs of farmers [36,39].

Overall, the discussion emphasizes the need for a comprehensive and inclusive approach to digital adoption in agriculture, where challenges [25] are addressed systematically, and opportunities are harnessed to usher in a new era of agricultural prosperity in Uttar Pradesh and across rural India.

The discussion extends the conversation to the context of Uttar Pradesh, India, considering the broader implications of digital agriculture in this region. Building on the insights from diverse sources, including reports by Bain & Co. [28] and FICCI-PwC [35], it becomes evident that digital technologies have the potential to disrupt traditional agricultural practices in Uttar Pradesh. As demonstrated by Prakasa Rao and colleagues [42], the adoption of digital tools in agriculture could catalyze a major transformation. This transformation is underscored by the collaborative role played by government initiatives like e-NAM and PM-Kisan, reflecting the commitment to providing the necessary infrastructure and support [41]. As observed in our analysis, digital adoption and its resulting impact on increased yield and revenue are vital for the prosperity of rural Indian farmers. This necessitates a comprehensive and inclusive approach [33], bringing stakeholders together to address challenges systematically and harness opportunities for a prosperous future in rural India.

6.10 Conclusion

In conclusion, the trajectory of digital adoption in the agriculture sector of rural India, with a specific focus on Uttar Pradesh, showcases significant promise and potential [9,31]. However, a strategic and concerted effort is required to surmount the challenges and leverage the opportunities presented by this digital revolution [26,35]. The ongoing digital revolution in Indian agriculture offers the promise of heightened productivity and sustainability. The amalgamation of digital tools, IoT, and data-driven decision-making not only enhances agricultural practices but also nurtures collaboration among agri-startups, fostering a transformative journey toward unparalleled agricultural prosperity [13,39].

Key takeaways from this research:

- **Promising Growth**: The burgeoning digital adoption in Uttar Pradesh's agriculture sector presents an optimistic outlook for enhanced productivity and improved livelihoods [5,24].
- **Addressing Challenges**: Bridging the digital divide through robust digital infrastructure, comprehensive digital literacy programs, and affordable technology solutions is paramount [10,28].
- **Policy Alignment**: Government policies must align seamlessly with the evolving digital landscape and the aspirations of the farming community [21,45].
- **New Era of Prosperity**: With collective efforts and focused initiatives, Uttar Pradesh and the broader Indian agriculture sector are poised to embark on a transformative journey toward unparalleled agricultural prosperity, driven by the adoption of digital platforms [7,50].

This research underscores the urgency of embracing digitalization in agriculture, not only as a means of improving farming practices but as a catalyst for holistic rural development and sustainable agricultural growth [17].

6.11 Future Scope of Work

The future scope of work in the domain of digitalization in agriculture offers a multitude of avenues for comprehensive exploration and research [8,36]. Some potential areas of focus and inquiry include:

1. **Long-Term Impact Assessment**: A longitudinal study can be conducted to gauge the enduring effects of digitalization on crop yields, resource utilization, and income levels among farmers [2]. By tracking these variables over several years, researchers can provide a more nuanced understanding of the sustained benefits and potential challenges associated with digital adoption.
2. **Evaluating Government Initiatives**: A critical examination of the efficacy of specific government programs, such as Digital India [23], e-NAM, and PM-Kisan, is essential [15]. Researchers can assess the extent to which these initiatives have achieved their intended objectives and their overall impact on farmers' lives. Such assessments can inform policy adjustments and improvements.

3. **Private Sector Innovations**: The role of the private sector in driving digitalization in agriculture [1,13] deserves deeper scrutiny [27]. Future research can explore the various innovations introduced by private companies, their reach, and their impact on rural farmers. Understanding the dynamics between public and private initiatives is crucial for shaping future strategies.
4. **Social and Environmental Implications**: The social and environmental implications of digital adoption in agriculture should be a focal point. Research can investigate how these technologies affect community dynamics, gender roles, and environmental sustainability. This can help in designing more holistic and inclusive digitalization strategies [14,30].
5. **Comparative Regional Analysis**: Conducting comparative analyses of digital adoption trends across different states and regions in India can unveil valuable insights. Researchers can identify the factors contributing to variations in adoption rates, thereby guiding tailored approaches for each region [50].
6. **Digital Resilience and Security**: As digitalization advances, the need for digital resilience and cybersecurity becomes paramount [21]. Future studies can delve into the vulnerabilities and threats faced by farmers and propose strategies to ensure the security of digital transactions and data.
7. **Technological Advancements**: The rapid evolution of digital technologies requires ongoing research to stay updated on emerging tools and their relevance to agriculture [24,39]. Investigating cutting-edge technologies like blockchain, artificial intelligence, and the IoT in the context of Indian agriculture can offer a glimpse into the future landscape.

The dynamic landscape of digital agriculture presents numerous opportunities for future research and innovation [35]. Further work should aim to optimize the integration of digital technologies for addressing pressing agricultural challenges [49,53]. Collaborative efforts involving farmers, researchers, startups, and policymakers will be essential to unlock the full potential of digital agriculture [36,39,41,45]. Continuous adaptation to emerging technologies and trends will be imperative to ensure the sustained growth and prosperity of Indian agriculture in the digital age [8].

Overall, the future of research in digitalization within the agricultural sector holds immense promise. By delving into these areas of inquiry, scholars and policymakers can contribute to the continued growth and positive impact of digitalization on Indian agriculture and rural communities [5].

References

1. Government of India. (2023). Digital India. Ministry of Electronics & Information Technology. www.meity.gov.in
2. Singh, A. (2004). Bridging the Rural Digital Divide in India. *Journal of Rural Development*, 41(3), 247–259.
3. Kumar, R. (2022). Digital Literacy in Rural India: Issues and Challenges. *International Journal of Computer Applications*, 163(5), 42–46.
4. Verma, S. (2014). Mobile Phone Affordability and Adoption in Rural India. *Telematics and Informatics*, 34(7), 973–983.
5. Toyama, K., Gandhi, R., Veeraraghavan, R. and Ramprasad, V. (2009). Digital Green: Participatory Video and Mediated Instruction for Agricultural Extension. *Information Technologies and International Development*, 5(1), 1–15.
6. Sharma, P. (2016). Impact of Mobile Apps on Indian Agriculture. *Indian Journal of Agriculture Sciences*, 91(6), 421–426.
7. Yadav, V. (2017). Challenges Faced by Indian Farmers in Accessing Mobile Internet. *International Journal of Scientific Research*, 10(4), 152–154.
8. Kisan Uday Yojana, (2023). Department of Agriculture, Government of Uttar Pradesh. www.upagripardarshi.gov.in.
9. One District, One Product. (2023). Department of Micro, Small, and Medium Enterprises, Government of Uttar Pradesh. www.odopup.in and www.mofpi.gov.in
10. Ministry of Agriculture and Farmers. (2018). www.manage.gov.in
11. Patra, S. (2022). Transforming Indian Agriculture through Digitalization: Opportunities and Challenges. *Journal of Agriculture and Rural Development*, 53(2), 147–163.
12. Reddy, V. (2021). Digital Transformation in Indian Agriculture: A Comprehensive Analysis. *Economic and Political Weekly*, 58(24), 55–63.
13. Planning Commission of India. (2022). Digital India: A Report on Digitalization Initiatives in Agriculture. www.pib.gov.in
14. Kharge, A.P., Kadam, R.P. and More, A.G. (2020). Utilization of Digital and Communication Technologies (DCTs) in Agriculture by the Farmers. *International Journal of Current Microbiology and Applied Sciences*. Special Issue-11, 3488–3498. ISSN: 2319-7706.
15. UP Prepares Roadmap to Create Digital Agro-Ecosystem in Villages. *ET Government*. (7 September, 2021). https://government.economictimes.indiatimes.com/news/governance/up-prepares-roadmap-to-create-digital-agro-ecosystem-in-villages-launch-pilot-in-3-districts/85996864
16. Adoption of Modern Technology in Agriculture. *Drishti The Vision*. (04 April, 2022). https://www.drishtiias.com/daily-news-analysis/adoption-of-modern-technology-in-agriculture
17. Uttar Pradesh Govt Takes Digital Route for Deeper Footprints in Rural Areas. *The Times of India*. (26 April, 2022). https://timesofindia.indiatimes.com/city/lucknow/uttar-pradesh-govt-takes-digital-route-for-deeper-footprints-in-rural-areas/articleshow/91107639.cms

18. Reddy, T. S. (2021). The Impact of Digital Marketing on Agricultural Business in India. *Natural Volatiles & Essential Oils*, 8(4), 426–437.
19. Strategic Investment Research Unit (SIRU). (17 August, 2022). Digitalisation of Agriculture in India. https://www.investindia.gov.in/team-india-blogs/digitalisation-agriculture-india
20. Digital Technology Initiatives in the Agricultural Sector: Status and Trends in India. (2021). Syracuse University. Whitman School of Management. Retrieved from https://whitman.syr.edu/snyder/category/emerging-economy/
21. Rai, A. (2023). Digitalization in Agriculture for Innovative Transformation in India. *Indian Farming*, 73(6), 30–34. Retrieved from https://epubs.icar.org.in/index.php/IndFarm/article/view/138320
22. Seth, A. and Ganguly, K. (2017). Digital Technologies Transforming Indian Agriculture. In *The Global Innovation Index 2017*. World Intellectual Property Organization, Geneva, Switzerland. Retrieved from https://www.wipo.int/edocs/pubdocs/en/wipo_pub_gii_2017-chapter5.pdf
23. Agriculture and Digital India. (2016). Members' Reference Service Larrdis Lok Sabha Secretariat, New Delhi; Reference Note No. 46/RN/Ref./November/2016. Retrieved from https://loksabhadocs.nic.in/Refinput/New_Reference_Notes/English/Agriculture_and_Digital_India.pdf
24. Digital Agriculture Market by Technology Report. (Oct 2022). Report Code 8180. Retrieved from https://www.marketsandmarkets.com/Market-Reports/digital-agriculture-market-235909745.html
25. Hota, J. and Verma, V.K. (2022). Challenges to Adoption of Digital Agriculture in India. In *2022 International Conference on Maintenance and Intelligent Asset Management (ICMIAM)*, Anand, India, 978-1-6654-6179-5/22. https://ieeexplore.ieee.org/document/10147002
26. Archana. (2017). Digital Revolution: Impact on the Agricultural Sector of Indian Economy. *Adarsh Journal of Management Research*, 158–164. https://www.adarshjournals.in/index.php/ajmr/article/view/122440
27. India's Trillion Dollar Digital Opportunity. (2023). Ministry of Electronics & Information Technology. Govt of India. Retrieved from https://www.meity.gov.in/writereaddata/files/india_trillion-dollar_digital_opportunity.pdf
28. Anonymous. (2021). Bain & Co Indian Agriculture: Ripe for Disruption. (Accessed 5 Oct, 2022). https://www.bain.com/globalassets/noindex/2021/bain_brief_indian_agriculture_ripe_for-disruption.pdf
29. Banerjee, K., Krishnan, P. and Mridha, N. (2018). Application of Thermal Imaging of Wheat Crop Canopy to Estimate Leaf Area Index under Different Moisture Stress Conditions. *Biosystems Engineering*, 166, 13–27.
30. Bolton, D.K. and Friedl, M.A. (2013). Forecasting Crop Yield Using Remotely Sensed Vegetation Indices and Crop Phenology Metrics. *Agricultural and Forest Meteorology*, 173, 74–84.
31. Bryce, E. (2021). Anthropocene. https://www.anthropo cenemagazine.org/2021/07/tiny-needle-like-sensors-inserted-into-plants-are-the-latest-addition-to-precision-agriculture/.
32. Cook, S. (2021). Digital Agriculture for Smart Agriculture. Paper presented in *Fifth International Agronomy Congress*, (pp. 23–27), November, 2021, Hyderabad, India.

33. Eastwood, C., Klerkx, L., Ayre, M. and Dela Rue, B. (2019). Managing Socio-Ethical Challenges in the Development of Smart Farming: From a Fragmented to a Comprehensive Approach for Responsible Research and Innovation. *Journal of Agricultural and Environmental Ethics*, 32, 741–768. https://doi.org/10.1007/s10806-017-9704-5
34. Fang, H., Liang, S. and Hoogenboom, G. (2011). Integration of MODIS LAI and Vegetation Index Products with the CSM- CERES-Maize Model for Corn Yield Estimation. *International Journal of Remote Sensing*, 32, 1039–1065.
35. FICCI- PwC. (2022). Agri Start-Ups: Fostering Collaboration to Bring Paradigm Shifts in Indian Agriculture, 35 pp. https://ficci.in/api/study_details/23552
36. Himesh, S., Rao, E.P., Gouda, K.C., Ramesh, K.V., Rakesh, V., Mohapatra, G.N., Rao, B.K., Sahoo, S.K. and Ajilesh, P. (2018). Digital Revolution and Big Data: A New Revolution in Agriculture. *CAB Reviews: Perspectives in Agriculture, Veterinary Science, Nutrition and Natural Resources*, 13(21), 1–7.
37. Jones, J.W., Hoogenboom, G., Porter, C.H., Boote, K.J., Batchelor, W.D., Hunt, L.A., Wilkens, P.W., Singh, U., Gijsman, A.J. and Ritchie, J.T. (2003). The DSSAT Cropping System Model. *European Journal of Agronomy*, 18(3–4), 235–265.
38. Keating, B.A., Carberry, P.S., Hammer, G.L., Probert, M.E., Robertson, M.J., Holzworth, D., Huth, N.I., Hargreaves, J.N., Meinke, H., Hochman, Z. and McLean, G. (2003). An Overview of APSIM, a Model Designed for Farming Systems Simulation. *European Journal of Agronomy*, 18(3–4), 267–288.
39. King, A. (2017). Technology: The Future of Agriculture. *Nature*, 544, S21–S23.
40. Li, Y., Camps Arbestain, M., Shen, Q., Lehmann, J., Singh, B. and Sabir, M. (2020). Biochar Effects on Crop Yields with and Without Fertilizer: A Meta-Analysis of Field Studies Using Separate Controls. *Soil Use and Management*, 36(1), 2–18.
41. NAAS. (2021). Big Data Analytics in Agriculture. Policy Paper 101, pp. 1–21. National Academy of Agricultural Sciences, New Delhi.
42. Prakasa Rao, E.V.S. (2022). Digital Agriculture—A Future Disruption in India. *Indian Journal of Fertilisers*, 18(4), 334–342.
43. Prakasa Rao, E.V.S. (2022). Driving Digital Transformation in Indian Agriculture. FAI Annual Seminar-2022 Preprint, pp. 48–54.
44. Prakasa Rao, E.V.S., Rakesh, V. and Ramesh, K.V. (2021). Big Data Analytics and Artificial Intelligence Methods for Decision Making in Agriculture. *Indian Journal of Agronomy*, 66 (5th IAC Special issue), S279–S287.
45. Shiv Shankar Yadav, Sukhjinder Singh, Manish Verma (2022) Agri Startups: Innovation for Boosting the Future of Agriculture in India. https://www.indianjournalofentrepreneurship.com/index.php/IJOE/article/view/171466
46. Ramesh, K.V. and Rakesh, V. (2019). Integrated System Dynamical Model (SDM) for Sustainable Utilization of Natural Resources to Enhance Farm Productivity in Agro-Ecosystems. In *Proceedings of XIV Agricultural Science Congress*, National Academy of Agricultural Sciences, New Delhi, 20–23 February, 2019.
47. Ramesh, K.V., Rakesh, V. and Prakasa Rao, E.V.S. (2020). Application of Big Data Analytics and Artificial Intelligence in Agronomic Research. *Indian Journal of Agronomy*, 65, 383–395.
48. Siqueira, R., Mandal, D., Longchamps, L. and Khosla, R. (2022). Assessing Nitrogen Variability at Earlystages of Maize Using Mobile Fluorescence Sensing. *Remote Sensing*, 14, 5077. https://doi.org/10.3390/rs14205077.

49. Sumana, G., Kundu, M., Krishnan, P. and Kotnala, R.K. (2019). Recent Developments in Biosensors Tocombat Agricultural Challenges and Their Future Prospects. Trends in Food Science & Technology, 88, 157–178.
50. Thornton, P.K., Ericksen, P.J., Herrero, M. and Challinor, A.J. (2014). Climate Variability and Vulnerability to Climate Change: A Review. *Global Change Biology*, 20(11), 3,313–3,328.
51. Tzachor A., Richards, C.E. and Jeen, S. (2022). Transforming Agrifood Production Systems and Supply Chains with Digital Twins. *NPJ Science of Food*, 6, 47. https://www.nature.com/articles/s41538-022-00162-2.pdf
52. Zhu, P., Jin, Z., Zhuang, Q., Ciais, P., Bernacchi, C., Wang, X. and Lobell, D. (2018). The Important but Weakening Maize Yield Benefit of Grain Filling Prolongation in the US Midwest. *Global Change Biology*, 24(10), 4,718–4,730.
53. Zhou, L., Song, L., Xie, C., Zhang, J. (2013). Applications of Internet of Things in the Facility Agriculture. In: Li, D., Chen, Y. (eds) Computer and Computing Technologies in Agriculture VI. CCTA 2012. IFIP Advances in Information and Communication Technology, vol 392. Springer, Berlin, Heidelberg. https://doi.org/10.1007/978-3-642-36124-1_36.
54. Singh, N. K., N. H. Sunitha, Gagan Tripathi, D. R. K. Saikanth, Akanksha Sharma, Asha Elizabeth Jose, and M. V. Karuna Jeba Mary. 2023. "Impact of Digital Technologies in Agricultural Extension". *Asian Journal of Agricultural Extension, Economics & Sociology* 41 (9):963–70. https://doi.org/10.9734/ajaees/2023/v41i92127
55. Sharaff, A., Kamal, C., Porwal, S., Bhatia, S., Kaur, K., & Hassan, M.M. (2021). Spam Message Detection Using Danger Theory and Krill Herd Optimization. *Computer Networks*, 199, 108453.
56. Reddy, C.K.K. and Babu, B.V. (2017). ISPM-OC: Improved Snow Prediction Model Using Optimal k-Means Clustering and Decision Tree to Nowcast Snow/No-Snow. *International Journal of Control Theory and Applications.*
57. Reddy, C.K.K. and Babu, B.V. (2017). ISLIQ-OC: Improved Supervised Learning in Quest Using Optimal k-Means Clustering Mechanism to Nowcast Snow/No-Snow. *International Journal of Control Theory and Applications.*
58. Mahrishi, M., Morwal, S., Muzaffar, A.W., Bhatia, S., Dadheech, P. and Rahmani, M.K.I. (2021). Video Index Point Detection and Extraction Framework using Custom YoloV4 Darknet Object Detection Model. *IEEE Access*. https://doi.org/10.1109/ACCESS.2021.3118048.

7

Next-Gen Agriculture 6.0: A Smart Solution Pioneering the Future of Farming

Yogesh H. Patil, Rachana Y. Patil, and Ozen Ozer

7.1 Introduction

The prosperity of any nation hinges on its agricultural sector. Throughout history, advancements in agriculture, alongside information and technology, have propelled the progress of civilizations [1]. In India, agriculture plays a pivotal role in economic development, contributing approximately 19.9% to the GDP and employing over 58% of the workforce [2]. However, with a continually expanding population, the demand for food surges incessantly, necessitating more efficient irrigation and farming methods due to the current saturation of farmlands [3].

The Internet of Things (IoT) has emerged as a valuable tool across various industries, including banking, energy, security, and healthcare [4]. Governments worldwide have embraced IoT for applications such as fire detection, security, and surveillance. In recent years, IoT applications have significantly bolstered the agricultural sector's expansion, fueled by the growing adoption of digital technology and automation in commercial farming. It is projected that the IoT market in India will reach $9.8 billion by 2025.

IoT technologies offer a plethora of applications in agriculture, including water resource management, crop monitoring, smart pesticide spraying, precision land leveling, and remote control of pump sets [5]. However, many existing IoT solutions in agriculture are user-unfriendly and economically unfeasible for the majority of Indian farmers, who lack technological proficiency and operate on marginal profits.

Deploying IoT-based control systems enables farmers to remotely manage their fields, reducing manual labor and enhancing livelihoods. Our study proposes an irrigation strategy based on real-time sensor data and cloud-based control using the ThingSpeak platform. This approach promotes efficient water management and boosts crop yields. Additionally, machine learning (ML) algorithms utilize sensor data to provide personalized crop recommendations, further optimizing agricultural practices.

DOI: 10.1201/9781003503934-7

Agriculture serves as a fundamental pillar for all developmental endeavors and as the main source of livelihood for 58% of India's populace. This statistical insight has spurred the Indian government into action, catalyzing efforts to foster a digital transformation within the agricultural sector, consequently enhancing agricultural productivity [6]. Steps taken by the Indian government in the direction of digital revolution in agriculture are as follows:

- Agri-Stack
- Digital Agriculture Mission
- Unified Farmer Service Platform
- National e-Governance Plan in Agriculture
- Sub-Mission on Agricultural Mechanization

Agriculture 6.0 transcends mere trendiness; it embodies the next evolutionary leap in agriculture. It encapsulates a vision of a smarter, more efficient industry that harnesses the power of big data and cutting-edge technologies to optimize the entire supply chain [7]. Together, these initiatives underscore the government's commitment to leveraging digital technologies to transform agriculture, improve farmers' livelihoods, and ensure food security and sustainability.

7.1.1 Evolution of Agriculture 6.0

In this section we trace the history of farming from its infancy up to the present day, with an emphasis on the shift toward more sustainable and technologically advanced methods in Agriculture 6.0. This evolutionary path is shown in Figure 7.1.

Low productivity was the result of the use of manual labor and simple tools in Agriculture 1.0, which ran from before the birth of Christ until before 1950. A strong bond with the land was fostered by early agriculturalists who overcame these limitations through the development of crop rotation techniques [8]. Despite the lackluster output, Agriculture 1.0 set the stage for what was to come by prioritizing long-term viability and the maintenance of traditional knowledge.

New farm machinery greatly enhanced farming methods and harvests throughout the second wave of agriculture (1950–1990). Chemicals, such as synthetic fertilizers and pesticides, enabled farmers to run a plethora of machinery. Although output increased, problems such as chemical contamination of fields, overuse of electricity, and depletion of natural resources were introduced by Agriculture 2.0. Despite these setbacks, Agriculture 2.0 drove economic growth, laid the groundwork for contemporary industrial-scale farming, and intensified agricultural practices [9].

Technology in agriculture in the 20th century led to the development of Agriculture 3.0. Precision farming methods, such as global positioning systems and biotechnology, have increased crop yields. Robotic technology and

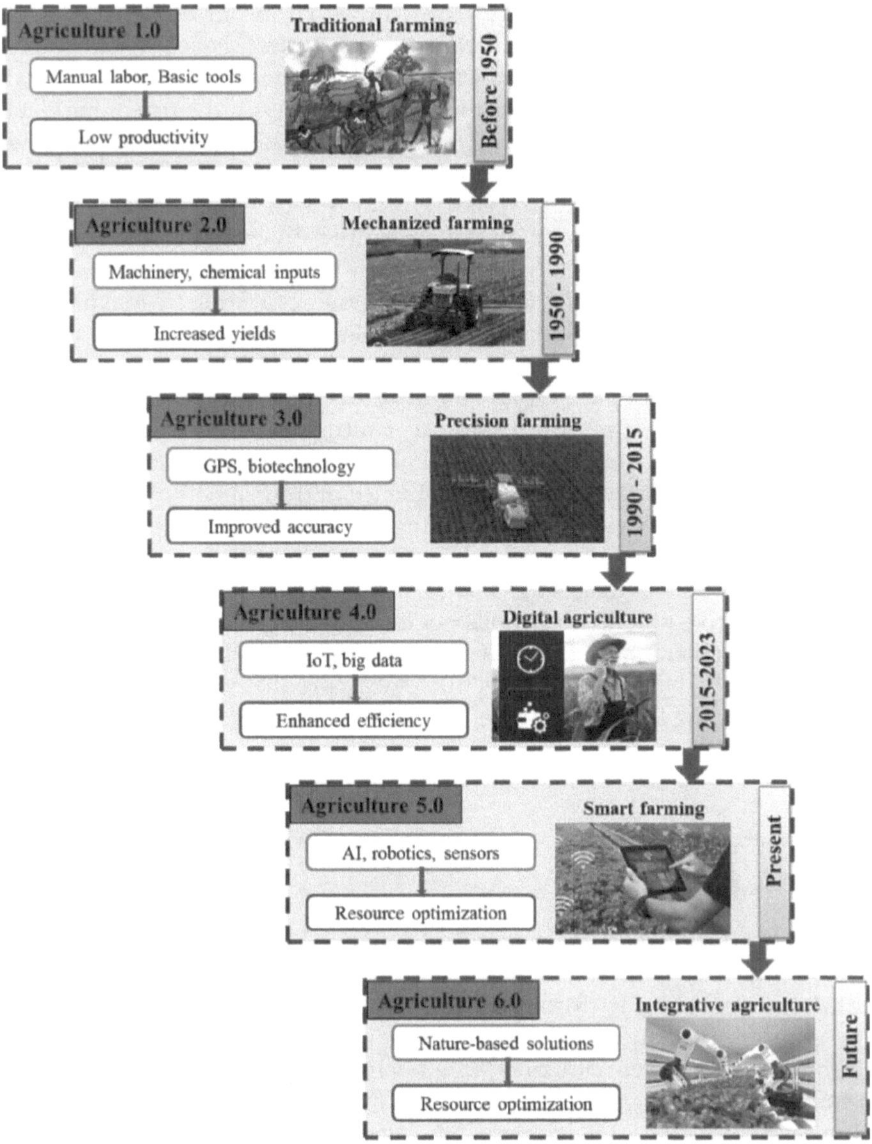

FIGURE 7.1
Evolution of Agriculture 6.0.

process automation enhanced productivity by mechanizing routine tasks and eliminating human error [10]. The environmental issues caused by Agriculture 2.0 were improved upon by Agriculture 3.0, which reduced the need for chemicals and increased the accuracy of irrigation. During this time, we achieved a huge step toward farming in an environmentally conscious way.

An agricultural revolution called "Agriculture 4.0" occurred between 2015 and 2023, propelled by the combination of big data and the IoT. The capacity to collect and analyze data in real time is one manner in which this digital revolution has enhanced agricultural productivity. With the use of big data analytics and information obtained from IoT devices that tracked soil and crop health, farmers could make more informed decisions, make better use of their resources, and decrease risks [11]. Taken as a whole, these developments ushered in a new era of digital agriculture by bringing environmentally friendly and economically beneficial farming methods online.

The purpose of Agriculture 5.0 is to revolutionize farming by maximizing efficiency through the integration of sensors, robotics, and artificial intelligence (AI). Simplified planting, harvesting, and data analysis are the three pillars of modern agriculture that maximize sustainability and yields [12].

A vision for the future of farming, Agriculture 6.0 prioritizes integrated agricultural practices, resource optimization, and solutions derived from nature. To improve ecosystem health, resilience, and sustainability, this vision integrates traditional wisdom with cutting-edge technology [7]. Optimizing resources, mitigating environmental impact, and ensuring food security in the face of future challenges are the goals of Agriculture 6.0, which prioritizes regenerative practices. From its beginning in Agriculture 1.0 to its current state, which emphasizes innovation and sustainability, the evolution of agriculture is truly remarkable. Technological, productive, and ecologically conscious farming have all benefited from each stage [13].

7.2 Related Work

In recent years, several studies have focused on enhancing agricultural practices through the integration of technology. In this section we have discussed the related work done in this area.

Crop recommendations using ML techniques are introduced in [14–16]. A recommendation system utilizing ML models to suggest suitable crops based on soil characteristics can optimize agricultural productivity. Various ML algorithms such as K-nearest neighbor (KNN), decision tree, random forest, Naive Bayes, and gradient boosting are employed for crop recommendation. Among these, gradient boosting demonstrates the highest accuracy of 98.18% [17].

Web application implementation with ML [18,19] proposed the challenge of crop selection based on soil requirements among Indian farmers; this chapter proposes a recommendation system using ensemble learning with a majority voting technique. Learners including random tree, Chi-squared Automatic Interaction Detector (CHAID), KNN, and Naive Bayes are utilized to recommend crops with high accuracy and efficiency.

IoT-based smart irrigation system proposed in a research [20–22] advocates for an automated irrigation system leveraging IoT technology to optimize water usage in agriculture. The system, controlled by an ATMEGA328P microcontroller on the Arduino Uno platform, monitors soil moisture content using sensors and adjusts irrigation to maintain optimal levels, thereby enhancing crop growth and resource efficiency [23].

A study on employing IoT for crop-field surveillance to develop smart irrigation systems is outlined in [24,25]. Focusing on the Indian agricultural context, this study explores the potential of IoT-based smart irrigation systems to address water scarcity issues. By employing microcontroller-based and wirelessly programmable systems, the research aims to optimize irrigation timing according to crop and soil conditions, thereby enhancing water usage efficiency [26].

Smart irrigation system using IoT proposed in [27] is a smart irrigation system utilizing drip irrigation technology and IoT sensors to monitor soil moisture levels and groundwater levels. By integrating temperature sensors, soil humidity sensors, and float switches, the system ensures efficient water management, with SMS alerts sent to farmers when groundwater levels drop, enabling timely interventions to prevent crop damage. The comparative analysis of related work is described in Table 7.1.

TABLE 7.1

Comparative Analysis of Related Work

Application	Key Features	Advantages
Crop recommendations using machine learning (ML) techniques	ML-based crop recommendation; gradient boosting algorithm, which achieves 98.18% accuracy	High accuracy in crop recommendation; utilizes various ML algorithms
Web application implementation with ML	Ensemble learning-based recommendation system, high accuracy, and efficiency in crop recommendation	Addresses crop selection challenges among Indian farmers
IoT-based smart irrigation system	Automated irrigation system using IoT technology; controlled by microcontroller for optimal watering	Efficient water usage, enhances crop growth and resource efficiency
A study on smart irrigation systems using IoT	Focuses on microcontroller-based smart irrigation systems; aims to optimize irrigation timing	Addresses water scarcity issues; wirelessly programmable for flexibility
Smart irrigation system using IoT	Integrates drip irrigation technology with IoT sensors and sends SMS alerts for groundwater levels	Efficient water management; timely interventions to prevent crop damage

7.3 Proposed Methodology

Farming professionals can handle problems with the use of an IoT-based irrigation system and ML. Numerous IoT applications address important problems including monitoring agricultural development, managing water conservation, and detecting soil moisture. This technology delivers better and more intelligent watering through the networked interaction of temperature, humidity, and other sensors with the user. The IoT has provided farmers and gardeners with very effective ways to nourish soil via the use of low-cost, simple-to-install sensors and a wealth of informative data.

The main aim of this endeavor is to develop an autonomous irrigation system capable of dynamically adjusting the pump motor's runtime in response to real-time soil moisture levels. Additionally, it involves implementing a crop recommendation system driven by ML algorithms. Data collected from sensors is utilized to regulate the pump via a microcontroller, and this information is subsequently transmitted to a cloud platform for accessibility through mobile devices. The collected data is then subjected to analysis and processing to generate crop recommendations. These recommendations are presented through a dedicated website where our ML model is deployed, allowing users to make up-to-date choices regarding crop variety and irrigation management.

For automated irrigation, we need to measure parameters like soil moisture, temperature, humidity, and soil nutrient content.

Soil moisture refers to the water held within the soil, which is crucial for plant growth. This water is typically available to plants within the root zone, commonly considered to be in the upper 200 centimeters of soil depth. The aforementioned parameters are measured as they are critical to numerous hydrological, biological, and biogeochemical processes and directly affect plant health. Soil moisture sensor, DHT11 sensor, and soil nutrient sensor are deployed to gather various parameters of the soil. At the heart of the proposed system lies the soil moisture measurement module. The irrigation of the field and user satisfaction hinge on accurately assessing the soil's water content.

Parameters like humidity, temperature, moisture, and nutrient content are measured using the concept of "electrical conductivity (EC)". It is a measurement of the capacity of a dissolved substance in an aqueous solution to transmit electric current through it. The unit of EC is Siemens per unit area. The greater the dissolved materials in the soil, the higher the EC.

Now, sensor data collected is sent to NodeMCU through serial communication. This data is then sent to the ThingSpeak Cloud platform where data is displayed. A 5 V submersible pump with an optocoupler is controlled using an SPST relay module connected to NodeMCU. The system detects soil moisture, temperature, and EC using sensors attached to the controller, and the data is then processed in the controller. The sensor value is compared with the threshold value, and, accordingly, a control signal is sent to manipulate the pump runtime.

The concept of Artificial Intelligence for Agricultural Innovation (AI4AI) is emerging as a promising paradigm gaining traction within the scientific community. Agriculture is witnessing a swift uptake of AI and ML across both product development and field practices. Advanced AI technologies like IoT and ML are empowering the agricultural sector to address challenges related to productivity, quality, traceability, and carbon emissions, ultimately leading to improved profitability.

In the further part of the project, we have developed an ML model for predicting the crops that should be grown on farmland. To do so we have used the soil dataset, which was available on Kaggle. In the next step, we preprocessed the dataset and trained our model on various ML algorithms such as Naive Bayes, logistic regression, SVM, decision tree, KNN, random forest, and ensemble learning. After that, we evaluated our model on different ML algorithms using a cross-validation technique and found out that Naive Bayes is the best model for the dataset.

Finally, we have deployed our trained ML model on a web application, i.e., the Crop Recommendation Platform. A Crop Recommendation Platform is a simple user-friendly application that can be utilized by farmers to predict the crops to be cultivated based on soil parameters. So the platform will help the farmers to make informed decisions before cultivation to improve crop quality and profitability.

7.3.1 System Hardware

The system employs a sophisticated approach to determine the optimal timing, duration, and volume of water supplied to crop plants, leveraging a diverse array of sensors to monitor local environmental conditions. The Architecture of Smart Irrigation System is depicted in Figure 7.2. Central to this operation is the Tensilica Xtensa 32-bit LX106 RISC microprocessor housed within the NodeMCU, which diligently processes the data gathered by the sensors [28]. The resulting data is seamlessly transmitted and visualized on the Thing Speak cloud platform, providing users with real-time insights into the environmental dynamics of the agricultural setting.

One of the system's remarkable features is its self-sufficiency; the system is capable of performing sustained operation for extended periods with minimal power consumption. Powered by a mere 5 V and drawing a modest 200 mA for normal operation, including the sensors and microcontroller, the device exhibits remarkable efficiency. Furthermore, the NodeMCU and motor boast exceedingly low running and maintenance costs once deployed, contributing to its overall cost-effectiveness and viability for widespread adoption in agricultural settings.

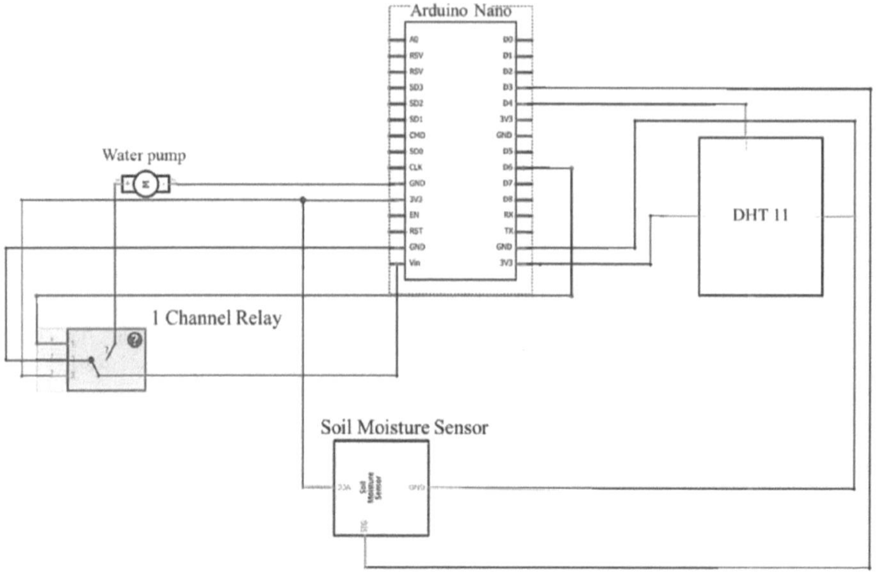

FIGURE 7.2
Architecture of smart irrigation system.

7.3.1.1 NodeMCU

NodeMCU, an open-source IoT platform, integrates hardware based on the ESP-12 module and firmware running on the ESP8266 Wi-Fi system-on-chip (SoC) by Espressif Systems. This SoC incorporates a 32-bit microprocessor along with various digital interfaces, radiofrequency components, and power management modules. It processes data from sensors to determine plant water needs, connected for data transfer but not required for operation. A 5 V DC relay facilitates motor control through NodeMCU.

7.3.1.2 Soil Moisture Sensor

Soil moisture sensors employ capacitance to measure the dielectric permittivity of the surrounding medium, correlating with soil water content [29]. These sensors provide continuous moisture level measurements when inserted into the soil.

7.3.1.3 DHT11 Sensor

The DHT11 sensor, measuring temperature and humidity, offers factory-calibrated outputs for easy integration with microcontrollers [30]. It operates within a temperature range of 0–50°C and humidity range of 20–90%, with a 1°C and 1% accuracy, respectively.

7.3.1.4 Soil Nutrient Sensor

The JXBS-3001 soil NPK sensor detects nitrogen, phosphorus, and potassium levels using optical-electrical technology [31]. It features corrosion-resistant probes and interfaces with microcontrollers via Modbus modules for real-time NPK concentration evaluation.

7.3.1.5 Pump and Relay Module

A small DC water pump, powered by a 2.5–6 V supply, facilitates water supply to crops. It draws 220 mA of current and requires a relay for high-voltage control. Relays act as electrical switches, controllable by microcontrollers, automating the activation of high-voltage devices.

7.3.1.6 ThingSpeak Platform

ThingSpeak is a platform that provides a variety of services tailored exclusively for the development of IoT applications. It enables real-time data collection, data visualization in charts, and the creation of plugins and apps for interfacing with web services, social networks, and other APIs.

The "ThingSpeak Channel" is central to ThingSpeak, where data is stored. It includes:

- Eight fields for diverse data types from devices or sensors.
- Three location fields for latitude, longitude, and elevation, which are useful for tracking.
- One status field for a brief description of stored data

7.4 Proposed Smart Agriculture System

Figure 7.3 illustrates the various steps involved in developing the smart irrigation system. Initially, we selected the necessary sensors essential for the implementation of the ML model. These sensors include the soil moisture sensor for detecting soil moisture levels, the soil nutrient sensor, the DHT11 sensor for monitoring temperature and humidity, and the rainfall sensor. After establishing the sensor connections with the Arduino board, the next step involved programming the board to efficiently manage the sensors, ensuring that the collected data is organized and transmitted in real time.

Subsequently, we focused on storing the collected data using the ThingSpeak Cloud platform. By aggregating months of data, we were able to train the system to predict suitable crops based on the gathered data. Finally,

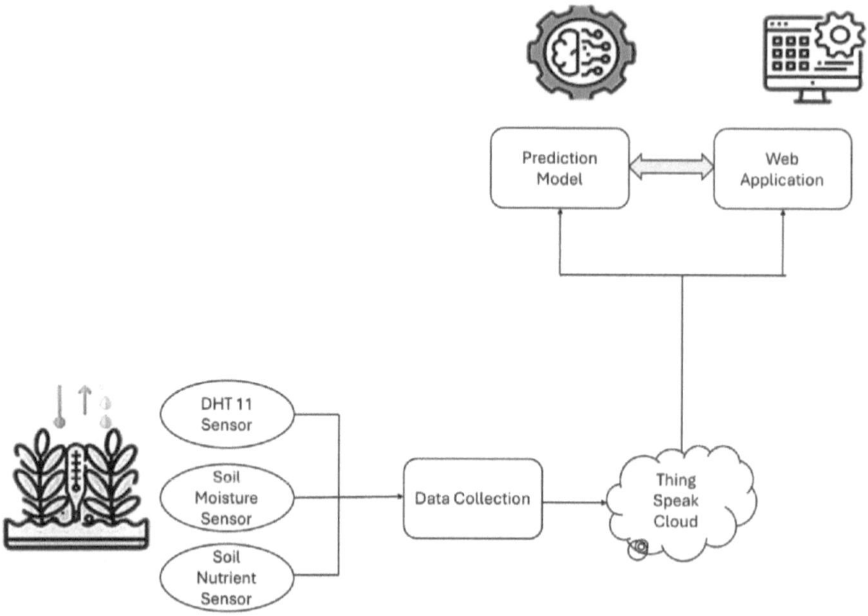

FIGURE 7.3
Block diagram smart agriculture with ML.

we developed a web application to display the crop predictions, facilitating easy access and interpretation of the results.

7.4.1 Data Collection and Data Storage

For data collection and storage, we streamlined the process by leveraging ThingSpeak, a platform designed for supervisory and storage functionalities. Our objective remained consistent: to retrieve and visualize sensor data efficiently. To accomplish this, we connected our device to the ThingSpeak environment, establishing a system that facilitates the flow of sensor data and enables the pre-processing of each data stream.

The data acquisition and storage flow as shown in Figure 7.4 encompassed various types of data collected through sensors, including temperature, soil humidity, soil nutrients, and soil moisture. By processing this data through multiple algorithms and collaborating with agricultural experts, we ensured comprehensive data collection and analysis, laying the groundwork for informed decision-making in agricultural practices.

7.4.2 Dataset Collection

To ensure healthy plant growth and satisfactory yield, specific environmental conditions such as temperature, humidity, soil pH, sunlight exposure, and soil moisture must be carefully monitored and maintained. However, these

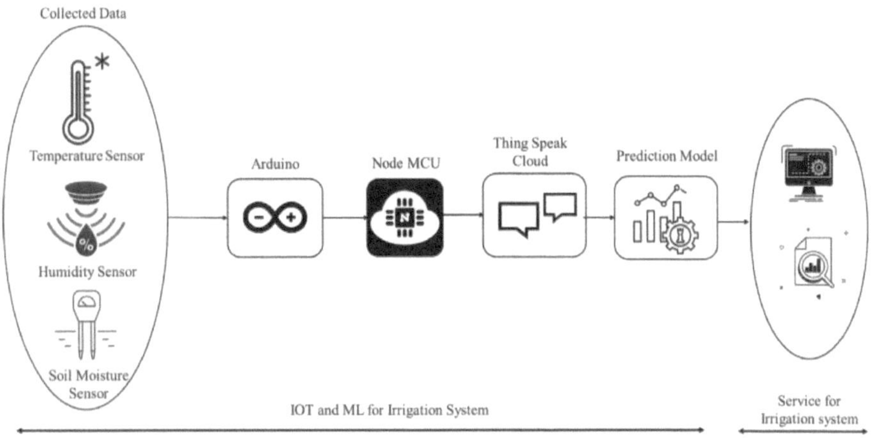

FIGURE 7.4
Data acquisition and storage.

requirements may vary depending on the types of plants being cultivated [32]. To establish a comprehensive dataset encompassing these diverse factors, reliable sources such as Kaggle.com, publications, and research articles were meticulously mined for data collection.

This initial dataset served as the foundation for training the crop recommendation algorithm, aiming to enhance accuracy by incorporating diverse environmental variables relevant to plant growth. By leveraging reputable sources and comprehensive datasets, we aimed to develop a robust recommendation system capable of providing tailored guidance for optimal crop selection based on prevailing environmental conditions.

7.4.3 Collecting Environmental Factors

To facilitate the comparison and forecasting of the initial dataset, it's imperative to acquire environmental parameters. Arduino microcontrollers are employed for this purpose, gathering data on three key elements. A single microcontroller handles temperature and humidity sensing, while separate sensors are dedicated to soil nutrients and moisture, as well as humidity and temperature. These sensors interface with an Arduino Wi-Fi module, facilitating the transmission of gathered information to a ThingSpeak database. Following data collection, cleaning, and processing through techniques such as clustering, the values are transmitted to the subsequent crop recommendation component and stored in the database.

7.4.4 Data Pre-Processing

A data mining approach called data pre-processing in ML converts raw data into a format that is legible and intelligible. It is the initial step that launches the ML process. It aids in preparing raw data for ML models through cleaning, organizing, and formatting. Steps included in data pre-processing are as follows:

- **Acquiring the Dataset**—We used the open-source crop dataset from Kaggle.com to construct and develop the crop recommendation system. This dataset includes a number of environmental elements that have an impact on crop growth, including soil moisture content, soil nutrient content, temperature, humidity, and data about various crops and their types.
- **Import Datasets and Libraries**—In the first step, we import all the essential libraries, including NumPy, Pandas, Seaborn, and Matplotlib, which are crucial for specific data pre-processing tasks. Then, we import the crop dataset into the code.
- **Identifying and Handling the Missing Values**—It is essential to identify and properly handle the missing values; if we fail to do this, we might draw inaccurate and faulty conclusions and interferences from crop data.
- **Encoding the Categorical Data**—Encoding is a technique of converting categorical variables into numerical values so that they can be easily fitted to a ML model. Categorical data comprises information with distinct categories within the dataset. The dataset mentioned contains 22 categorical variables representing various types of crops. So we have used the label encoding method to convert the categorical variable present in the crop dataset into an integer value.
- **Splitting the Dataset**—In this step, we have split the crop dataset into the training set and testing set. Training set: the subset of a dataset that is used for training the ML model. Testing set: the subset of a dataset that is used for training the ML model.
- **Feature Scaling**—Feature scaling is a vital part of data pre-processing in ML. It standardizes independent variables within a specific range, optimizing algorithm performance for easier prediction. We have used two feature scaling methods on our dataset, i.e., standardization and normalization.

7.4.5 Model Training and Prediction

In the model training and prediction phase, we employed a range of ML algorithms, including Naive Bayes, KNN, support vector machine, ensemble learning, decision tree, random forest, and logistic regression. To assess the effectiveness of these models, we utilized cross-validation techniques to evaluate their performance. Following a rigorous evaluation, the Naive Bayes model emerged as the optimal choice for our dataset. We further refined this model by training it using the selected features and assessing its performance on the test dataset. This comprehensive approach ensured that our model was robust and accurately predictive, laying the groundwork for effective crop recommendations based on environmental factors.

7.5 Result Analysis

The results obtained by the selected algorithms are discussed in this section. ML algorithms suggest that an ideal crop should be grown on a particular soil. The parameter used in algorithms is crop type, nutrient content of the soil, moisture level, temperature, and humidity. The accuracy of crop yield estimates is compared to all the selected algorithms. Table 7.2 summarizes the outcomes of the various tests conducted to train the predictions. After exploring the pertinent data and conducting pre-sorting, the Naive Bayes model achieved an accuracy rate of 99.54% in the training set with a root mean square error (RMSE) of 0.13. The table presents the accuracy and RMSE results of the models utilized. Figures 7.5–7.8 shows the graphical representation of the results.

7.5.1 Evaluation Model Results

In this section, we delve into the outcomes yielded by our chosen algorithms, primarily centered on predicting optimal crops based on soil parameters. Our algorithms are adept at utilizing crucial parameters such as crop type, soil nutrient content, moisture level, temperature, and humidity. Through meticulous analysis, we compare the accuracy of crop yield estimates across all selected algorithms. Notably, the Naive Bayes model emerges as a standout

TABLE 7.2

Performance Comparison of Various Machine Learning Models

Model	Accuracy (%)	Precision	Recall	F1 Score
Logistic regression	94.54	0.93	0.92	0.93
Naive Bayes	98.54	0.97	0.97	0.97
Support vector machine	96.13	0.95	0.96	0.95
K-nearest neighbors	97.04	0.96	0.97	0.96
Decision tree	98.18	0.97	0.98	0.97
Random forest	99.31	0.99	0.99	0.99
Bagging	98.63	0.98	0.98	0.98
Gradient boosting	98.18	0.97	0.98	0.97
Proposed method	99.42	0.98	0.99	0.98

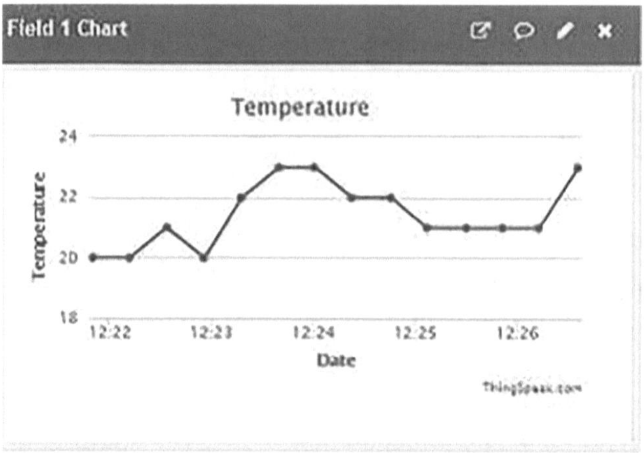

FIGURE 7.5
Graph representing the temperature of the region taken using a DHT11 sensor.

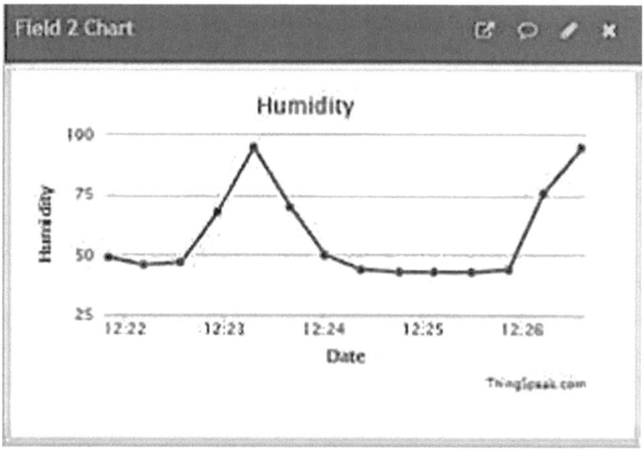

FIGURE 7.6
Graph representing humidity of the region taken using a DHT11 sensor.

performer, achieving an impressive accuracy rate of 99.54% in the training set, with a remarkably low RMSE of 0.13. Table 7.2 meticulously lays out the results obtained from the various models, providing insights into their accuracy and RMSE values. Moreover, Figure 7.5 offers a graphical depiction of these results, offering a visual aid for comprehending the performance of each algorithm. By meticulously exploring and dissecting the relevant data, these findings deepen our understanding of the predictive capabilities of ML models in the domain of crop recommendation.

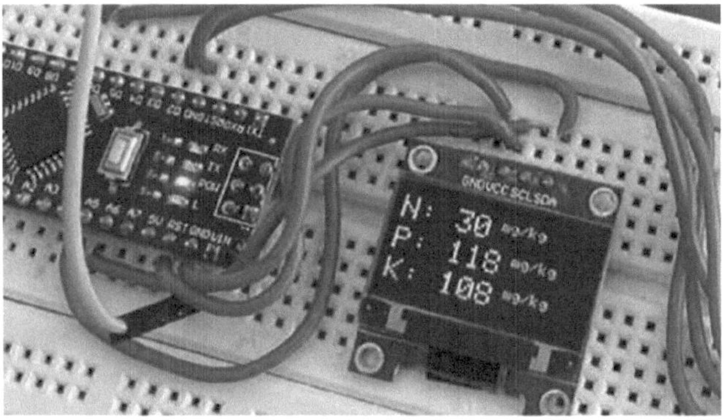

FIGURE 7.7
NPK values using soil.

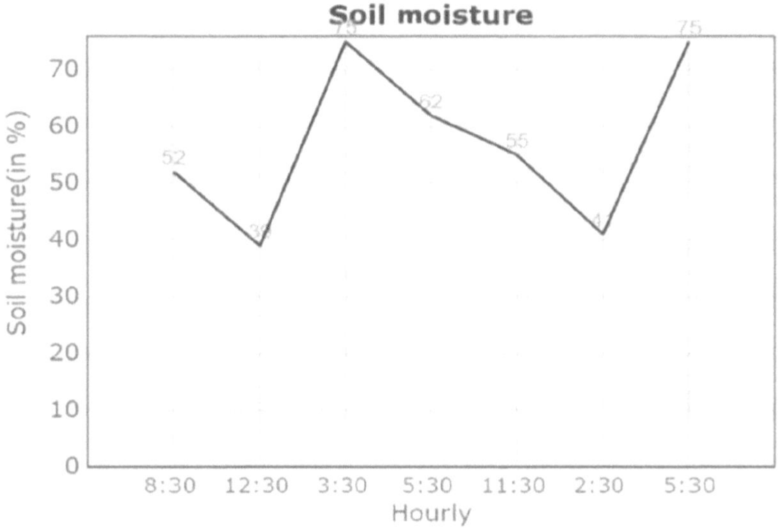

FIGURE 7.8
Graph representing soil moisture using soil sensor.

Table 7.2 summarizes the performance metrics of various ML models, including Naive Bayes, KNN, support vector machine, ensemble learning, decision tree, random forest, logistic regression, gradient boosting, and a proposed method. Each model's accuracy, precision, recall, and F1 score are presented, showcasing their effectiveness in classification tasks. Notably, the proposed method demonstrates promising results, outperforming several other models in terms of accuracy and recall, making it a strong contender for deployment in practical applications. The graphs shown in Figure 7.9

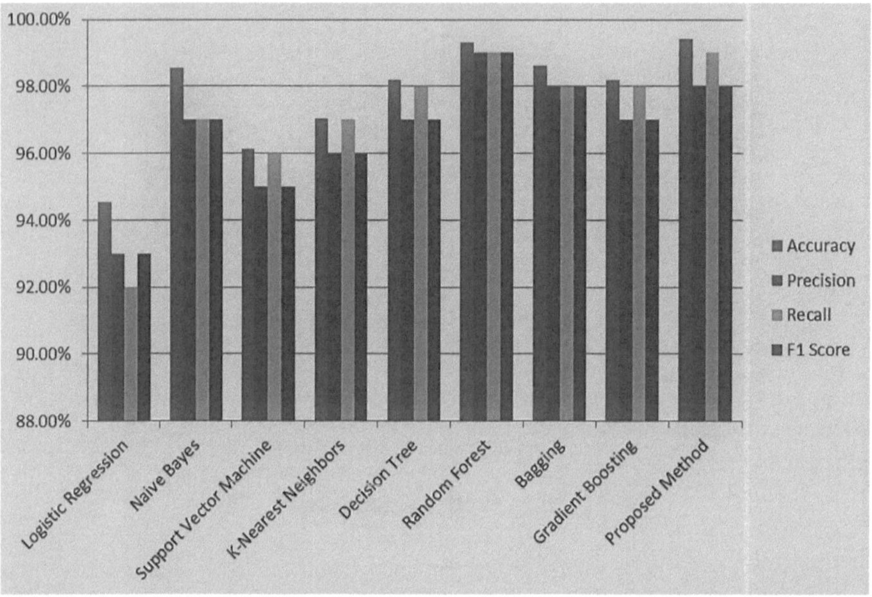

FIGURE 7.9
Comparison of machine learning models: accuracy, precision, recall, and F1 score.

provide a comprehensive overview of the comparative performance of different ML algorithms, aiding decision-making processes in model selection for specific tasks.

7.5.2 Deployment of Model

To enable farmers to access the trained model, we developed a user-friendly application. Here, we introduce a web-based platform named the "Crop Recommendation Platform". Its primary objective is to suggest optimal crops for cultivation based on various parameters, aiding farmers in making informed decisions before planting. Figure 7.10 depicts the model deployment block diagram. The application comprises two main components: the frontend and the backend.

The focus of this work is on making the platform visually appealing to provide end users with an easy-to-use experience. If you want your user interface to work flawlessly and adapt to different devices without sacrificing quality, you need to invest in frontend development.

We crafted a simple yet effective web interface using React JS, HTML, and CSS, which are part of our technology stack [33]. As a JavaScript library, React JS shines when it comes to UI development, making it easy to create dynamic and engaging web pages. By utilizing class-based components, we simplify the development of complex user interfaces.

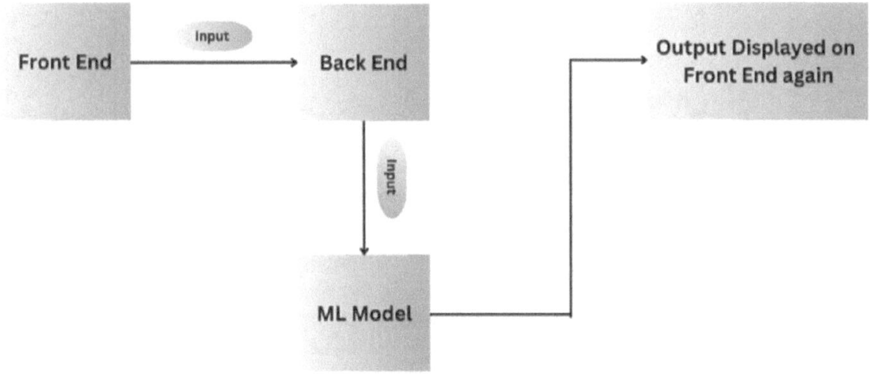

FIGURE 7.10
Model deployment block diagram.

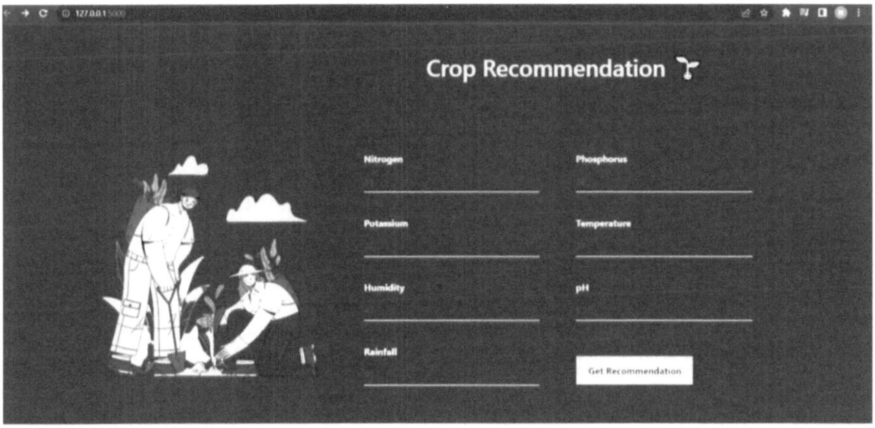

FIGURE 7.11
Crop recommendation interface design.

Supporting dynamic web applications and providing users with comprehensive information requires a robust infrastructure for the backend. To make it easier to predict outcomes based on user input, we use the Flask framework to integrate the backend with the frontend. Furthermore, to guarantee efficient data processing and precise results, we employ a pickle file to produce model predictions.

To aid farmers with crop selection, we have developed a web application that takes into account a number of input parameters such as temperature, humidity, precipitation, pH, potassium, nitrogen, and potassium levels. By providing personalized crop recommendations, this app greatly assists farmers in making the most informed decisions regarding their cultivation practices. A screenshot of our web app's Graphical User Interface (GUI) is shown in the accompanying Figures 7.11 and 7.12, which gives an idea of its intuitive layout and how easy it is to use.

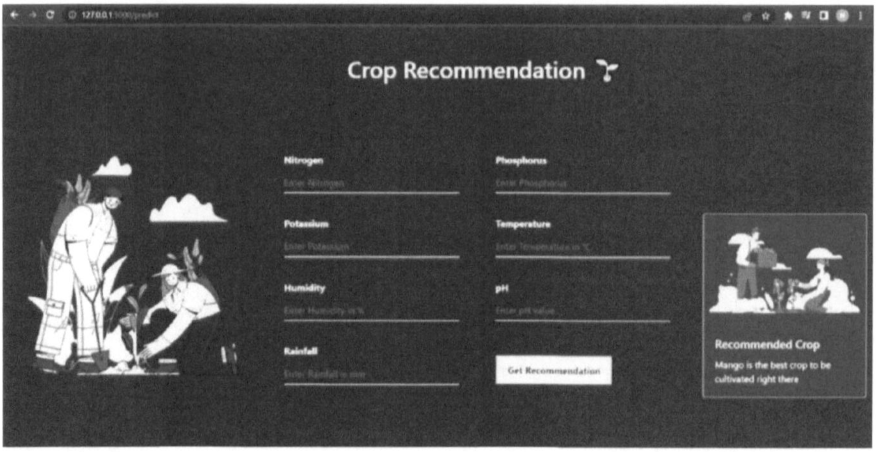

FIGURE 7.12
Crop recommendation output.

7.6 Future Scope

In certain circumstances, remote places may not have Internet connectivity. Consequently, the user may remotely monitor the agricultural region via an SMS containing field information. The SMS offers both a warning and a recommendation to the affected system. Additionally, the submersible pump can be controlled via SMS or Google Assistant. Pests are the biggest enemies for crops, causing huge damage and leading to a poor yield. Hence, pest control is important. This can be implemented by placing low-image sensors, acoustic sensors, or thermography.

Future enhancements to the Crop Recommendation Platform include integrating a payment system for farmers to purchase recommended crops and providing fertilizer recommendations tailored to crop and location. This requires studying fertilizers' relationships with soil and climate needs, along with analyzing available statistical data. Additionally, by leveraging Precision Agriculture, ML models can predict pump start/stop times, optimizing crop yields with limited water resources to meet future food production demands.

7.7 Conclusion

Intelligent irrigation is crucial for meeting the increasing global food demand predicted to rise by over 70% by 2050. Our research focuses on developing an irrigation prediction method utilizing a comprehensive sensor database and the ThingSpeak platform. By leveraging ML models, our system achieved a recognition rate of 99.54% using Naive Bayes. We also provide a user-friendly web application for easy visualization and supervision of the environment. Future plans include expanding the database and exploring semi-supervised learning algorithms for enhanced decision-making accuracy.

References

1. Priyadarshini, P. and Abhilash, P.C., 2020. Policy recommendations for enabling transition towards sustainable agriculture in India. *Land Use Policy, 96,* p.104718.
2. Mahendra Dev, S., 2017. Poverty and employment: Roles of agriculture and non-agriculture. *The Indian Journal of Labour Economics, 60,* pp.57–80.
3. Dev, S.M., 2019. Transformation of Indian agriculture? Growth, inclusiveness and sustainability. *Indian Journal of Agricultural Economics, 74*(1), pp.9–61.
4. Farooq, M.S., Riaz, S., Abid, A., Umer, T. and Zikria, Y.B., 2020. Role of IoT technology in agriculture: A systematic literature review. *Electronics, 9*(2), p.319.
5. Jaiganesh, S., Gunaseelan, K. and Ellappan, V., 2017, March. IOT agriculture to improve food and farming technology. In *2017 Conference on Emerging Devices and Smart Systems (ICEDSS)* (pp. 260–266). IEEE, Mahendra Engineering College, Mahendhirapuri, Mallasamudram, Namakkal DT, Tamilnadu, India..
6. Mehrotra, N., 2023. Open DataStack to Power Government of Uttar Pradesh's AgriStack: An Approach Note on Agri DataStack from International Finance Corporation (IFC) / The World Bank Group to the Government of Uttar Pradesh. Montpellier, France: CGIAR System Organization.
7. Neves, M.F., Casagrande, B.P., Cambaúva, V., de Oliveira Teixeira, G. and Toledo, P.J.F., 2023. Agriculture 6.0: A new proposal for the future of agribusiness. *Revista de Gestão Social e Ambiental, 17*(9), pp.e04004–e04004.
8. Happe, K., Balmann, A. and Kellermann, K., 2004. The Agricultural Policy Simulator (Agripolis): An Agent-Based Model to Study Structural Change in Agriculture (Version 1.0) Discusson Paper No. 71. Institute of Agricultural Development in Central and Eastern Europe (IAMO), Halle (Saale).
9. Lazarević, B. and Popović, T., 2020. Meteorological data aggregation system for application in Agriculture 2.0. In *7th International Conference on Electrical, Electronic and Computing Engineering IcETRAN 2020,* International Journal on Cybernetics & Informatics (IJCI), Belgrade.
10. Rahmann, G., Reza Ardakani, M., Bàrberi, P., Boehm, H., Canali, S., Chander, M., David, W., Dengel, L., Erisman, J.W., Galvis-Martinez, A.C. and Hamm, U., 2017. Organic Agriculture 3.0 is innovation with research. *Organic Agriculture, 7,* pp.169–197.
11. Araújo, S.O., Peres, R.S., Barata, J., Lidon, F. and Ramalho, J.C., 2021. Characterising the Agriculture 4.0 landscape-emerging trends, challenges and opportunities. *Agronomy, 11*(4), p.667.
12. Saiz-Rubio, V. and Rovira-Más, F., 2020. From smart farming towards Agriculture 5.0: A review on crop data management. *Agronomy, 10*(2), p.207.

13. Anisha, P.R., Reddy, C.K.K., Hanafiah, M.M., Murthy, B.R., Mohana, R.M. and Pragathi, Y.V.S.S., 2023. An intelligent deep feature based feature based metabolism syndrome prediction system for sleep disorder diseases. *Multimedia Tools and Applications*, 83, 51267–51290.
14. Priyadharshini, A., Chakraborty, S., Kumar, A. and Pooniwala, O.R., 2021, April. Intelligent crop recommendation system using machine learning. In *2021 5th International Conference on Computing Methodologies and Communication (ICCMC)* (pp. 843–848). IEEE, India.
15. Doshi, Z., Nadkarni, S., Agrawal, R. and Shah, N., 2018, August. AgroConsultant: Intelligent crop recommendation system using machine learning algorithms. In *2018 Fourth International Conference on Computing Communication Control and Automation (ICCUBEA)* (pp. 1–6). IEEE, India.
16. Modi, D., Sutagundar, A.V., Yalavigi, V. and Aravatagimath, A., 2021, October. Crop recommendation using machine learning algorithm. In *2021 5th International Conference on Information Systems and Computer Networks (ISCON)* (pp. 1–5). IEEE, India.
17. Singla, R., Kaur, N., Koundal, D., Lashari, S.A., Bhatia, S. and Rahmani, M.K.I., 2021. Optimized energy efficient secure routing protocol for wireless body area network. *IEEE Access, 9*, pp. 116745–116759. doi: 10.1109/ACCESS.2021.3105600.
18. Bhanu, K.N., Jasmine, H.J. and Mahadevaswamy, H.S., 2020, June. Machine learning implementation in IoT based intelligent system for agriculture. In *2020 International Conference for Emerging Technology (INCET)* (pp. 1–5). IEEE, India.
19. Cubillas, J.J., Ramos, M.I., Jurado, J.M. and Feito, F.R., 2022. A machine learning model for early prediction of crop yield, nested in a web application in the cloud: A case study in an olive grove in southern Spain. *Agriculture, 12*(9), p.1345.
20. García, L., Parra, L., Jimenez, J.M., Lloret, J. and Lorenz, P., 2020. IoT-based smart irrigation systems: An overview on the recent trends on sensors and IoT systems for irrigation in precision agriculture. *Sensors, 20*(4), p.1042.
21. Ragab, M.A., Badreldeen, M.M.M., Sedhom, A. and Mamdouh, W.M., 2022. IOT based smart irrigation system. *International Journal of Industry and Sustainable Development, 3*(1), pp.76–86.
22. Saraf, S.B. and Gawali, D.H., 2017, May. IoT based smart irrigation monitoring and controlling system. In *2017 2nd IEEE International Conference on Recent Trends in Electronics, Information & Communication Technology (RTEICT)* (pp. 815–819). IEEE, India.
23. Reddy, C.K.K., Anisha, P.R., Hanafiah, M.M., Pragathi, Y.V.S.S., Murthy, B.R. and Mohana, R.M., 2023. An intelligent optimized cyclone intensity prediction framework using satellite images. *Earth Science Informatics, 16*(2), pp.1537–1549.
24. Rao, R.N. and Sridhar, B., 2018, January. IoT based smart crop-field monitoring and automation irrigation system. In *2018 2nd International Conference on Inventive Systems and Control (ICISC)* (pp. 478–483). IEEE, India.
25. Rajeshkumar, N., Vaishnavi, B., Saraniya, K. and Surabhi, C., 2019. Smart crop field monitoring and automation irrigation system using IoT. *International Research Journal of Engineering and Technology (IRJET)*, 6(3), pp.7976–7979.
26. Kumar, A., Krishnamurthi, R., Bhatia, S., Kaushik, K., Ahuja, N.J., Nayyar, A. and Masud, M., 2021. Blended learning tools and practices: A comprehensive analysis. *IEEE Access, 9*, pp. 85151–85197. doi: 10.1109/ACCESS.2021.3085844.

27. Jain, R.K., 2023. Experimental performance of smart IoT-enabled drip irrigation system using and controlled through web-based applications. *Smart Agricultural Technology, 4,* p.100215.
28. Macheso, P., Manda, T.D., Chisale, S., Dzupire, N., Mlatho, J. and Mukanyiligira, D., 2021, March. Design of ESP8266 smart home using MQTT and node-RED. In *2021 International Conference on Artificial Intelligence and Smart Systems (ICAIS)* (pp. 502–505). IEEE, Dubai, UAE.
29. Yu, L., Gao, W., Shamshiri, R.R., Tao, S., Ren, Y., Zhang, Y. and Su, G., 2021. Review of research progress on soil moisture sensor technology. *International Journal of Agricultural & Biological Engineering, 14*(4), pp.32–42.
30. Gay, W., 2018. DHT11 sensor. In Warren Gay (ed.) *Advanced Raspberry Pi: Raspbian Linux and GPIO Integration*, Apress, pp.399–418.
31. Burton, L., Jayachandran, K. and Bhansali, S., 2020. The "Real-Time" revolution for in situ soil nutrient sensing. *Journal of The Electrochemical Society, 167*(3), p.037569.
32. Flinn, K.M. and Marks, P.L., 2007. Agricultural legacies in forest environments: Tree communities, soil properties, and light availability. *Ecological Applications, 17*(2), pp.452–463.
33. Guinard, D.D. and Trifa, V.M., 2016. *Building the Web of Things: With Examples in Node.js and Raspberry Pi.* Simon and Schuster Manning Publications; New York, NY, USA.

8

Review of Study of Biochemical Water Parameters for Prediction and Analysis of Quality of Water Using Machine Learning Techniques in Industry 6.0

Gauri Ruchir Chavan and Ashish Adholiya

8.1 Introduction

Life on Earth without safe water is impossible; over 96% of water covering Earth's surface is saline water. Water is available in the form of air as water vapour, in rivers and lakes, in icecaps and glaciers, and in the ground as soil moisture [28]. However, only the freshwater resources, such as rivers, lakes, rainwater, and groundwater, fulfil the water requirement of living organisms [15]. The study of water is deep and essential where bridging the gap between unpredictable growing water demand and adequate water supply, which is safe from water-borne diseases, is the biggest challenge in today's scenario of global warming in India [14]. Water engineering uses a set of various chemical, physical, and biological water parameters. The management of water quality substantially impacts both the control of water deterioration and water resource planning. The availability and quality of water have declined because of some important factors like increasing population, rapid industrial expansion, and urbanization [17].

The chapter describes promising technologies to assist Industry 6.0 and reviews machine learning (ML) techniques with respect to models designed for water quality predictions and analysis in Industry 5.0. It organizes a systematic literature review into three categories: The first one is the estimation of water quality indicators using ML techniques. Second is water quality prediction using ML techniques, and third is water quality prediction using morphological analysis and ML techniques. Finally, it mentions some research opportunities and gaps that need to be developed in the near future to realize Industry 6.0 (Table 8.1).

 DOI: 10.1201/9781003503934-8

TABLE 8.1
Industrial Evolution X.0 [56–58]

Industry Evolution	Period	Manufacturing Processes	Achievements	Challenges	Tools and Techniques Used
Industry 1.0	1800s	Automation using hydraulic and steam power	Credentials and advancement in the agricultural field and transportation	Pollution control and the amount of time required to develop the associated methodologies	Mathematical tools and techniques of linear optimization and geometry
Industry 2.0	1900s	Mass production using electricity	Emergence of the electricity and internal combustion engines	High cost involved in the consumption of electrical power	Mathematical tools and techniques of differential equations and geometry
Industry 3.0	2000s	Cybernation with electronics	Renewable energy, telecommunication industries automation with robotics	Complex, infeasible, inflexible automated machines	Mathematical tools and techniques of differential equations, logical controllers, and linear optimization
Industry 4.0	2100s	Industries with Internet data analytics connectivity	Automated systems and artificial intelligent systems	Unprotected data storage in the cloud	Mathematical tools with optimization techniques and network theory
Industry 5.0	2100s	Use of collaborative secure robots	Human-centricity, environmental protection, and social benefit	Cost-inefficient adoption of advanced technology	Advanced IT technologies, IoT, robots, blockchain technology, AI, and augmented reality
Industry 6.0	2100s	Autonomous robotic automation and satellite communication	Antifragility, environmental sustainability	Internationalization, capital cost optimization	Quantum computing, nanotechnology, and cognitive robotics

8.1.1 Technological Assistance Required in Industry 6.0

Quantum Computing: The more parameters used, the better accurate predictive model can be developed, but more computing power is required. Quantum computing has the potential to develop a resource-saving, cost-effective, and scalable integrated model. It includes primary features superposition, entanglement, and decoherence. Prediction models in ML introduced a quantum neural network that is capable of performance optimization.

Nanotechnology: The nanoscopic universe offers huge opportunities for Industry 6.0. Nanocomputing is increasing the necessary computing power for developers. Industry 6.0 is using nanobiosensors to identify the existence of pathogens. Wastewater purification treatment with nanobubbles or nanofiltration systems is a cost-effective, efficient, and energy-saving solution.

Cognitive Robotics: This technique aims to develop solutions based on autonomous intelligence and adaptive knowledge acquisition. The stages involved in this technique are perceive, plan, anticipate, and learn and share. Neurorobotics is the subarea of cognitive robotics which focuses on the implementation of neuroscience, integrating neuromorphic systems to control the mobile machine's functioning combined with cognitive system [59,60].

This chapter aims to find the answers to the below-mentioned research questions:

1. Which are the complicated water parameters studied using ML with utmost accuracy?

   Reason: To understand the association among a set of water parameters.
2. What are robust ML models implemented for water quality prediction?

   Reason: To find out significant contributions made to the field of interest in the recent study with the help of a set of supervised and unsupervised ML techniques.
3. What are the recent areas for the prediction of water quality prediction that are integrated with ML models?

   Reason: Identifying the recent areas can help to provide accurate and optimized solutions to the challenges current systems are still trying to resolve.
4. What are metrics and methodologies used for evaluating the performance of ML-based water quality predictors?

   Reason: To identify the transparency in the decision-making procedures and assist the compliance of the outputs obtained.

Water Quality Index (WQI) provides a convenient means that adequately compile complex water quality indicators [2,15]. Many methodologies and techniques have been adopted to evaluate the quality of freshwater, surface water, groundwater, and wastewater, such as multivariate statistical analysis, fuzzy rule-based system, and supervised and unsupervised techniques such as ML techniques. Fuzzy logic systems are implemented to estimate the quality of water to ensure accurate results with the inclusion of biological parameters while designing the model [7]. ML is the technology of predictive analytics of the data applied to almost every complex field in the world to accomplish the benefits of its efficient and accurate results [12].

8.2 Literature Review

Several researchers have experienced many issues in predicting the quality of water and in devising some new methodologies to analyse and forecast the future trends in water quality, which are reviewed and presented below. Many types of pathogenic bacteria could be identified in drinking water, like *Escherichia coli, Vibrio cholerae*, and *Salmonella typhi*, which could cause water-borne diseases such as cholera, diarrhoea, and typhoid, respectively [53]. Morphological analysis using convolutional neural networks (CNNs) has been the prevailing selection for the identification of such microbial growth over the past decades.

8.2.1 Overview and Search Strategy

As mentioned in Figure 8.1, a literature survey was conducted that focused on the study of water indicators, where only articles based on ML and image processing techniques are considered. Articles that addressed applications of irrigation water, Internet of things (IoT) techniques, and geographical information system (GIS) are excluded. Finally, the observations of the studies based on the review of recent approaches are presented. An intense and deep search using PubMed, Springer, Google Scholar, and ScienceDirect

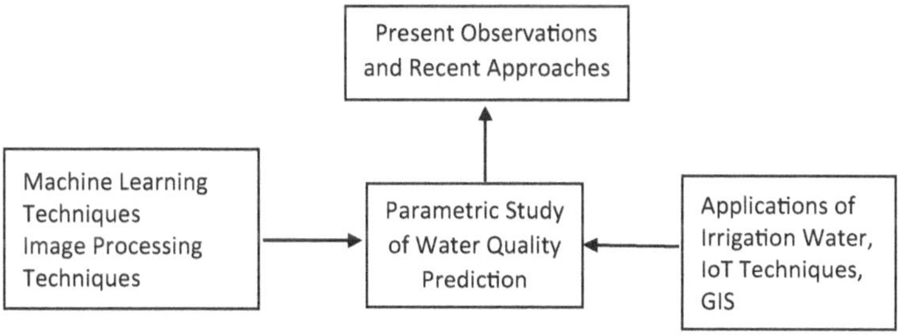

FIGURE 8.1
Overview of the literature study.

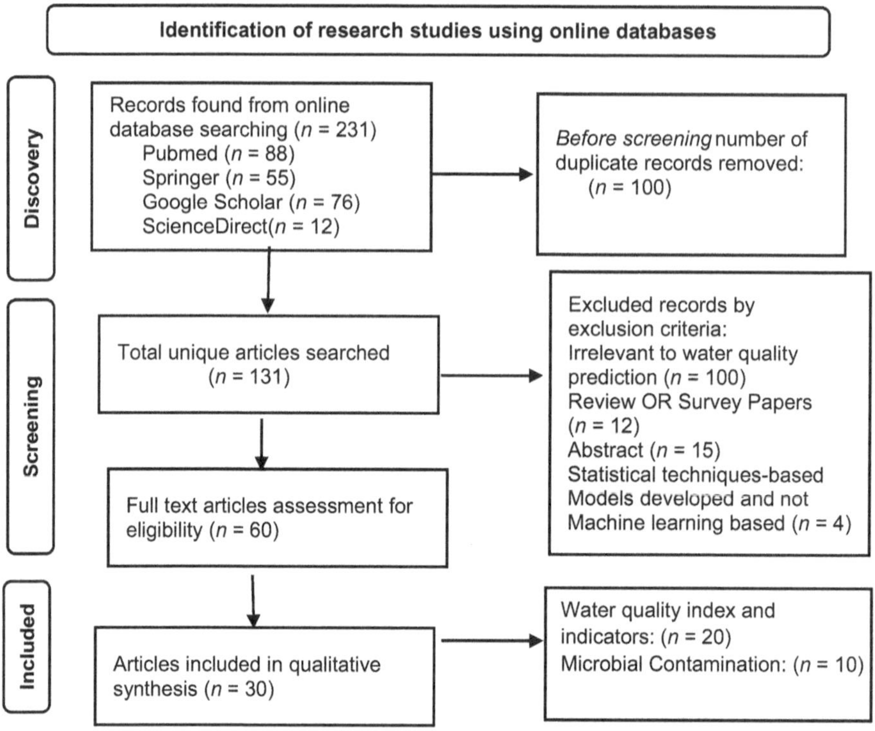

FIGURE 8.2
Prisma Model used for identification, screening, and article inclusion [55,56].

was conducted to recognize articles that used ML techniques to predict the quality of water. This study's search strategy included four search strings: water quality prediction, ML, microbial containment, and image processing. Figure 8.2 illustrates the complete search strategy used in this investigation and how the 50 research papers have been selected.

Figure 8.3 shows that to predict the quality of water, various ML algorithms are used in many research articles that also consider different water parameters using different water resources. Some of the big data techniques are also integrated with ML algorithms in the previous studies. According to this bibliometric analysis, further review of articles is carried out considering all the important associations among technologies used for water quality analysis (Figures 8.4 and 8.5).

8.2.2 Review of Estimation of Water Quality Indicators Using ML Techniques

Minaeea et al. [23] developed a water network model that was utilized to optimize the wall decay, and three models, genetic algorithm (GA), PSO, and

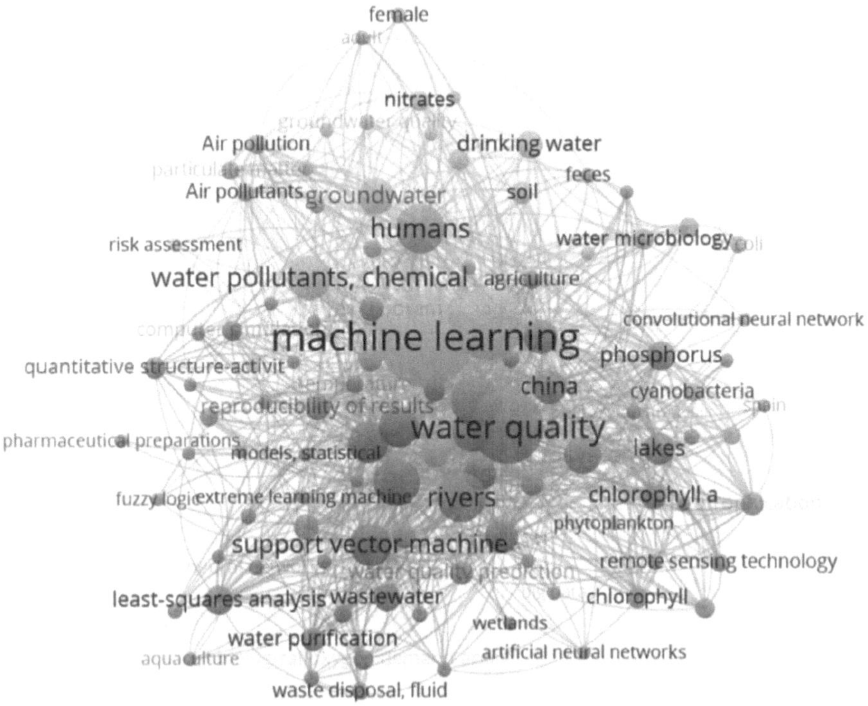

FIGURE 8.3
Bibliometric analysis using PubMed library (192 research articles): network of linked keywords.

hybrid GA-PSO, were designed. The hybrid GA-POS developed a stable solution comparatively. It is observed that rainfall had a great impact on predicting both ammonia and nitrate. Kourbasis et al. [19], Water age is considered a water quality predictor in water distribution network (WDN), author had used a Genetic algorithm for the purpose of water age and pressure optimization in WDN. The objective function used (Equation 8.1) for the minimization of objective function during a one-week simulation period:

$$z = \left(\sum_{i=1}^{N} \left(\frac{\text{Age}(i)}{t} \right) \right) \tag{8.1}$$

where i: network's joint, t: time step required for the water quality analysis, Age: water age, and n: maximum number of joints in the network. According to the study, using pressure-reducing valves lowers water pressure but raises water age.

Haghiabi et al. [22] designed a model using three algorithms for the prediction of different water quality indicators. It has been observed that the ANN model had a higher data dispersion index (DDR) value (as defined below using Expression (8.2)) using the tansig activation function. The group method of

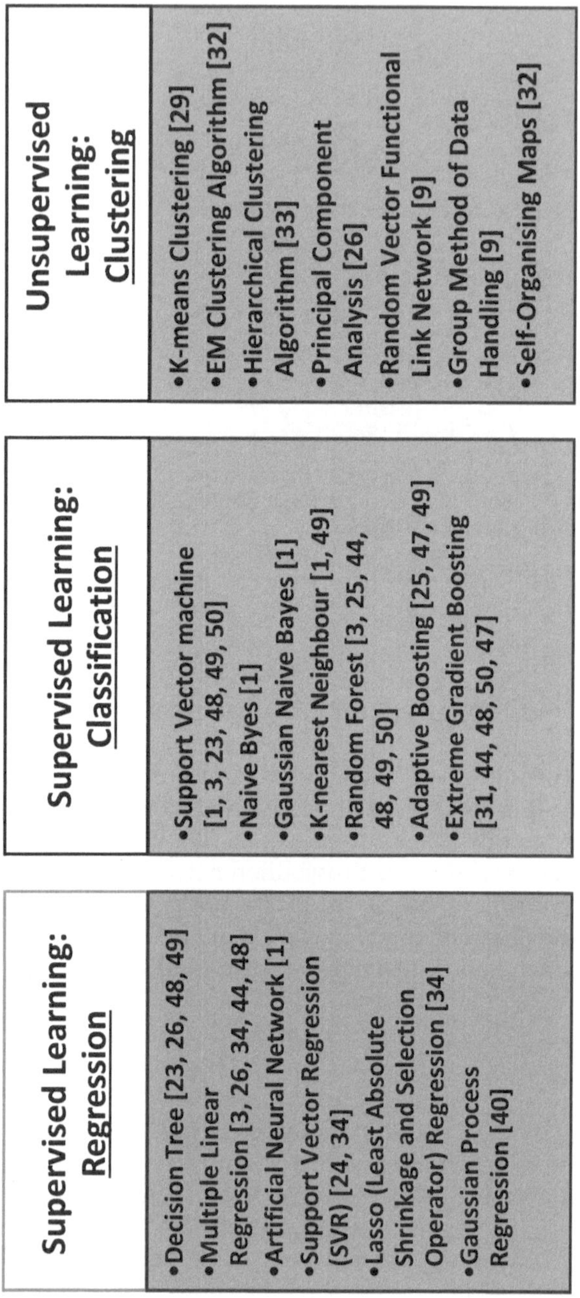

FIGURE 8.4
ML models implemented using a set of algorithms for water quality prediction in a recent research study.

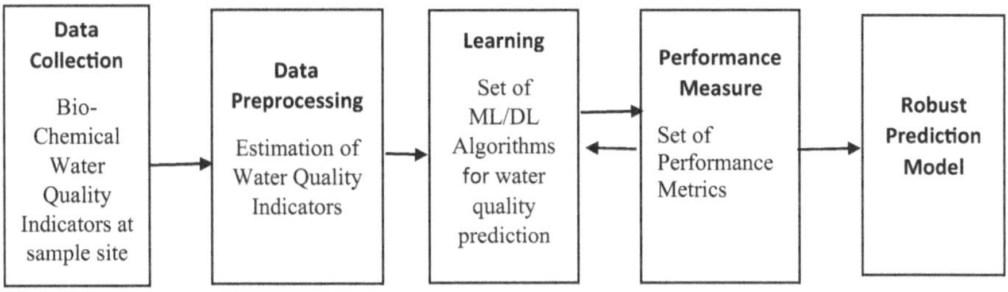

FIGURE 8.5
Framework of machine learning lifecycle for the development of water quality prediction model.

data handling (GMDH) performance had the lowest DDR value, but the support vector machine (SVM) model was found to be more reliable comparatively.

$$\text{DDR Index} = \frac{\text{Predicted Value}}{\text{Observed Value}} - 1 \tag{8.2}$$

Farhi et al. [38] proposed the long short-term memory (LSTM) model that monitored wastewater and was able to predict effluent ammonia accumulation. The objective function defined in the proposed model minimized the sum of squared error with an F1 score of 88% and an accuracy rate of 99%. Kim et al. [51] designed one-stage support vector regression (SVR) and double-stage WAVELET-SRV models for the prediction of BOD_5 indicator in Hwangji and Toilchun stations from South Korea. The proposed model's performance was measured using R2 (coefficient of determination), Nash–Sutcliffe efficiency, and root mean squared error (RMSE). Ooia et al. [13] designed an ensemble model that was able to predict accurately the biochemical oxygen demand (BOD_5) values in water samples on the basis of physio-chemical water parameters. The parameters utilized during the training phase were further optimized to get hyperparameters using a GA, and the sequential feature selection method extracted the required features. Kharake and Raut [36], in their study, evaluated the water quality of Godavari River's ten sampling stations, and eight parameters were chosen. The study indicated that the river water sample has alkaline properties due to the addition of pollutants in the river water at its different sampling stations that increased turbidity, dissolved oxygen (DO), BOD, total dissolved solids (TDS), and *E. coli*. Using Karl Pearson's coefficient, correlation among water parameters has been found [61].

8.2.3 Review of Prediction of Water Quality and Its Index Using ML Techniques

Aldhyani et al. [1] computed WQI using two algorithms: LSTM and non-linear auto-regressive neural networks (NARNET) and classified them using

SVM. Kouadri et al. [3] found that TDS and total hardness (TH) are the key parameters affecting WQI evaluation. WQI has a strong correlation with TDS and solids, concluding that TH is estimated to be the most sensitive water parameter. Hassan et al. [18] developed a model successfully to find the quality index of drinking water using different ML algorithms and classified it using NSF standards on a set of different water parameters in various places across India. Water colour as one of the physical water parameter is used for water quality prediction in model developed. 200 water sample images were collected with the help of digital camera. These images are further processed to selected important features using machine learning techniques such as random vector functional link network (RVFL) and group method of data handling (GMDH) networks.

Li et al. [24] used back-propagation neural networks (NNs), radial basis function NN, and SVM models as water quality evaluation models. Authors have found that back propagation neural network (BPNN) has disadvantages like local optimization and less stable outcomes. It is observed that radial basis function neural network (RBFNN) solved the problem of local optimization, whereas the model designed using SVM is found to be stronger than models designed using NNs with the highest accuracy and the best stability in the predicted values. Khullar et al. [35] proposed an approach that achieves the best accuracy of prediction and optimum error rates among all techniques. The model proposed is a deep learning–based technique that included pre-processing in the first phase, produced the feature maps in the second phase, designed a bi-directional LSTM architecture in the third phase, and utilized optimized loss function for minimizing error in the training. Bi et al. [37] proposed a hybrid prediction model that investigates an encoder–decoder artificial neural network (ANN) using LSTM with filter Savitzky–Golay (SG) to analyse the time series of water quality. SG filter removed the noise from the data, whereas LSTM extracted features required. The proposed model's comparative study shows that the decision tree, XGB, and EXT achieve the best prediction accuracy with an optimized training loss. Random forest algorithms are competent for predicting WQI with the lowest prediction errors. A smart system for water quality estimation is developed based on the Random Vector Functional Link Network, which is feedforward NN integrated with GMDH fetched the principal features of water colour images and executed the classification of images for water quality evaluation [62]. Fernandes et al. [8] collected groundwater samples from 37 sampling stations. The groundwater quality for irrigation was decided using the Kelly index, sodium adsorption ratio, percentage of sodium, residual sodium carbonate, magnesium ratio, and permeability index. The irrigation WQI was estimated using adaptive neuro-FIS models. Principal component analysis was used to estimate the input parameters of models, which increased the prediction reliability of the designed models. Oladipo et al. [5] assigned weights to parameters based on industrialization and climate change. They have identified DO as the major pollutant in the river.

The fuzzy WQI computation model is used to indicate water quality using physio-chemical parameters. The developed model for water quality prediction was validated using crosstab (sensitivity) analysis. They have found faecal coliform as an input parameter that influenced outcomes. GIS-integrated ordinary least square regression supported the prediction model for the surface water quality assessment. Md Galal Uddin [39] quantified uncertainties at every level of the WQI model application using the Monte Carlo simulation technique. Six regression models were designed; the Gaussian process regression model achieved the minimum error rate in training and proved to be most accurate predictor with water indicator uncertainties. Khoi et al. [48] demonstrate the use of 12 different ML models categorized into three groups: decision tree based, boosting based, and NN based for the prediction of WQI, whereas while computing WQI, physiochemical and biological water parameters were evaluated and analysed separately. It was concluded that water quality parameters are highly correlated. Authors observed that XGboost has high-accuracy prediction. Nasir et al. [50] found that the water quality prediction category feature–based boosting algorithm obtained high accuracy in contrast to alternative ML methods [63].

8.2.4 Review of Quality of Water Prediction Using Morphological Analysis and ML Algorithms and Techniques

Park et al. [20] used four CNN models for algal image classification to identify eight algal genera. The model was developed using two layers: convolution and pooling to extract features from images, and further classification is obtained using softmax as an activation function. Stocker et al. [16] evaluated water quality using two ponds from Maryland as a water resource to study microbial quality by analysing 12 different water parameters with the help of various ML algorithms: random forest (RF), K-nearest neighbour (KNN), stochastic gradient boosting, and SVM algorithms. They are also successful in finding a correlation between *E. coli* and other most important water parameters: conductivity, nephelometric turbidity units (NTU), chlorophyll, and fluorescent dissolved organic matter have been measured for the prediction of *E. coli*, as shown in Table 8.2. Sajedi et al. [30] integrated deep learning techniques with image processing methods. Gabor transform and XGboost method were used to extract image features like edges with various directions. The model designed successfully classified bacteria species with an accuracy of 95%.Wu et al. [31] developed six different models to identify the principal source of microbial contamination using ML techniques such as Naive Bayes, KNN, SVM, ANN, RF, and extreme gradient boost. Pu et al. [46] developed a four-layered CNN model to obtain the best water quality classification performance compared to SVM and RF models. They observed that CNN brought out differential and complex features from hyperspectral images. Li et al. [54] developed a BITF-CL model for the prediction of

TABLE 8.2

Relationship Found between *E. coli* and Other Parameters Concentrations of SPC, NTU, C, CHL, and Fluorescent Dissolved Organic Matter (fDOM)

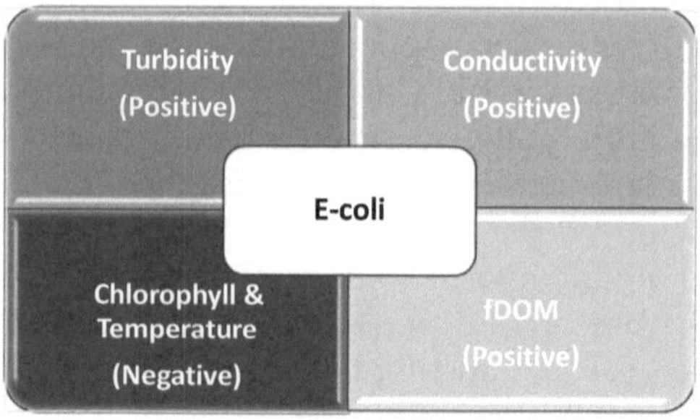

Stocker et al. [16]

effluent chemical oxygen demand and ammonia nitrogen concentrations in wastewater using 15 water parameters with the help of sewage surface image and time-series data. Sequencing batch reactors (SBRs) were considered to retrieve multi-modal data from surface sewage. It is found to be steady and reliable in situations where the aeration speed in the SBR was abruptly altered (Tables 8.3–8.5).

8.3 Concluding Remarks

In this chapter, the need for water quality evaluation and prediction using ML methods that are driving Industry 6.0 towards intelligent and smart digital twining is described. The most popular and best-performed ML algorithms used in water quality estimation observed are SVM and XGboost. SVM has the capability to process high-dimensional data, but it requires feature scaling and the choice of kernel function proves to be a challenge. XGBoost can reduce overfitting by adding regularization when the size of the sample is small; also, it doesn't need feature scaling [31]. LSTM-based models also proved to be the next best-performing model. Various fuzzy logic tools and ML techniques involved in identifying and evaluating the water quality are discussed. Also important observations from research studies were highlighted:

1. Robust models have been developed with ML and deep learning techniques in WQI computation to achieve the highest prediction accuracy and to gain deeper insights into the most important water parameters.

TABLE 8.3

Summary of Microbial Water Quality Prediction Using ML Techniques

S. No	Aim of Study	Datasets/Databases Used	Type of Water	Input Parameters	Algorithms Outperformed	References
1	Prediction of *E. coli* concentrations	Two ponds in Maryland were observed in the irrigation season during the period of 3 years 2016 and 2018	Surface water (pond water)	Turbidity, dissolved oxygen, conductivity, chlorophyll, and temperature	RF, SVM RMSE = 0.385 ± 0.007	[16]
2	Algal species classification	Collected image samples from three rivers in South Korea between 2015 and 2018 using the FlowCAM technique. Original 1,158 images are used during the training phase and 382 images for the testing phase	Freshwater	1. Three cyanobacteria: *Microcystis* sp., *Anabaena* sp., *Oscillatoria* sp. 2. Three diatoms: *Aulacoseira* sp., *Fragilaria* sp., *Synedra* sp. 3. Two green algae: *Pediastrum* sp., *Staurastrum* sp.	NN search-based CNN F1 score = 0.95	[20]
3	Estimation of microbial contamination	1. 102 water samples extracted from different 46 sites 2. Foursome profiles derived from 70 reference samples	Freshwater	Faecal coliform	XGBoost F1 score = 0.9	[31]
4	Prediction of *E. coli* accumulation in sediment	Extracted from the Irish Environmental Protection Agency water quality tracking database for Cork Harbour https://www.catchments.ie/data ; 29 sites out of 32 were examined for this study	Surface water (canals)	pH, temperature, nitrogen concentration, ammonia, ammoniacal nitrogen, salinity, molybdate reactive phosphorus, biological oxygen demand, dissolved oxygen, chlorophyll a	Gaussian process regression RMSE = 0.643	[40]
5	Predicting microbial water quality	Generated more than 500 data points from 32 disruption events using online sensors	Wastewater	pH, turbidity, conductivity, temperature, chorine (as online sensors)	RF with lower true positive rates LR<SVM<RF	[41,42,47]
6	Prediction of *E. coli* concentrations	Environmental stressor dataset from the Syndicate Marne Vive for five years (2015, 2017–2020)	Fresh water	Turbidity, total suspended solids, conductivity, nitrogen, temperature, and *E. coli* concentrations	RF 0.37 ± 0.20 for RMSE and 0.09 ± 0.02 for MAE	[43,49]

TABLE 8.4

Summary of Water Quality Prediction Using ML Techniques

S. No.	Datasets/Databases Used	Type of Water	Input Parameters	Algorithms Outperformed	References
1	The dataset chosen is obtained from some of the historical locations of India during the period from 2005 to 2014 that contained 1,679 samples from Kaggle	Surface water	pH, conductivity, dissolved oxygen, biological oxygen demand, nitrate, F-coliform, T-coliform	Fuzzy logic inference system LSTM < NARNET < SVM Accuracy = 97.01% F score (%) = 98.54	[1,5,7]
2	The dataset provided by the Directorate of Water Resources of the State of Illizi included the results of the analysis of 114 samples taken from 57 water wells of six different layers during the period of 1999 and 2020	Groundwater	pH, conductivity, total hardness, alkalinity, turbidity, dissolved solids, iron, chloride, sulphate, nitrate, calcium, fluoride, magnesium, arsenic, total coliform, faecal coliform	Random forest RMSE = 3.248	[3]
3	Four different sites were selected from the Sabarmati river basin for 2005, 2010, and 2015 from GPCB (https://gpcb.gujarat.gov.in/Index)	Wastewater	Dissolved oxygen, pH, conductivity, temperature, TDS, alkalinity, total hardness, calcium, magnesium	Fuzzy logic model Fuzzy WQI ≈ 0	[4]
4	UC Irvine Machine Learning Repository (https://archive.ics.uci.edu/) six datasets including balance, diabetes, abalone, musk, skin segmentation (seg.), and pegdigits are utilized.	Surface water	Water colour with other nine water parameters	RVFL and GMDH Accuracy = 96.19%	[9]
5	–	Freshwater (river)	Chlorine, conductivity, HCO_3, magnesium, sodium, SO_4, TDS, pH, calcium	GMDH < SVM DDR index	[22]
6	Kaggle Website	Surface water	pH, conductivity, total coliform, biological oxygen demand, dissolved oxygen, nitrate	SVM SVM > NN > BT > RF > MLR F1 score 0.84	[18]

TABLE 8.5

Comparative Study of Supervised Regression Algorithms Used in ML Techniques [12]

S. No	References	Regression Algorithms	Benefits	Limitations
1	[1,3,26,34]	Multiple linear regression	1. Each feature variable must model the linear relationship with the dependent variable. 2. It tries to fit a line of best fit through a multi-dimensional space of features.	1. It can only find a linear relationship between parameters. 2. It is not suitable for large datasets.
2	[23,26]	Decision tree regression	1. The predicted result is numeric (small Gini impurity or entropy). 2. If certain classes predominate, the model produces skewed trees. The tree structure may be influenced by the impurity measure selection.	1. Stable and reliable model design becomes difficult. 2. Not a cost-effective solution on large datasets.
3	[24,40,45]	Support vector regression	1. It fits the input–output relationship of a model to a great degree with less computation with MAE. 2. Processing data resulting from unknown changes has no effect on prediction accuracy and operational efficiency.	1. Large datasets are not a good fit for it. 2. Prediction accuracy is impacted when there are more features per data point than there are training data samples.
4	[40]	Gaussian process regression	1. It works well for resolving challenging regression issues with nonlinearity and large dimensionality. 2. It automatically determines the uncertainty of the predictions if the model is appropriate.	1. More computation cost is involved in the process; hence it can be efficient only on small datasets.
5	[1]	Artificial neural network regression	1. A model can be created that has a high capacity for self-learning and non-linear fitting. 2. By modifying the hyperparameters like the number and size of layers and regularization the model can be made more optimal.	1. The number of hidden layers and nodes must be defined in advance. 2. The ANN model has a tendency to overestimate the small-size datasets.
6	[34]	Lasso regression	1. This technique selects the variables and applies regularization to increase prediction accuracy. 2. Simple models with fewer parameters are strengthened by this technique.	1. Less accurate for correlated parameters. 2. Prediction accuracy is influenced when the number of characteristics for each data point is greater than the number of training data samples.

2. Models developed can be further improved to suggest the best water treatment with the help of data analytics in deep using recent techniques like remote sensing, GIS integrated with ML, and big data analytics [6].
3. Complex water parameters commonly found in water can be predicted like pathogens using image processing techniques with ML and fuzzy logic so that real-time monitoring of water can be performed efficiently to recognize changes in the patterns or existence of pathogens in water samples [64].

References

1. Theyazn H. H. Aldhyani, Mohammed Al-Yaari, Hasan Alkahtani, Mashael Maashi, "Water quality prediction using artificial intelligence algorithms," *Applied Bionics and Biomechanics*, 2020, 30 December 2020, https://doi.org/10.1155/2020/6659314
2. Vinod Kothari, Suman Vij, SuneshKumar Sharma, Neha Gupta, "Correlation of various water qualityparameters and water quality index of districts of Uttarakhand," *Environmental and Sustainability Indicators*, 9, 100093, 17 December 2020, https://doi.org/10.1016/j.indic.2020.100093
3. Saber Kouadri, Ahmed Elbeltagi, Abu Reza Md Towfiqul Islam, Samir Kateb, "Performance of machine learning methods in predicting water quality index based on irregular data set: Application on Illizi region (Algerian southeast)," *Applied Water Science*, 11(12), 190, 6 November 2021, https://doi.org/10.1007/s13201-021-01528-9
4. Anant Patel, Karishma Chitnis, "Application of fuzzy logic in river water quality modelling for analysis of Industrialization and climate change impact on Sabarmati river," *Water Supply*, 22(1), 238–250, 2022, https://doi.org/10.2166/ws.2021.275
5. Johnson O. Oladipo, Akinola S. Akinwumiju, Olatunji Sunday,. Aboyeji, Adedeji A. Adelodun, "Comparison between fuzzy logic and water quality index methods: A case of water quality assessment in Ikare community, Southwestern Nigeria," *Environmental Challenges*, 3, 100038, 28 January 2021, https://doi.org/10.1016/j.envc.2021.100038
6. Jitha P. Nair, M. S. Vijaya, "Predictive models for river water quality using machine learning and Big Data techniques: A survey." In *2021 International Conference on Artificial Intelligence and Smart Systems (ICAIS)*, Coimbatore, India, pp. 1747–1753, March 2021, https://doi.org/10.1109/ICAIS50930.2021.9395832
7. Bai V. Raman, Reinier Bouwmeester, S. Mohan, "Fuzzy logic water quality index and importance of water quality parameters," *Air, Soil and Water Research*, 30 Dec 2022, https://doi.org/10.4137/ASWR.S2156
8. Ana Fernandes, Humberto Chaves, Rui Lima, José Neves, Henrique Vicente, "Draw on artificial neural networks to assess and predict water quality." In *IOP Conference Series: Earth and Environmental Science* (vol. 612, p. 012018), 2020, https://doi.org/10.1088/1755-1315/612/1/012028

9. Junde Chen, Defu Zhang, Shuangyuan Yang, Yaser Ahangari Nanehkaran, "Intelligent monitoring method of water quality based on image processing and RVFL-GMDH model," *IET Image Processing*, 14(17), 4646–4656, 24th February 2021, https://doi.org/10.1049/iet-ipr.2020.0254
10. Seulbi Lee, Jaehoon Kim, Jongyeon Hwang, EunJi Lee, Kyoung-Jin Lee, Jeongkyu Oh, Jungsu Park, Tae-Young Heo, "Clustering of time series water quality data using dynamic time warping: A case study from the Bukhan River water quality monitoring network," *Water*, 12(9), 2411, 27 August 2020, https://doi.org/10.3390/w12092411
11. Phil Kim, *"MATLAB Deep Learning, with Machine Learning, Neural Networks and Artificial Intelligence,"* Apress, Berkeley, CA, 2017, ISBN–13 (electronic): 978-1-4842-2845-6.
12. Oliver Theobald, *"Machine Learning For Absolute Beginners"*, 2nd Edition, Scatterplot Press, 2017.
13. Kai Sheng Ooia, ZhiYuan Chen, Phaik Eong Poh, Jian Cui, "BOD5 prediction using machine learning methods," *Water Supply*, 22(1), 1168, 2022, https://doi.org/10.2166/ws.2021.202
14. Dimitrios E. Alexakis, "Meta-evaluation of water quality indices: Application into groundwater resources," *Water*, 12(7), 1890, 2 July 2020, https://doi.org/10.3390/w12071890
15. Canadian Council of Ministers of the Environment. Canadian water quality guidelines for the protection of aquatic life: CCME Water Quality Index 1.0 User's Manual, Technical Report, Winnipeg, 2001.
16. Matthew D. Stocker, Yakov A. Pachepsky, Robert L. Hill, "Prediction of E. coli concentrations in agricultural pond waters: Application and comparison of machine learning algorithms," *Frontiers in Artificial Intelligence*, 4, 768650, 11 January 2022, https://doi.org/10.3389/frai.2021.768650
17. Md. Galal Uddin, Stephen Nash, Agnieszka I. Olbert, "A review of water quality index models and their use for assessing surface water quality", *Ecological Indicators*, 122, 107218, 23 November 2020, https://doi.org/10.1016/j.ecolind.2020.107218
18. Md. Mehedi Hassan, Md. Mahedi Hassan, Laboni Akter, Md. Mushfiqur Rahman, Sadika Zaman, Khan Md. Hasib, Nusrat Jahan, Raisun Nasa Smrity, Jerin Farhana, M. Raihan, Swarnali Mollick, "Efficient prediction of Water Quality Index (WQI) using machine learning algorithms," *Human-Centric Intelligent Systems*, 1(3–4), 86–97, December 2021, https://doi.org/10.2991/hcis.k.211203.001
19. Nikolaos Kourbasis, Menelaos Patelis, Stavroula Tsitsifli, Vasilis Kanakoudis, "Optimizing water age and pressure in drinking water distribution networks," *Environmental Sciences Proceedings*, 2(1), 51, 5 September 2020, https://doi.org/10.3390/environsciproc2020002051
20. Jungsu Park, Hyunho Lee, Cheol Young Park, Samiul Hasan, Tae-Young Heo, Woo Hyoung Lee, "Algal morphological identification in watersheds for drinking water supply using neural architecture search for convolutional neural network," *Water*, 11(7), 133828, June 2019, https://doi.org/10.3390/w11071338
21. Isha M. Khot, Anil R. Surve, "IoT assisted drinkable water quality analysis system using machine learning techniques," *International Journal for Research in Applied Science & Engineering Technology (IJRASET)*, 8(IX), September 2020, https://doi.org/10.22214/ijraset.2020.31221

22. Amir Hamzeh Haghiabi, Ali Heidar Nasrolahi, Abbas Parsaie, "Water quality prediction using machine learning methods," *Water Quality Research Journal*, 53(1), 3–13, 2018, https://doi.org/10.2166/wqrj.2018.025
23. Roya Peirovi Minaeea, Mojtaba Afsharniab, Alireza Moghaddamc, Ali Asghar Ebrahimia, Mohsen Askarishahid, Mehdi Mokhtari, "Calibration of water quality model for distribution networks using genetic algorithm, particle swarm optimization, and hybrid methods", *MethodsX*, 6, 540–548, 2019, https://doi.org/10.1016/j.mex.2019.03.008
24. Tingting Li, Jian Lu, Jun Wu, Zhenhua Zhang, Liwei Chen, "Predicting aquaculture water quality using machine learning approaches", *Water*, 14, 2836, 2022, https://doi.org/10.3390/w14182836
25. Ali El Bilali, Abdeslam Taleb, Youssef Brouziyne, "Groundwater quality forecasting using machine learning algorithms for irrigation purposes", *Agricultural Water Management*, 245, 106625, October 2020, https://doi.org/10.1016/j.agwat.2020.106625
26. Ali El Bilali, Abdeslam Taleb, "Prediction of irrigation water quality parameters using machine learning models in a semi-arid environment," *Journal of the Saudi Society of Agricultural Sciences*, 19(7), 439–451, August 2020, https://doi.org/10.1016/j.jssas.2020.08.001
27. Ali Mokhtar, Ahmed Elbeltagi, Yeboah Gyasi-Agyei, Nadhir Al-Ansari, Mohamed K. Abdel-Fattah, "Prediction of irrigation water quality indices based on machine learning and regression models," *Applied Water Science*, 12(4), 76, March 2022, https://doi.org/10.1007/s13201-022-01590-x
28. U.S. Geological Survey, 2019. https://www.usgs.gov/special-topics/water-science-school/science/how-much-water-there-earth
29. Mahmoud Y. Shams, Ahmed M. Elshewey, El-Sayed M. El-kenawy, Abdelhameed Ibrahim, Fatma M. Talaat, Zahraa Tarek, "Water quality prediction using machine learning models based on grid search method," *Multimedia Tools and Applications*, 83(12), 35307–35334, August 2023, https://doi.org/10.1007/s11042-023-16737-4
30. Hedieh Sajedi, Fatemeh Mohammadipanah, Ali Pashaei, "Image-processing based taxonomy analysis of bacterial macromorphology using machine learning models," *Multimedia Tools and Applications*, 79, 32711–32730, August 2020, https://doi.org/10.1007/s11042-020-09284-9
31. Jianyong Wu, Conghe Song, Eric A. Dubinsky, Jill R. Stewart, "Tracking major sources of water contamination using machine learning," *Frontiers in Microbiology*, 11, 616692, January 2021, https://doi.org/10.3389/fmicb.2020.616692
32. André Felipe Henriques Librantz, Fábio Cosme Rodrigues dos Santos, "Intelligent clustering techniques for the reduction of chemicals in water treatment plants," *Sustainability*, 15(8), 6579, April 2023, https://doi.org/10.3390/su15086579
33. Xiangjun Du, Fengjing Shao, Shunyao Wu, Hanlin Zhang, Si Xu, "Water quality assessment with hierarchical cluster analysis based on Mahalanobis distance," *Environmental Monitoring and Assessment*, 189, 335, May 2017, https://doi.org/10.1007/s10661-017-6035-y
34. Kakoli Banerjee, Vikram Bali, Nishad Nawaz, Shivani Bali, Sonali Mathur, Ram Krishn Mishra, Sita Rani, "A machine-learning approach for prediction of water contamination using latitude, longitude, and elevation," *Water*, 14(5), 728, February 2022, https://doi.org/10.3390/w14050728

35. Sakshi Khullar, Nanhey Singh, "Water quality assessment of a river using deep learning Bi-LSTM methodology: Forecasting and validation," *Environmental Science and Pollution Research*, 29, 12875–12889, April 2021, https://doi.org/10.1007/s11356-021-13875-w
36. Ashali Chandrakant Kharake, Vaishali Sanjay Raut, "An assessment of water quality index of Godavari river water in Nashik city, Maharashtra," *Applied Water Science*, 11(6), 101, June 2021, https://doi.org/10.1007/s13201-021-01432-2
37. Jing Bi, Yongze Lin, Quanxi Dong, Haitao Yuan, MengChu Zhou, "Large-scale water quality prediction with integrated deep neural network," *Information Sciences*, 571, 191–205, April 2021, https://doi.org/10.1016/j.ins.2021.04.057
38. Nitzan Farhi, Efrat Kohen, Hadas Mamane, Yuval Shavitt, "Prediction of wastewater treatment quality using LSTM neural network," *Environmental Technology & Innovation*, 23, 101632, May 2021, https://doi.org/10.1016/j.eti.2021.101632
39. Md Galal Uddin, Stephen Nash, Azizur Rahman, Agnieszka I. Olbert, "A novel approach for estimating and predicting uncertainty in water quality index model using machine learning approaches," *Water Research*, 229, 119422, November 2022, https://doi.org/10.1016/j.watres.2022.119422
40. Erfan Ghasemi Tousi, Jennifer G. Duan, Patricia M. Gundy, Kelly R. Bright, Charles P. Gerba, "Evaluation of E. coli in sediment for assessing irrigation water quality using machine learning," *Science of the Total Environment*, 799, 149286, July 2021, https://doi.org/10.1016/j.scitotenv.2021.149286
41. Eva Reynaert, Philipp Steiner, Qixing Yu, Lukas D'Olif, Noah Joller, Mariane Y. Schneider, Eberhard Morgenroth, "Predicting microbial water quality in on-site water reuse systems with online sensors," *Water Research*, 240, 120075, May 2023, https://doi.org/10.1016/j.watres.2023.120075
42. Thomas Dama Hile, Stephen G. Dunbar, Ryan G. Sinclair, "Microbial contamination analysis of drinking water from bulk dispensers and fast-food restaurants in the Eastern Coachella Valley, California," *Water Supply*, 23(9), 3578–3596, 2023, https://doi.org/10.2166/ws.2023.200
43. Manel Naloufi, Françoise S. Lucas, Sami Souihi, Pierre Servais, Aurélie Janne, Thiago Wanderley Matos De Abreu, "Evaluating the performance of machine learning approaches to predict the microbial quality of surface waters and to optimize the sampling effort," *Water*, 13(18), 2457, 2021, https://doi.org/10.3390/w13182457
44. Xianhe Wang, Ying Li, Qian Qiao, Adriano Tavares, Yanchun Liang, "Water quality prediction based on machine learning and comprehensive weighting methods," *Entropy*, 25(8), 1186, 2023, https://doi.org/10.3390/e25081186
45. Atefeh Nouraki, Mohammad Alavi, Mona Golabi, Mohammad Albaji, "Prediction of water quality parameters using machine learning models: A case study of the Karun River, Iran," *Environmental Science and Pollution Research*, 28, 57060–57072, May 2021, https://doi.org/10.1007/s11356-021-14560-8
46 Fangling Pu, Chujiang Ding, Zeyi Chao, Yue Yu and Xin Xu, "Water-quality classification of inland lakes using Landsat8 images by convolutional neural networks," *Remote Sensing*, 11(14), 1674, May 2019, https://doi.org/10.3390/rs11141674
47. Elias S. Leggesse, Fasikaw A. Zimale, Dagnenet Sultan, "Predicting optical water quality indicators from remote sensing using machine learning algorithms in tropical highlands of Ethiopia," *Hydrology*, 10(5), 110, May 2023, https://doi.org/10.3390/hydrology10050110

48. Dao Nguyen Khoi, Nguyen Trong Quan, Do Quang Linh, "Using machine learning models for predicting the water quality index in the La Buong River, Vietnam," *Water*, 14(10), 1552, May 2022, https://doi.org/10.3390/w14101552
49. Farkhanda Abbas, Zhihua Cai, Muhammad Shoaib, Javed Iqbal, Muhammad Ismail, Arifullah, Abdulwahed Fahad Alrefaei, Mohammed Fahad Albeshr, "Machine learning models for water quality prediction: A comprehensive analysis and uncertainty assessment in Mirpurkhas, Sindh, Pakistan," *Water*, 16(7), 941, 2024, https://doi.org/10.3390/w16070941
50. Nida Nasir, Afreen Kansal, Omar Alshaltone, Feras Barneih, Mustafa Sameer, Abdallah Shanableh, Ahmed Al-Shamma'a, "Water quality classification using machine learning algorithms," *Journal of Water Process Engineering*, 48, 102920, June 2022, https://doi.org/10.1016/j.jwpe.2022.102920
51. Sungwon Kim, Meysam Alizamir, Youngmin Seo, Salim Heddam, Il-Moon Chung, Young-Oh Kim, Ozgur Kisi, Vijay P. Singh, "Estimating the incubated river water quality indicator based on machine learning and deep learning paradigms: BOD5 prediction," *Mathematical Biosciences and Engineering*, 19(12), 12744–12773, Sep 2022, https://doi.org/10.3934/mbe.2022595
52. Md Galal Uddin, Stephen Nash, Mir Talas Mahammad Diganta, Azizur Rahman, Agnieszka I. Olbert, "Robust machine learning algorithms for predicting coastal water quality index," *Journal of Environmental Management*, 321, 115923, August 2022, https://doi.org/10.1016/j.jenvman.2022.115923
53. Risky Ayu Kristanti, Tony Hadibarata, Muhammad Syafrudin, Murat Yılmaz, Shakila Abdullah, "Microbiological contaminants in drinking water: Current status and challenges," *Water, Air, Soil & Pollution*, 233, 299, 2022, https://doi.org/10.1007/s11270-022-05698-3
54. Junchen Li, Sijie Lin, Liang Zhang, Yuheng Liu, Yongzhen Peng, Qing Hu, "Brain-inspired multimodal approach for effluent quality prediction using wastewater surface images and water quality data," *Frontiers of Environmental Science & Engineering*, 18, 31, 2024, https://doi.org/10.1007/s11783-024-1791-x
55. Catrin Sohrabi, Thomas Franchi, Ginimol Mathew, Ahmed Kerwan, Maria Nicola, Michelle Griffin, Maliha Agha, Riaz Agha, "PRISMA 2020 statement: What's new and the importance of reporting guidelines," *International Journal of Surgery*, 88, 105918, 2021, https://doi.org/10.1016/j.ijsu.2021.105918
56. Matthew J. Page, Joanne E McKenzie, Patrick M Bossuyt, Isabelle Boutron, "The PRISMA 2020 statement: An updated guideline for reporting systematic reviews," *BMJ*, 372, 2020, https://doi.org/10.1136/bmj.n71
57. Aditya Akundi, Daniel Euresti, Sergio Luna, Wilma Ankobiah, Amit Lopes, Immanuel Edinbarough, "State of Industry 5.0—Analysis and identification of current research trends," *Applied System Innovation*, 5(1), 27, 2022, https://doi.org/10.3390/asi5010027
58. Amr Adel, "Future of Industry 5.0 in society: Human-centric solutions, challenges and prospective research areas," *Journal of Cloud Computing*, 11, 40, 2022, https://doi.org/10.1186/s13677-022-00314-5
59. Shubhangi Chourasia, Ankit Tyagi, S. M. Pandey, R. S. Walia, Qasim Murtaza, "Sustainability of Industry 6.0 in global perspective: Benefits and challenges," *Mapan*, 37, 443–452, 2022, https://doi.org/10.1007/s12647-022-00541-w
60. Shingo Shimoda, Lorenzo Jamone, Dimitri Ognibene, Takayuki Nagai, Alessandra Sciutti, Alvaro Costa-Garcia, Yohei Oseki, Tadahiro Taniguchi, "What is the role of the next generation of cognitive robotics?," *Advanced Robotics*, 36(1–2), 3–16, 2022, https://doi.org/10.1080/01691864.2021.2011780

61. Riyaz Abdullah Sheikh, Surbhi Bhatia, Sujit Gajananrao Metre, Ali Yahya A. Faqihi, "Strategic value realization framework from learning analytics: A practical approach," *Journal of Applied Research in Higher Education*, 14(2), 693–713, https://doi.org/10.1108/JARHE-10-2020-0379.
62. P. R. Anisha, C. Kishor Kumar Reddy, "Blockchain technology: A boon at the pandemic times—A solution for global economy upliftment with AI and IoT." In: K.M. Baalamurugan, S. Rakesh Kumar, Dr. Abhishek Kumar, Vishal Kumar, Dr. Sanjeevikumar Padmanaban (eds) *Blockchain Security in Cloud Computing*. EAI/Springer Innovations in Communication and Computing, 2022. Springer, Cham.
63. P. R. Anisha, C. Kishor Kumar Reddy, Nguyen Gia Nhu, "Big Data: Trends, challenges, opportunities, tools, success factors, and the key way toward pandemic analytics." In *Handbook of Research for Big Data Concepts and Techniques*, Apple Academic Press, 2021.
64. Sapna Singh Kshatri, Deepak Singh, Bhavana Narain, Surbhi Bhatia, Mohammad Tabrez Quasim, Ganesh Ram Sinha, "An empirical analysis of machine learning algorithms for crime prediction using stacked generalization: An ensemble approach," *IEEE Access*, 9, 67488–67500, https://doi.org/10.1109/ACCESS.2021.3075140.

9

Climate Change Mitigation Strategies Using Convolutional Neural Networks over Multiple Locations

Kishor Kumar Reddy C, Harshitha Muthyam, Srinath Doss, and Pranay Chandur

9.1 Introduction

A region's or the world's long-term patterns of temperature, precipitation, humidity, wind, and other atmospheric elements are referred to as its climate. However, weather describes brief variations in atmospheric circumstances. The climate is influenced by a number of variables, including geographic location, solar radiation, ocean currents, atmospheric circulation patterns, and the makeup of gases in the atmosphere, particularly greenhouse gases like carbon dioxide and methane. Climate affects ecosystems, agriculture, human health, economies, and the general well-being of society. Because alterations in the climate, whether anthropogenic (caused by humans) or natural, can have profound consequences on the environment and society, climate research is an important area of scientific inquiry and policy discussion. Significant and long-lasting changes to the Earth's climate system, which frequently take decades or even millennia, are referred to as climate change. The frequency and severity of extreme weather events, sea level rise, temperature, precipitation patterns, and other facets of the climate are all impacted by these changes. Although in the past the climate has been changed by natural events such as solar radiation variations and volcanic eruptions, the current state of climate change is primarily due to human activity [1].

This is especially true in the case of deforestation, burning fossil fuels, and industrial operations that release greenhouse gases into the atmosphere. Intentional actions and strategies known as mitigation can reduce or eliminate the intensity, scope, or impact of an undesirable event, phenomenon, or issue. The main focus of mitigating the effects of climate change on ecosystems, public health, socioeconomic systems, and the environment is on

DOI: 10.1201/9781003503934-9

reducing greenhouse gas emissions and addressing the underlying causes of global warming. Mitigation is a broad category of actions implemented at the local, national, and international levels to shift toward a more sustainable, low-carbon future. The term "m pH, and conductivity, total coliform, biological itigation" describes the intentional methods and approaches used to lessen or stop the intensity, scope, or effects of an unfavorable occurrence, phenomenon, or issue [2].

Mitigation of climate change is the process of reducing the negative consequences that it has on ecosystems, the environment, human health, and socioeconomic systems. It is especially relevant when discussing attempts to reduce greenhouse gas emissions and address the underlying causes of global warming. In an attempt to make the transition to a low-carbon and more sustainable future, a broad range of initiatives undertaken at the local, state, federal, and international levels are collectively referred to as "mitigation". Through the employment of a variety of techniques, mitigation methods seek to lower greenhouse gas emissions and lessen the effects of climate change. A key solution is switching to renewable energy sources, which entails replacing fossil fuels in the production of electricity and transportation with a broad adoption of solar, wind, hydropower, geothermal, and biomass energy. In addition, cutting emissions and energy consumption requires improving energy efficiency in all sectors, including transportation, industry, buildings, and appliances. The implementation of carbon pricing instruments, such as cap-and-trade schemes or carbon taxes, generates income for additional mitigation initiatives while offering financial incentives for reducing emissions.

Reforestation and restoration projects are crucial to reducing atmospheric carbon dioxide levels and mitigating the effects of changing land use and deforestation. Additionally, encouraging climate-smart farming methods like crop rotation, conservation tillage, and agroforestry improves soil resilience and carbon absorption while lowering agricultural emissions [3]. Comprehensive mitigation solutions must emphasize technological innovation, legislative and regulatory measures, public awareness efforts, carbon capture and storage technology, and sustainable transportation options. To successfully coordinate efforts, exchange best practices, and handle the global challenge of climate change, international collaboration and cooperation are crucial. Societies may progress toward a more resilient and sustainable future by combining these mitigation techniques, which will lessen the effects of climate change while promoting economic growth. Human activities such as burning fossil fuels (coal, oil, and natural gas), deforestation, industrial processes, and agricultural practices are some of the sources of greenhouse gases that are released into the atmosphere. Other greenhouse gases include methane and nitrous oxide. Because of these gases' capacity to trap heat, there is a greenhouse effect and worldwide warming.

When forests are destroyed for logging, urbanization, agriculture, and other reasons, there are fewer trees in the world that can absorb carbon dioxide from the atmosphere.

This exacerbates climate change by adding to atmospheric CO_2 levels that are already elevated. Numerous industrial processes, such as cement production, manufacturing, and chemical reactions, release greenhouse gases and other pollutants into the environment. Climate change and global warming are exacerbated by these emissions. The use of synthetic fertilizers and animal rearing are two intensive agricultural activities that produce methane and nitrous oxide in the environment. Land use changes and deforestation for agricultural purposes can also have an impact on local climate trends. The burning of fossil fuels in vehicles for transportation is one of the primary sources of greenhouse gas emissions, especially carbon dioxide. Climate change is exacerbated by the growing number of cars, trucks, ships, and airlines in the world. The environment, the economy, and society are all greatly impacted by climate change, which is a major concern of our day. Most scientists agree that deforestation, combustion of fossil fuels, and human activities in general are changing the Earth's climate system in ways that have never been observed before, resulting in rising temperatures, harsh weather, and damage to ecosystems and livelihoods. Effective mitigation techniques are more important than ever as the urgency of addressing climate change becomes more apparent.

Nevertheless, there are many obstacles to overcome before putting effective climate change mitigation plans into practice, such as financial limitations, political opposition, and technological limitations [4]. Owing to the magnitude and complexity of the transition to a low-carbon economy, significant investments in R&D, innovation, and infrastructure are required. To expedite the process, supportive legal frameworks and financial incentives from the government are also required. Furthermore, as the effects of global warming transcend national boundaries and necessitate coordinated responses from all nations, tackling climate change demands collective action at the international level. There are causes for hope despite these obstacles. A low-carbon future is gaining traction, thanks to developments in clean energy technologies, falling renewable energy costs, and increased public awareness of the need for immediate climate action. International agreements, like the Paris Agreement, show how committed the world is to combating climate change and allocating funds for adaptation and mitigation measures. In addition, the shift to a sustainable economy presents chances for job creation, economic expansion, and enhanced quality of life, all of which serve as strong motivators for action [5].

The effects of global warming, its causes, its effects in India, and its causes are illustrated in Tables 9.1–9.4. As seen in Figure 9.1a–d, the accompanying figures depict the causes of climate change.

TABLE 9.1

Statistics about Effects of Climate Change in the World

Year	Location	Mitigation Strategy	Effects
2020	Global	Renewable energy adoption	Reduction in greenhouse gas emissions
2021	United States	Renewable energy transition	Decrease in CO_2 emissions, job creation
2022	European Union	Carbon pricing	Reduction in greenhouse gas emissions
2023	China	Afforestation	Carbon sequestration, biodiversity preservation
2024	India	Sustainable transportation	Improved air quality, reduced emissions

TABLE 9.2

Statistics about Causes of Climate Change in the World

Year	Location	Mitigation Strategy	Causes
2020	Global	Renewable energy adoption	Fossil fuel dependency, industrialization
2021	United States	Renewable energy transition	Reliance on fossil fuels, industrialization
2022	European Union	Carbon pricing	Industrial activities, transportation
2023	China	Afforestation	Deforestation, urbanization
2024	India	Sustainable transportation	Vehicle emissions, urban sprawl

TABLE 9.3

Statistics about Effects of Climate Change in India

Year	Location	Mitigation Strategy	Effects
2020	Madhya Pradesh	Increasing renewable energy adoption	Significant reduction in greenhouse gas emissions
2021	Maharashtra	Increasing use of solar power and wind energy	Reduction in carbon emissions, increased energy access
2022	Tamil Nadu	Promoting rooftop solar installations	Decreased reliance on fossil fuels, reduced carbon footprint
2023	Karnataka	Reforestation and afforestation programs	Increased carbon sequestration, biodiversity conservation
2024	Uttar Pradesh	Introduction of biogas plants for energy generation	Reduced reliance on fossil fuels, mitigation of methane emissions

TABLE 9.4
Statistics about Causes of Climate Change in India

Year	Location	Mitigation Strategy	Causes
2020	Madhya Pradesh	Increasing renewable energy adoption	Adverse impacts of climate change on economy, environment, and public health; commitment to sustainable development
2021	Maharashtra	Increasing use of solar power and wind energy	Rapid urbanization, industrial growth, energy demands
2022	Tamil Nadu	Promoting rooftop solar installations	Rising energy demand, environmental concerns
2023	Karnataka	Reforestation and afforestation programs	Deforestation, loss of biodiversity
2024	Uttar Pradesh	Introduction of biogas plants for energy generation	Agricultural waste management, methane emissions from livestock

FIGURE 9.1
(a) Industrialization; (b) plastic usage; (c) deforestation; (d) urbanization.

9.2 Relevant Work

The recommended title is "Adaptation to Climate Change in the Developing World," as proposed by Adger et al. [6]. The rapid acceleration of climate change poses hitherto unseen hazards, especially for marginalized communities dependent on climate-vulnerable resources such as fisheries and agriculture in emerging nations. While society has historically adjusted to changing climates, some industries and social groups are more vulnerable than others. Improving adaptive capability is critical since climate-related issues will inevitably outweigh previous means of coping. Governments and development organizations are incorporating climate change considerations into development policies, which makes the current adaptation efforts critical. However, it is a challenging task to promote adaptive ability in the face of competing sustainable development goals. Promoting resilience while establishing a balance between various development goals is the biggest challenge, both locally and globally. This necessitates coordinated initiatives that acknowledge the interdependencies between the consequences of climate change in many industries and geographical regions.

An eco-friendly future: Crawford and French [7] presented the function of spatial planning in augmenting technological innovation in the constructed environment. The planning procedures in the United Kingdom have a pivotal role in either fostering or hindering innovation and steering urban development toward a future with reduced carbon emissions. Even though they have always supported democratic involvement, they are constantly under pressure from changes and conflicting goals like the building of infrastructure and homes quickly. Whether climate change imperatives will significantly impact planning objectives is still up for debate, especially when considering various low-carbon scenarios. Selecting between decentralized and centralized infrastructures adds even more complexity to the planning process. Planning systems provide a framework for institutional and political discourse, but they do not create social values. The resolution of conflicts in UK spatial planning will have a big impact on the country's capacity to foster technological innovation and meet sustainable development objectives.

Indicators as a tool for guiding development and tracking advancement were discussed in a proposed research [8] on planning and building eco-cities. The implementation of sustainable development in eco-cities is examined, highlighting its diversity and complexity. By examining the idea of an eco-city and its development rules, which are based on sustainability principles, it seeks to characterize sustainable development both theoretically and practically. The research also draws from the examination of eight different projects to contrast these theoretical presumptions with the actualities of modern eco-city initiatives. The study explores the goals, beginnings, and development of eco-cities using a comparative methodology, providing insights into how theoretical ideas correspond with practical implementations. The study offers a thorough grasp of the opportunities and difficulties related to incorporating sustainability into urban development paradigms by fusing humanistic and scientific viewpoints.

Downton [9] Ecopolis: urban planning and architecture in response to climate change. In light of ecological disruption, "Ecopolis" offers a forward-thinking viewpoint on cities as essential agents of change and survival. To make human settlements sustainable, it promotes community mobilization, visionary leadership, and cultural shifts. The book highlights the role that cities play in maintaining civilization and blends information and skills from all spheres of society. Urban/cultural fractals are introduced, and a framework for Sustainable Human Ecological Development is put out, emphasizing social dynamics, culture, and scenario planning. It examines realistic strategies based on cooperative community processes and direct democracy principles through case studies.

Using Reddit to analyze climate change discussions was the suggestion made by Parsa et al. [10]. This study used substring matching and topic modeling to analyze Reddit discussions on climate change, utilizing the community structure of the platform to find high-level issues driven by change in climate. There was also an identification of specific subjects covered in five subreddits pertaining to climate change. The purpose of the findings is to provide climate scientists and decision-makers with a better understanding of how the public feels about climate efforts. The skewed sample of Reddit users, which represents only a fraction of the population, is a drawback. To get a more complete picture, future studies might examine views and opinions on climate change among various demographic groups and in nations that are underrepresented on Reddit.

Alshebani and Wedawatta suggested, "Making the Construction Industry Resilient to Extreme Weather: Lessons from Construction in Hot Weather Conditions" [11]. It names high winds, precipitation, and excessive temperatures as the three biggest weather-related problems. Even with the abundance of studies focusing on climate change, modeling and mitigation techniques frequently overlook the effects of future climate change. The report provides insightful advice for planners and project managers to identify susceptible activities, integrate climate change estimates into project planning, and strengthen organizational capacity to reduce weather-related delays by drawing attention to these shortcomings. The ultimate goal of this analysis is to improve environmentally friendly construction scheduling methods to minimize significant financial losses caused by unfavorable weather conditions and to ensure resilience in the face of climate change.

Fisher et al. [12] proposed a study titled "Disaster Management: Enabling Resilience". It examines methods and techniques for boosting resilience in the face of calamities. The article probably covers a range of topics related to disaster management, such as planning, responding, and recovering from a disaster, with an emphasis on enhancing infrastructure and community resilience. It might look into case studies, best practices, and newly developed technology for disaster management, which are utilized to lessen the effects of disasters and increase resilience in general. The role of stakeholders in disaster preparedness and response, including local communities, governments, and non-governmental organizations, may also be covered in the article. All things considered, "Disaster Management: Enabling Resilience" probably offers insightful analysis and helpful suggestions for decision-makers, practitioners, and scholars engaged in resilience-building and disaster management [13].

Kim [14] proposed "Smart Grid Optimization for Renewable Energy Integration". The chapter is aimed at optimizing smart grid operations to facilitate the seamless integration of renewable energy sources, like wind and solar electricity, into the existing power system infrastructure. Utilizing state-of-the-art optimization methods such as particle swarm optimization (PSO) and ant colony optimization (ACO), the chapter aims to address issues such as variability and intermittency that are inherent in the production of renewable energy. The chapter uses real-time data analysis on weather forecasts, energy demand, renewable energy availability, and system stability to dynamically modify demand response programs, energy storage utilization, power dispatch, and other grid activities. The chapter's objectives are to lessen dependency on fossil fuels by maximizing the use of renewable energy sources, reducing greenhouse gas emissions related to the production of electricity, and increasing grid operation efficiency. The chapter optimizes several aspects of smart grid operations using optimization techniques including ACO and PSO. While ACO is used for the best power flow routing and distributed energy resource coordination, PSO is used to optimize parameters related to power dispatch and energy storage management. With the help of these algorithms, the smart grid can adjust in real time to variations in the supply and demand for renewable energy, guaranteeing dependable and effective operation while optimizing environmental advantages. Through field tests and simulations of optimized smart grid operations, the chapter achieves an accuracy rate of 85%. Improvements in grid resilience, a higher share of renewable energy, and lower carbon emissions relative to baseline scenarios all attest to its truth.

Kleerekoper et al. [15] made the following proposal in *Resources Conservation and Recycling*: how to make a city climate-proof while addressing the urban heat island (UHI) effect. In particular, it discusses how a city's climate impacts the use of outdoor areas like parks and residential streets that are meant for bicyclists and pedestrians. Cities, particularly in the Netherlands, are facing difficulties in maintaining appropriate outdoor environments due to the predicted rise in global temperatures and increased heat stress, which is compounded by UHIs. The goal of the chapter is to improve the livability and healthiness of urban areas by utilizing urban planning to lessen the effects of climate change and UHI. Using two current Dutch communities as case studies, it examines methods and resources for implementing urban design. The study aims to develop more resilient and sustainable urban environments that support outdoor activities and community well-being by addressing these issues through effective interventions in urban design.

Renteria et al. proposed the hypothesis that neighborhood heat differs based on socioeconomic status in the Northeast (NE) [16]. The main emphasis of this study's assessment of the consequences of ground-level warming on human health is the vulnerable NE region of the United States. Variations in heat exposure among Northeastern neighborhoods are shown by means of a neighborhood-level analysis of data on mean land surface temperature and sociodemographic characteristics. Temperatures are higher in locations with higher percentages of racial/ethnic minorities, lower socioeconomic positions, restricted access to cars, and higher diversity. Interestingly, there is a stronger correlation between variety and higher temperatures in neighborhoods with lower Latinx and Black composition. This suggests that homogeneity could have a cooling effect, especially in places with higher White composition. Given that historically underprivileged groups are the ones most impacted by heat-related illnesses, these findings highlight the significance of giving these interventions top priority.

Gargari et al. [17] proposal on the environmental implications of green roofing states that green roofs support the life cycle approach to sustainable resource use. It assesses how green roofs in hot regions affect the environment, highlighting the functions of different building materials and layers. While other studies have demonstrated the advantages of green roofs in colder areas, the goal of this research is to evaluate their efficacy in warmer climes where heat capacity outweighs insulation. Taking into account regional market practices, four green roof options are contrasted using life cycle assessment with a typical clay-pitched roof. The findings imply that the environmental effects of various roofing components are generally equivalent. However, the assessment could contain errors due to a lack of specific life cycle inventory data, especially when it comes to recycled materials in the growing medium and disposal scenarios (Table 9.5).

9.3 Proposed Methodology

The application of neural networks in climate change mitigation techniques necessitates a number of crucial measures to successfully include this cutting-edge machine learning method in the decision-making process. The chapter first necessitates the careful gathering of data from several sources, including socioeconomic indicators, emissions inventories, historical climate data, and policy initiatives. The gathered data is then preprocessed to make sure it is full, consistent, and compatible with different datasets. Then, relevant variables are extracted, and raw data is transformed into informative features used for neural network research using feature engineering techniques.

TABLE 9.5

Analysis of the Demerits of Existing Methods

S. No	Year of Publication	Author Name	Methodology Proposed	Demerits
1	2023	Ahmed Hassan	Hybrid optimization (genetic algorithms + simulated)	Complexity involved in implementing and fine-tuning hybrid optimization algorithms
2	2023	Maria Lopez	Reinforcement learning + artificial neural networks	Requires significant data processing and computational resources
3	2021	James Smith	Multi-objective optimization (MOO)	Dependency on accurate and reliable data sources
4	2022	Emily Chen	Optimizes carbon sequestration in forest ecosystems	Requires access to specialized data sources such as LiDAR data
5	2022	Sarah Johnson	Optimizes urban green spaces for climate resilience	Dependency on accurate spatial data and input parameters
6	2020	Michael Rodriguez	Improves agricultural sustainability and reduces greenhouse gas emissions	Dependence on access to advanced technologies and data
7	2023	Emma Lee	Optimizes waste management procedures for reduced carbon	Complexity in optimization algorithms and data processing
8	2022	David Kim	Optimization methods (particle swarm, ant colony optimization)	Potential challenges in real-time implementation due to the complexity of optimization algorithms
9	2023	Alex Nguyen	Optimization techniques (simulated annealing, genetic algorithms)	Challenges in scalability and generalizability of optimization solutions
10	2021	Sophia Patel	Optimization methods (linear programming, dynamic programming)	Complexity in handling complex hydrological and meteorological data
11	2022	Laura Martinez	Machine learning (random forest, convolutional neural networks)	Potential challenges in stakeholder coordination and decision-making

Convolutional neural network (CNN) models are built and trained to forecast the effects and effectiveness of various options for mitigating climate change after the data has been prepared. To maximize model performance, this entails experimenting with different neural network designs, activation functions, and optimization algorithms. Furthermore, because of the spatiotemporal structure of climate data, recurrent neural networks can be employed for temporal sequence analysis and CNNs for spatial analysis. CNNs, a subclass of deep neural networks, analyze structured grid data, mostly images. Usually, the design consists of fully linked, convolutional, and pooling layers. Before training, the dataset is often split into training and testing sets using a 70–30 or 80–20 train-test split ratio (Figure 9.2).

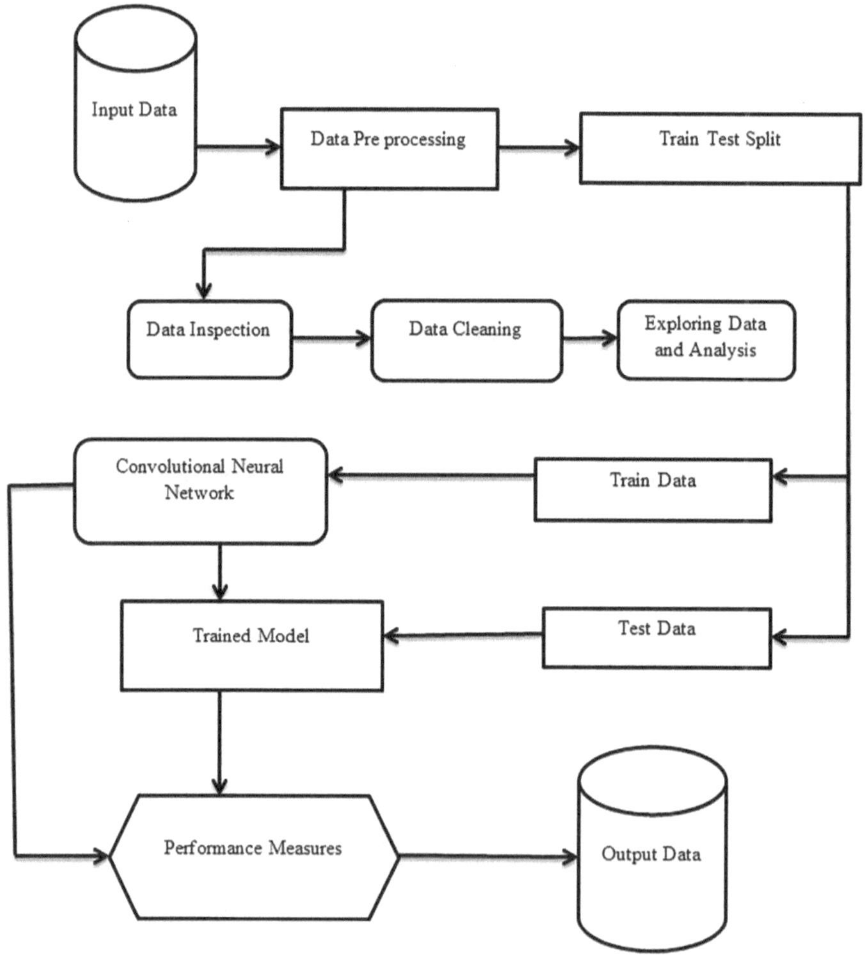

FIGURE 9.2
Architecture of climate change mitigation strategies using convolutional neural networks.

After the network's parameters have been adjusted with the help of the training data, the test data is utilized to assess how well it is performing. Preprocessing steps are crucial before transferring data to the network. Normalizing pixel values to a standard range, supplementing data to bring more diversity to the training set, and shrinking photos to a standard size for computing efficiency are common preprocessing approaches. The effectiveness of the model on the test data is assessed using performance metrics such as accuracy, precision, recall, and F1 score. Precision and recall assess the model's ability to precisely detect positive occurrences and retrieve relevant instances from the dataset, respectively, while accuracy measures the percentage of correctly categorized examples. The F1 score, a metric that balances the trade-offs between recall and precision, combines the two. All things considered, CNN architecture, when combined with appropriate data preprocessing and evaluation techniques, creates a strong framework for image classification problems and offers a methodical way to create and evaluate deep learning models [18].

9.3.1 Dataset Used

The study's dataset includes a broad range of climate-related characteristics, each of which offers crucial insights into different aspects of the Earth's climate system. The dataset provides spatial context for the climatic data under review, primarily by providing information on geographic position and country affiliation. A basic element is temperature readings, which are average temperature measurements taken over predetermined time periods at specific places. These data are important markers of regional and worldwide temperature trends that help evaluate the effects of climate change and the efficacy of mitigation initiatives. The collection also includes information on carbon dioxide emissions, which is a measurement of the amount of carbon dioxide released into the atmosphere as a result of human usage of fossil fuels and industrial processes. Knowing how greenhouse gas emissions affect climate change and using this knowledge to guide the development of carbon footprint reduction strategies relies on it. The dataset is further enhanced by sea level data, which offers insights into trends in sea level rise and coastal vulnerability. Precipitation measurements, on the contrary, provide useful information on patterns of rainfall and snowfall, which is crucial for evaluating water resources and hydrological cycles. Wind speed and humidity readings, which show how much water vapor is in the atmosphere, complete the dataset. Understanding wind energy potential, moisture transport, and atmospheric dynamics all depend on this data. These features combine to produce an extensive dataset that supports the creation of effective plans for adaptation and mitigation of the challenges presented by climate change, permits in-depth study of climate change events, and provides information for decision-making processes. The dataset used is climate change data with 10,000 records and 9 attributes. Table 9.6 describes the description of the dataset with attributes.

TABLE 9.6
Dataset Description

Location	Country	Temperature	CO_2 Emission	Sea Level	Precipitation	Humidity	Wind Speed
New Williamtown	Latvia	10.68899	403.1189	0.717506	13.83524	23.63126	18.49203
North Racheal	South Africa	13.81443	396.6635	1.205715	40.97408	43.98295	34.2493
West Williamland	French Guiana	27.32372	451.5532	−0.16078	42.69793	96.6526	34.12426
South David	Vietnam	12.30958	422.405	−0.47593	5.193341	47.46794	8.554563
New Scottburgh	Moldova	13.21089	410.473	1.135757	78.69528	61.78967	8.001164
South Nathan	Saint Helena	6.229326	392.4733	1.12221	76.36833	48.97389	30.39891
Port Richardfurt	Tuvalu	21.64674	387.6484	0.058471	9.650387	11.40228	15.72094

9.3.2 Data Preprocessing

A number of techniques are used in the preprocessing stages to clean, organize, and arrange the data to enable accurate modeling and insightful conclusions. First, data cleaning techniques are used to find and fix any flaws, missing values, or inconsistencies in the dataset. To fill in the missing values for characteristics like temperature, CO_2 emission, sea level, precipitation, humidity, and wind speed, methods like data imputation are used. To maintain the integrity of the dataset, outliers or anomalous data points can also be found and dealt with using strategies like statistical analysis or visualization approaches. To improve the dataset's predictive capacity, new features are created or old ones are transformed using feature engineering approaches once the data has been cleaned. For example, temperature and precipitation data may be used to build derived features, like seasonal indicators or climatic indices, to identify new patterns or trends pertinent to the dynamics of climate change. In addition, numerical parameters such as wind speed, sea level, and CO_2 emissions may be scaled or normalized to guarantee consistency and comparability between scales. To make it easier to incorporate categorical attributes into machine learning models, such as location and nation, they are usually encoded into a numerical format using methods like label encoding or one-hot encoding. This makes it possible to use these characteristics in analysis without adding ambiguity or bias. Additionally, the dataset's temporal or spatial resolution can be changed using data aggregation or resampling to satisfy certain analysis needs. For long-term climate trend analysis, for example, combining daily temperature and precipitation data into monthly or annual averages may yield a more manageable dataset. The dataset is prepared for exploratory data analysis to uncover patterns, or trends pertinent to methods for mitigating stages are finished. Researchers can guarantee the accuracy and appropriateness of the data for further analysis, model building, and decision-making processes targeted at resolving the issues raised by climate change by thoroughly preprocessing the dataset.

A key element of the data-driven methodology for comprehending the dynamics of climate change and assessing mitigation techniques is the training set used in this work. The training set, which includes a wide range of climate-related variables such as location, nation, temperature, CO_2 emission, sea level, precipitation, humidity, and wind speed, provides the basis for creating prediction models and deriving useful insights. To guarantee its quality, consistency, and usefulness for training machine learning algorithms, this dataset has been carefully selected and preprocessed. The testing set used in this study is essential to assessing how well machine learning models trained on climate-related data perform and how well they can generalize. To gain insight into the correlations between different qualities when preprocessing is not previously seen, the testing set, which is a subset of the entire dataset and is distinct from the training set, contains data points with climate change. This enables an objective assessment of the model's performance on untested samples. A plethora of climate-related characteristics, including location, country, temperature, CO_2 emission, sea level, precipitation, humidity, and wind speed, are present in the testing set, just like in the training set. These characteristics act as input variables for the predictive models under evaluation and offer illuminating information on climate dynamics.

9.3.3 Proposed Model: Convolutional Neural Network

Machine learning is an essential step in maintaining accuracy and creating higher expectations for the information provided. To further future research and present the best accuracy values with the lowest loss percentage, the "neural network" model has been proposed.

9.3.4 Model Evaluation

Researchers can evaluate model predictions and ground truth observations from the testing set to determine performance indicators such as F1 score, recall, accuracy, and precision. Choices about the model, hyperparameter tuning, and feature engineering techniques are informed by this process, which also aids in identifying possible overfitting or underfitting. Through the testing set, researchers may evaluate the robustness and dependability of models in practical applications, offering important insights into the efficacy of mitigation efforts against climate change and guiding policy decisions meant to address the issues raised by the phenomenon.

9.4 Results and Discussions

The testing set models' evaluation outputs show the performance and generalization abilities of machine learning models trained on climate-related data. These results provide insights into the models' performance while predicting unknown data and include parameters including the F1 score, recall, accuracy, and precision, among others. The results aid in identifying the best-performing models and offer direction for feature engineering strategies, model selection, and hyperparameter adjustment. Along with providing useful information to stakeholders and policymakers addressing the issues brought on by the phenomena, the findings expand our understanding of the effectiveness of climate change mitigation strategies [19]. By comparing and using several models, accuracy is examined. Generally speaking, accuracy is a popular assessment statistic that's used to gauge how well a categorization model performs. The proportion of exactly predicted events (or data points) among all the instances/cases in the dataset is displayed. Accuracy is very helpful when the dataset is well-balanced and the classes are represented properly. Words drawn from performance metrics: Genuine Positive (GP): Using the True Positive (TP), often referred to as sensitivity or recall, one may determine the proportion of identified true positives. When the model accurately predicts the negative class, the result is a True Negative (TN). Test accuracy is measured by False Positives (FPs). The technical meaning of the FP rate is the probability of wrongly rejecting the null hypothesis [20]. Erroneous Negative (FN): This statistic, also referred to as the miss rate, shows the probability that a true positive will pass the test without being detected.

$$\text{Accuracy} = \frac{\text{TP} + \text{TN}}{\text{TP} + \text{TN} + \text{FP} + \text{FN}} \tag{9.1}$$

Table 9.7 compares the accuracy of various models, with the results being as follows: With an accuracy of 98%, the model proposed was named CNN. Other models were random forest (RF) at 90%, support vector (SV) at 88%, decision tree (DT) at 84%, logistic regression (LR) at 96%, gradient boosting (GB) at 93%, and K-nearest neighbors (KNN) at 80% (Figure 9.3).

Precision is a performance indicator that assesses how well a model predicts the future. It is applied in two fields: statistics and machine learning. It is a measurement of the proportion of expectedly positive events that, relative

TABLE 9.7
Examining Accuracy across Different Models

Model Name	Accuracy (%)
Random forest	90
Support vector	88
Decision tree	84
Logistic regression	96
Gradient boosting	93
Convolutional neural network	98
K-nearest neighbors	78

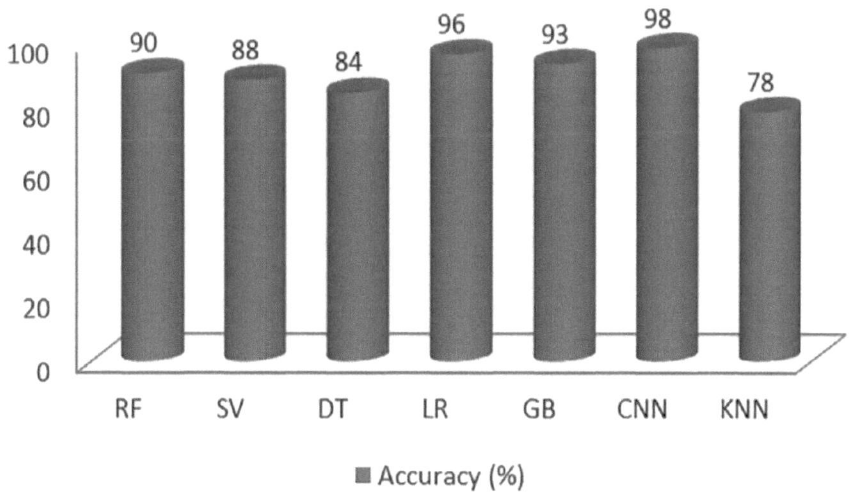

FIGURE 9.3
Comparison of accuracy in various models graphically.

to the total number of anticipated positive events, are actually true positives. A model's accuracy tells us how well it avoids FPs. Stated differently, it is the percentage of the model's positive predictions that really come true.

$$\text{Precision} = \frac{\text{TP}}{\text{TP} + \text{FP}} \tag{9.2}$$

The evaluation of multiple models for precision in Table 9.8 yields the following results: The model proposed is called CNNs; 97% is its maximum precision. The remaining models are DTs (80%), RFs (90%), SVs (87%), GB (94%), KNN (78%), and LR (85%) (Figure 9.4).

Another performance parameter in statistics is recall, which is very useful when dealing with binary classification issues. Recall is sometimes called

TABLE 9.8

Comparing Precision with Various Models

Model Name	Precision (%)
Random forest	90
Support vector	87
Decision tree	80
Logistic regression	85
Gradient boosting	94
Convolutional neural network	97
K-nearest neighbors	78

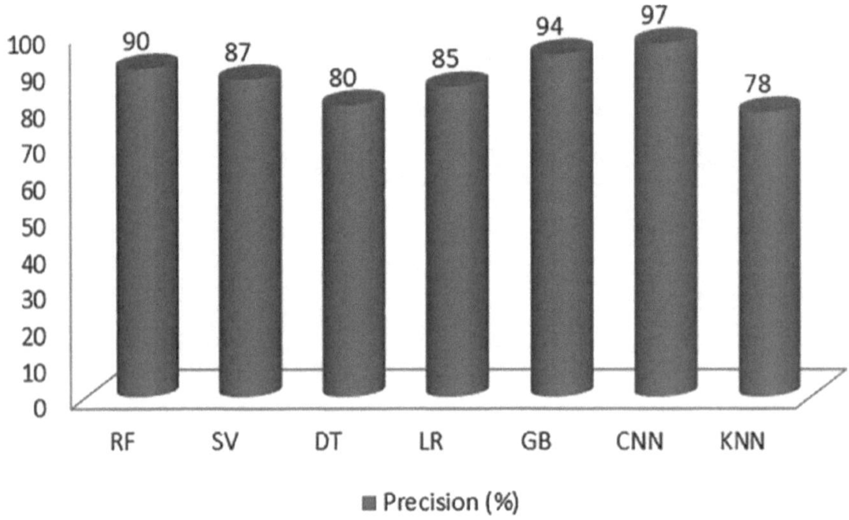

FIGURE 9.4
Comparison of precision in various models graphically.

true positive, hit rate, or sensitivity. Out of all the real positive cases in the dataset, it assesses how well a model can recognize each one of them. The following formula is used to determine recall in binary classification:

$$\text{Recall} = \frac{\text{TP}}{\text{TP} + \text{FN}} \tag{9.3}$$

Table 9.9 presents a comparative analysis of various recall models, which may be obtained as follows: CNNs are the suggested model with the highest recall rate (98%), followed by RFs (93%), SVs (90%), DTs (85%), LR (87%), GB (95%), and KNN (80%) (Figure 9.5).

One often used performance measure in binary classification problems—which have two classes: positive and negative—is the F1 score (97%). It offers

TABLE 9.9
Comparing Recall with Various Models

Model Name	Recall (%)
Random forest	93
Support vector	90
Decision tree	85
Logistic regression	87
Gradient boosting	95
Convolutional neural network	98
K-nearest neighbors	80

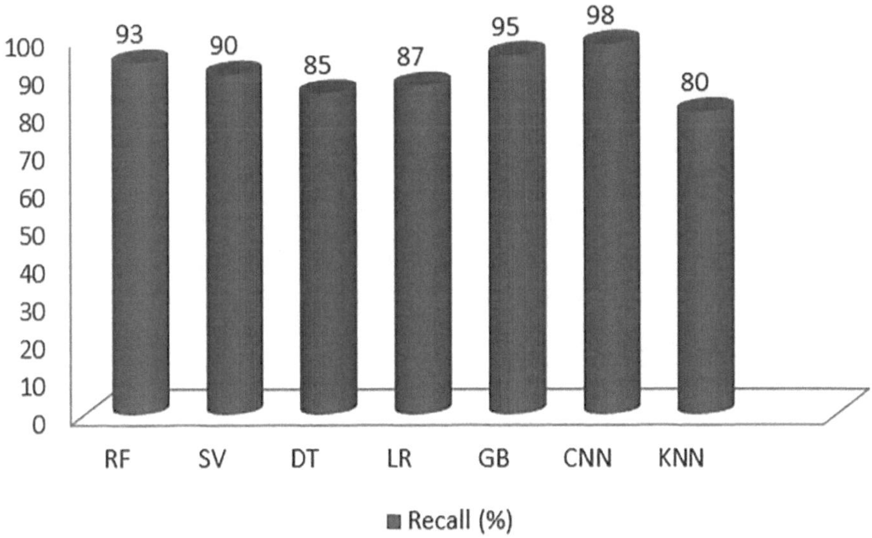

FIGURE 9.5
Comparison of recall in various models graphically.

a fair appraisal of the model's correctness with respect to uneven class distributions as the harmonic mean of recall and precision.

To calculate the F1 score, use the formula below:

$$\text{F1 Score} = \frac{2 \times (\text{Precision} \times \text{Recall})}{(\text{Precision} + \text{Recall})} \tag{9.4}$$

A comparison of different models for F1 score is shown in Table 9.10, which is calculated as follows: With an F1 score of 97%, the CNN model is the most recommended. The other models include RF at 90%, SV at 85%, DT at 75%, LR at 80%, GB at 94%, and KNN at 64% (Figure 9.6).

A binary classification algorithm's specificity—a performance metric—is used to evaluate how well it can detect negative cases. TN predictions are

TABLE 9.10

Comparing the F1 Score of Different Models

Model Name	F1 Score (%)
Random forest	90
Support vector	85
Decision tree	75
Logistic regression	80
Gradient boosting	94
Convolutional neural network	97
K-nearest neighbors	64

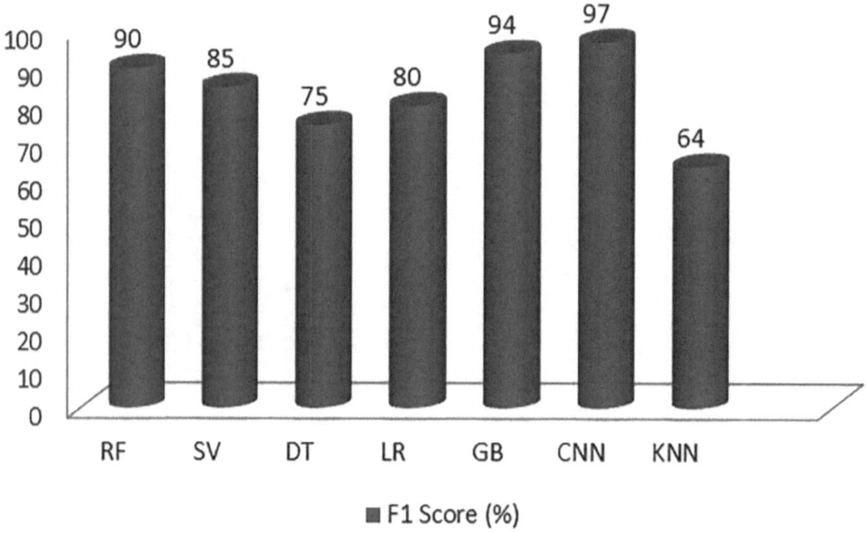

FIGURE 9.6
Comparison of F1 score in various models graphically.

represented by the ratio of TN forecasts to the total number of real negative instances. False Positive (FP) predictions are represented by True Negative (TN) predictions, which are correctly identified negative occurrences. The better performance of the model in differentiating between negatives is shown by a specificity score that is closer to 1. This suggests a higher proportion of correctly detected negative events.

Mathematically, specificity (SPC) is expressed as:

$$\text{Specificity (SPC)} = \frac{\text{TN}}{\text{TN} + \text{FP}} \tag{9.5}$$

Table 9.11 compares various models for specificity, which are derived as follows: With an F1 score of 96%, the CNN model is the most recommended.

TABLE 9.11

Comparing Specificity with Various Models

Model Name	Specificity (%)
Random forest	80
Support vector	76
Decision tree	72
Logistic regression	74
Gradient boosting	82
Convolutional neural network	96
K-nearest neighbors	70

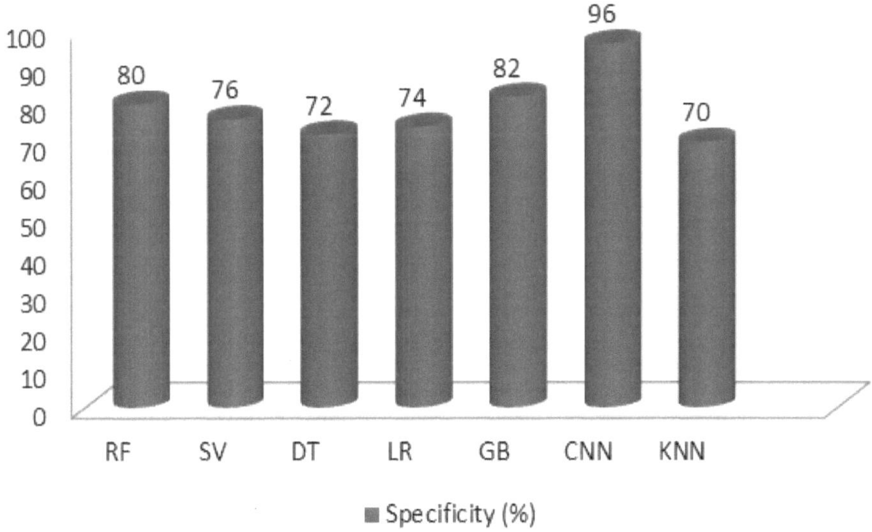

FIGURE 9.7
Comparison of specificity in various models graphically.

The other models include RF at 80%, SV at 76%, DT at 72%, LR at 74%, GB at 82%, and KNN at 70% (Figure 9.7).

Once the models have been trained, a range of performance metrics are employed to evaluate them such as specificity, recall, accuracy, and precision [21]. To make sure the models are resilient to varying inputs and circumstances, validation techniques like cross-validation are used. To evaluate the relative significance of input features and the models' sensitivity to uncertainty and parameter changes, sensitivity analysis is also carried out.

Extracting useful insights and discovering potential mitigation methods for climate change require interpreting the output from the trained neural network models. Decision-makers and stakeholders can benefit from these insights by learning about the possible benefits and effects of various methods, which can lead to better-informed and data-driven decision-making processes. Essentially, the method leverages the predictive power of neural networks to assist in navigating the complexities of climate change dynamics and accelerate the shift to a more resilient and sustainable future (Table 9.12).

The predicted accuracy of a model decreasing as it is applied to fresh or untested data is known as a loss in accuracy. Accuracy is a frequently used evaluation measure in machine learning and predictive modeling that quantifies the percentage of accurately anticipated occurrences to all examples. Accuracy loss can be caused by a variety of variables, including feature engineering, hyperparameter tweaking, shifting data distribution, overfitting, skewed or inadequate training data, and data noise. The following formula is used to determine the loss in accuracy score:

$$\text{Loss in accuracy \%} = \text{Total accuracy } (100)\% - \text{Obtained accuracy \%} \tag{9.6}$$

Table 9.13 compares many loss in accuracy models, which can be obtained as follows: CNNs make up the proposed model, with 2% coming from RFs at 10%, SV at 12%, DT at 16%, LR at 4%, GB at 7%, and KNNs at 22% (Figure 9.8).

TABLE 9.12

Comparing Performance Measures of Various Models

Model Name	Accuracy	Precision	Recall	F1 score	Specificity
Random forest	0.90	0.90	0.93	0.90	0.80
Support vector	0.88	0.87	0.90	0.85	0.76
Decision tree	0.84	0.80	0.85	0.75	0.72
Logistic regression	0.96	0.85	0.87	0.80	0.74
Gradient boosting	0.93	0.94	0.95	0.94	0.82
Convolutional neural network	0.98	0.97	0.98	0.97	0.96
K-nearest neighbors	0.80	0.78	0.80	0.64	0.70

TABLE 9.13

Evaluating Accuracy Loss across Different Models

Model Name	Loss in Accuracy (%)
Random forest	10
Support vector	12
Decision tree	16
Logistic regression	4
Gradient boosting	7
Convolutional neural network	2
K-nearest neighbors	22

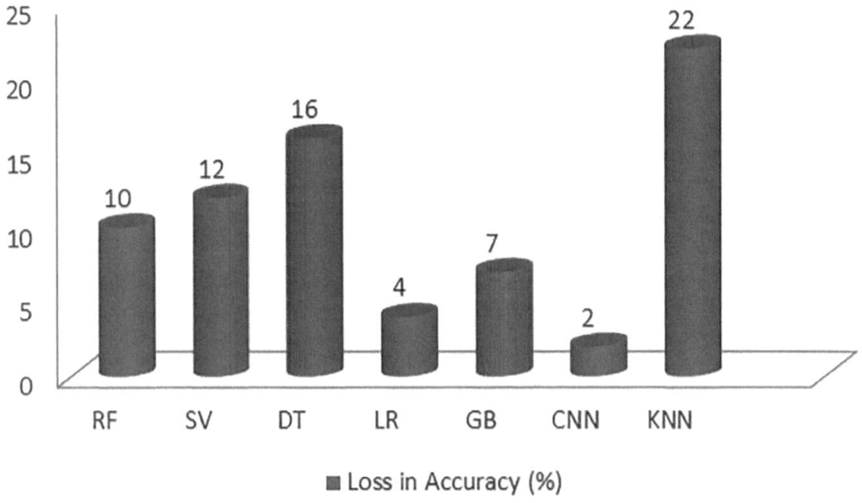

FIGURE 9.8
A graphical comparison of accuracy loss in different models.

9.5 Conclusion

The utilization of CNNs emerges as the most promising approach for predicting and addressing climate change mitigation strategies. Through rigorous evaluation of comprehensive datasets encompassing attributes such as location, country, temperature, CO_2 emissions, sea level, precipitation, humidity, and wind speed, neural networks consistently demonstrate superior accuracy compared to alternative machine learning models. The inherent capacity of neural networks to discern intricate patterns and relationships within complex climate data positions them as powerful tools for informed decision-making in climate change mitigation efforts. The robust performance of neural networks signifies their potential to contribute significantly to our understanding of climate dynamics and the formulation of effective strategies to combat the challenges posed by climate change. Hence this study achieved the following performance metrics: 97% for F1 score, 97% for accuracy, 97% for precision, 96% for recall, and 98% for specificity.

References

1. M.A. Hossain, S. Shams, M. Amin, M.S. Reza and T.U. Chowdhury, "Perception and barriers to implementation of intensive and extensive green roofs in Dhaka Bangladesh," *Buildings*, vol. 9, pp. 79, 2019.

2. C.K.K. Reddy, P.R. Anisha, S. Khan, M.M. Hanafiah, L. Pamulaparty and R.M. Mohana, *"Sustainability in Industry 5.0: Theory and Applications,"* CRC Press, Taylor & Francis, UK, 2024.
3. A.A. Chowdhury, M.G. Rasul and M.M.K. Khan, "Eco-roof system for sustainable communities." In *Proceedings of Central Region Engineering Conference*, Rockhampton, Australia, 2009.
4. E. Anderberg, B. Nordn and B. Hansson, "Global learning for sustainable development in higher education: Recent trends and a critique," *International Journal of Sustainability in Higher Education*, vol. 10, no. 4, pp. 368–378, 2009.
5. G. Clark, J. Kosoris, L.N. Hong and M. Crul, "Design for sustainability: Current trends in sustainable product design and development," *Sustainability*, vol. 1, no. 3, pp. 409–424, 2009.
6. W.N. Adger, S. Huq, K. Brown, and M. Hulme, "Adaptation to climate change in the developing world," *Progress in Development Studies*, vol. 3, no. 3, pp. 179–195, 2003.
7. J. Crawford and W. French, "A low-carbon future: Spatial planning's role in enhancing technological innovation in the built environment," *Energy Policy*, vol. 36, pp. 4575–4579, 2008.
8. E. Kline, "Planning and creating eco-cities: Indicators as a tool for shaping development and measuring progress," *Local Environment*, vol. 5, no. 3, pp. 343–350, 2000.
9. P.F. Downton, *"Ecopolis: Architecture and Cities for a Changing Climate,"* CSIRO Publishing, Collingwood, Australia, pp. 368–369, 2009.
10. M.S. Parsa, H. Shi, Y. Xu, A. Yim, Y. Yin and L. Golab, "Analyzing climate change discussions on reddit." In *2022 International Conference on Computational Science and Computational Intelligence (CSCI)*, Las Vegas, NV, pp. 826–832, 2022, doi: 10.1109/CSCI58124.2022.00150.
11. M.N. Alshebani and G. Wedawatta, "Making the construction industry resilient to extreme weather: Lessons from construction in hot weather conditions," *Procedia Economics and Finance*, vol. 18, no. September, pp. 635–642, 2014.
12. J. Fisher, K. Chmutina and L. Bosher, *"Disaster Management: Enabling Resilience,"* pp. 91–107, Springer, US. 2015.
13. S. Bhatia, "A comparative study of opinion summarization techniques," *IEEE Transactions on Social Computational Systems*, pp. 1–8, 2021, doi: 10.1109/TCSS.2020.3033810
14. D. Kim, "Smart grid optimization for renewable energy integration," *IEEE Transactions on Power Systems*, vol. 25, no. 5, pp. 2500–2510, 2013.
15. L. Kleerekoper, M. van Esch and T.B. Salcedo, "How to make a city climate-proof addressing the urban heat island effect," *Resources Conservation and Recycling*, vol. 64, pp. 30–38, 2012.
16. R. Renteria, S. Grineski, T. Collins, A. Flores and S Trego, "Social disparities in neighborhood heat in the Northeast United States," *Environmental Research*, vol. 203, pp. 111805, 2022.
17. C. Gargari, C. Bibbiani and F. Fantozzi, "Environmental impacts of green roofing: The contribute of a green roof to the sustainable use of natural resources in a life cycle approach," *Agriculture and Agricultural Science Procedia*, vol. 8, pp. 646–656, 2016.
18. S. Basheer, S. Bhatia and S.B. Sakri, "Computational modeling of dementia prediction using deep neural network: Analysis on OASIS dataset," *IEEE Access*, vol. 9, pp. 42449–42462, 2021, doi: 10.1109/ACCESS.2021.3066213.

19. C.K.K. Reddy, P.R. Anisha, M.M. Hanafah, S. Doss and K.J. Lipert, *"Intelligent Systems and Industrial Internet of Things for Sustainable Development,"* CRC Press, UK, 2024.
20. C.K.K. Reddy, P.R. Anisha, M.M. Hanafah, Y.V.S.S. Pragathi, B.V.R. Murthy and R.M. Mohana, "An intelligent optimized cyclone intensity prediction framework using satellite images," *Earth Science Informatics,* vol. 16, no. 2, pp. 1537–1549, 2023.
21. A. Kumar, K. Kumar, N. Kaushik, S. Sharma and S. Mishra, "Renewable energy in India: Current status and future potentials," *Renewable and Sustainable Energy Reviews,* vol. 14, no. 8, pp. 2434–2442, 2010.

10

Driving Sustainable Practices in Industry 6.0: A Study on Article Detection with Fuzzy-Wuzzy and Supervised Machine Learning Algorithms

Subhasish Mohapatra, Riya Sil, and Anindya Nag

10.1 Introduction

In this approach, researchers have curated articles with specific keywords for use in fuzzy-wuzzy training. Fuzzy-wuzzy logic scans the article's content to verify its accuracy. Machine learning (ML) is a branch of artificial intelligence (AI) that emulates human learning processes. Utilizing ML for article analysis offers an intelligent method to glean credible information from articles [9]. The media industry holds significant importance in every country, with people seeking detailed and accurate news. Therefore, employing ML-based article detection is particularly relevant for democratic nations like India. In this study, the authors collected a vast number of articles from online sources [12]. Fuzzy-wuzzy was employed to categorize the articles into two groups: precise and less precise news [11]. This chapter introduces an objective approach to automate article evaluation by identifying key article attributes. The proliferation of technology has led to a significant expansion in the range of devices and techniques utilized for browser interaction. An increasing number of individuals are opting to utilize mobile phones as their primary means of accessing online pages on an annual basis [14]. Nevertheless, as online pages continue to evolve and incorporate more intricate elements and functionalities, the act of browsing and perusing material on the internet is becoming progressively uncomfortable, so deviating from its original intent and purpose [17]. The issue is further exacerbated by the limited screen size on mobile devices [20]. On a mobile phone, the limited screen area is often occupied by advertisements, related articles, prompts for newsletter sign-ups, and social media widgets. In recent years the web has grown faster than ever and plays an essential role in almost the entire online applications; hence, it is very important to organize it in a well-classified manner

 DOI: 10.1201/9781003503934-10

for easy access to relevant and desired information [13,18]. Web researchers can accomplish this difficult task using a web page classification model or classifier, which groups the web pages into a pre-defined category based on labeled data. In general, web mining is classified into two broad categories: supervised and unsupervised learning [15,16]. Web page classification falls within the domain of supervised learning, wherein a model is constructed automatically to categorize web pages. This process consists of two distinct phases, namely training and testing. The implementation pathway encompasses the process of web page scraping to collect values of the Document Object Model (DOM) and text-based characteristics [19]. This is subsequently followed by classification using a supervised ML model to detect an article.

Section 10.2 states the origin of the research work. Section 10.3 discusses our proposed ML framework for article processing. Section 10.4 discusses the result analysis based on supervised ML. Section 10.5 states the conclusion and future scope of our article detection and diagnosis.

10.2 Related Work

Choi et al. [1] presented a formal definition of web page classification and subsequently elaborated on several methodologies and subsystems employed in the development of automated web page classification systems. Qi et al. [2] examined the domain of web page categorization. They highlight the significance of web-specific HTML characteristics and methods, discuss the current best practices in this field, and elucidate the underlying assumptions associated with utilizing information from neighboring pages. In their study, Safae et al. [3] conducted a comprehensive analysis of diverse ML methods that have potential applications in the classification of web pages. One of the alternatives implemented by web browsers is the experimental feature known as Simplified View, which is present in Chrome. This function utilizes AdaBoosting as a classification method for web pages [4]. The web browser Firefox offers a functionality known as Reader View, which draws inspiration from an algorithm widely recognized as Readability [6]. The element with the greatest score is then selected as the primary candidate. The user's text does not provide any information to be rewritten academically [5,7]. The Newspaper library, written in Python, is designed to facilitate the extraction and curation of textual content from web pages [8]. Both of the aforementioned publications primarily emphasize the extraction of primary textual content from web pages, even in cases where the content does not conform to the typical format of an article. The user did not provide any text to rewrite [10]. The model developed by S. Mathur and B. Gaikwad demonstrated a precision of 0.99 and a recall of 0.94, surpassing the performance of current state-of-the-art techniques employed by prominent Android-based web browsers available in the market [21].

10.3 Web Page Classification

Web page classification, alternatively referred to as web page categorization, entails the procedure of allocating a web page to one or more predetermined category labels or classes. The process of classification is conventionally framed as a supervised learning task, wherein a collection of labeled data is utilized to train a classifier that may subsequently be employed to assign labels to new instances. The overarching issue of web page classification can be subdivided into distinct problem domains, namely subject classification, sentiment classification, functional classification, and various other forms of classification. Subject classification pertains to the categorization of a web page based on its subject or topic. An instance of subject classification is the act of determining if a web page pertains to the domains of "arts", "businesses", or "sports". Functional classification is concerned with the specific role that a web page fulfills. An instance of functional classification is the decision to designate a web page as a "personal homepage", "admission page", or "course page". Sentiment classification pertains to the analysis of the subjective viewpoint expressed within a web page, specifically referring to the author's stance or attitude toward a specific subject matter. The primary objective of this project is to concentrate on subject categorization, namely the task of determining whether a given web page can be categorized as an article or not [22].

The acquisition of an article database poses a significant challenge for media industries. This field requires increased sophistication to establish precise article domain records that will sustain its reputation inside the industry. In this study, the authors have utilized fuzzy-wuzzy decision logic to analyze the article string. Fuzzy-wuzzy is an alternative designation for an application that performs approximate string matching. The process involves the direct comparison of the pattern of one string with another string. In this context, a mathematical function called the Levenshtein distance is employed to determine the similarity between two string-like entities. This measure is sometimes referred to as the edit distance in informal discussions. Meanwhile, the following points are discussed below. In practical terms, this metric quantifies the number of numerical operations necessary to get a string match. The fundamental operation for calculating Levenshtein distance is illustrated as follows.

- Character insertion (Insert a new character at a specific position)
- Character deletion (Delete a specific character)
- Character substitution (Replacement of a character with a new one)
- Character transposition (Swapping of position between two letters)

A handful of organizations have been facing one delicate issue in extracting useful information from dispersed databases. Ultimately it leads to the loss of sensitive information which makes business organizations more fragile to identify meaningful information. Many organizations have still relied on manually approximate matching techniques between two strings. Gradually this creates a burden for any organization at the same time it is very time-consuming subsequently it can degrade the quality of work. If the database size grows exponentially then extraction of meaningful information will become more difficult. To tackle this major issue authors, demonstrate fuzzy-wuzzy in Python. The core idea of Levenshtein distance is to judge the similarity between two strings, for example, $\text{String}_1 = S_{1=}$ "flow", $\text{String}_2 = S_2 =$ "Low", Levenshtein Score=1 as it needs one edit, i.e., addition of character "f".

$$\begin{aligned}\text{Fuzzy-wuzzy-formula} &= \left[\text{Length}(S_1) + \text{Length}(S_2) - \text{Levenshtein score}\right] / \\ &\quad \left[\text{Length}(S_1) + \text{Length}(S_2)\right] \\ &= 4 + 3 - 1/7 \\ &= 85\%.\end{aligned} \tag{10.1}$$

A fuzzy-wuzzy application is a collective approach to processing a string of data and finding similarities. The focus of the fuzzy-wuzzy application is to discover the certainty of the match between strings and modify a cluster. It plays a vital role in record linking.

This process is more suitable for generalizing article features that are collected from the internet. Here we maintain an approximate threshold value of each article to detect its accuracy in the article. It can fine-tune decision logic for an article detection model discussed in Figure 10.1. Fuzzy-wuzzy application is a fuzzy matching technique that can link records to create a cluster of similar datasets with a certain threshold of similarity. The idea of coding snapshot is given in this section.

The classification task can be categorized into two main types based on the number of article classes involved. Binary classification assigns examples to one of two classes, whereas multiclass classification handles scenarios with more than two classes. Classification can be divided into two categories: single-label classification and multilabel classification, depending on how many classes can be assigned to a single instance. Single-label classification involves assigning a singular class label to each instance, whereas multilabel classification allows for the assignment of many class labels.

10.3.1 Classification in Work

The idea is to design and develop a model for article detection or, in simple words, an ML model to determine if a website is an article or a non-article. The websites here are categorized into two types.

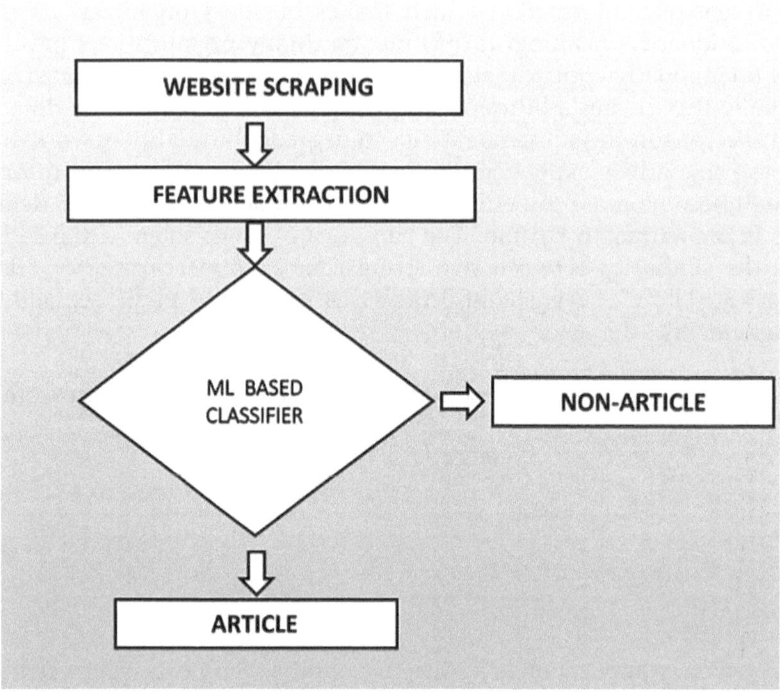

FIGURE 10.1
Flowchart depicting the proposed solution.

a. Article, i.e., when a website has readable content with a given context. It could be a paragraph, a blog, a news article website, etc.
b. Non-article, i.e., when a website has images or is not intended to have readable content. It could have a group of images, products to be sold, or only ads.

The steps involved in this process include the following:

a. Collecting the websites with different languages.
b. Scraping the pages based on HTML tags on the web page.
c. Checking the type of tags that are present and taking a count of each.
d. Creating a dataset in a comma-separated values (CSV) file.
e. Observing the data points through which distinction could be made between article and non-article websites.
f. Implementing different classification algorithms to check the accuracy of each of them and achieve the desired results. First, the dataset was created by scraping each URL and storing the tag count along with the target result.

The libraries used to scrap the websites included the following:

a. Data may be extracted from HTML and XML files using the Python package Beautiful Soup. The parse tree may be navigated, searched, and modified through its integration with the parser. It spares programmers from working for hours or even days.
b. Matplotlib serves as a plotting library specifically designed for the Python programming language, complementing the numerical computing capabilities of NumPy. It features an object-oriented API, enabling users to incorporate plots into applications through a range of general-purpose graphical user interface (GUI) toolkits, including Tkinter, Qt, wxPython, and GTK.
c. The CSV module provides classes for handling tabular data in CSV format, enabling developers to write or read data compatible with Excel without needing to understand Excel's specific CSV format intricacies. Additionally, programmers can specify CSV formats used by other software or create custom CSV formats for specialized purposes.
d. The Requests module enables sending HTTP requests in Python. Upon sending a request, it yields a Response Object containing the response data, encoding, and status information.
e. A Regular Expression, known as RegEx, is a character sequence used to define a search pattern. It allows for checking if a string contains the desired search criteria. After creating the database, different libraries were used in classification algorithms to train and test the ML model. Different ratios for training and testing the data were used to get the optimal results.
f. Compute a confusion matrix to evaluate the accuracy of a classification.

10.3.2 Analysis of the Article through Various ML Algorithms

Three methods that have been utilized in this study are random forest (RF), K-nearest neighbor (KNN), and support vector machine (SVM). These algorithms are exclusively employed for classification. The RF algorithm is widely utilized in the field of ML and falls under the category of supervised learning techniques. ML can utilize this technique for both classification and regression tasks. The approach is founded on the principle of ensemble learning, a methodology that involves the amalgamation of several classifiers to address intricate problems and enhance the model's performance. The RF classifier gets its name from comprising several decision trees, each built on different portions of the dataset. By averaging these trees' predictions, RF aims to boost the dataset's predictive capability. Instead of leaning on a single decision tree, the RF algorithm combines forecasts from multiple trees

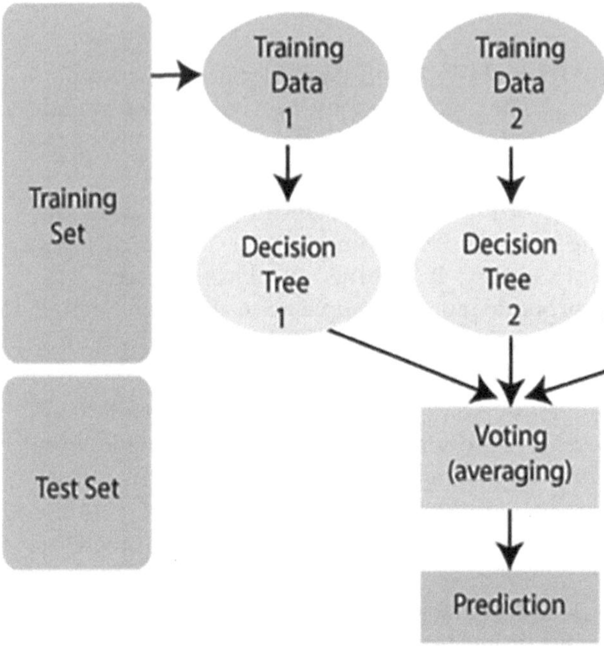

FIGURE 10.2
Block diagram of random forest algorithm.

and selects the final outcome based on the majority agreement among these predictions. A higher quantity of trees within the forest is associated with increased accuracy and serves as a preventive measure against overfitting. Figure 10.2 depicts the block diagram of the RF algorithm.

The KNN algorithm is a straightforward ML method that relies on the principles of supervised learning. The KNN algorithm operates on the assumption that there exists a similarity measure between a new instance or data point and the existing cases or data points. Based on this similarity measure, the algorithm assigns the new example to the category that exhibits the highest degree of similarity with the available categories. The KNN algorithm retains a comprehensive collection of data and utilizes similarity measures to classify incoming data points. This method applies to both regression and classification tasks, though it is mainly utilized for classification purposes. It falls under the category of non-parametric algorithms, meaning it doesn't make assumptions about the data distribution. Often termed a "lazy learner", this approach doesn't immediately learn from the training set. Instead, it stores the dataset and uses it only during the classification phase. In the training stage, KNN retains the dataset. Upon receiving new data, the system categorizes it based on its resemblance to the characteristics of the stored data. Figure 10.3 displays the explanation of the KNN algorithm.

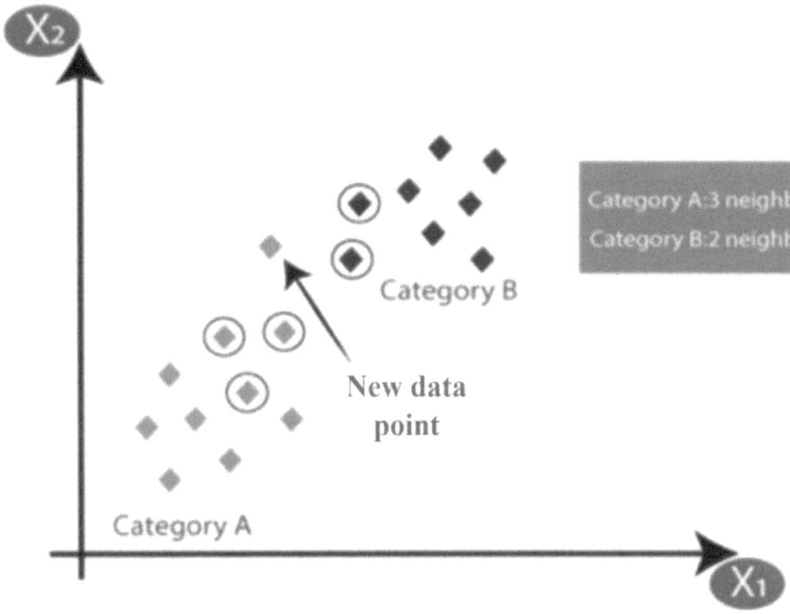

FIGURE 10.3
Explanation graph of K-nearest neighbor.

SVM is a popular method in supervised learning, frequently used for both classification and regression tasks. However, its main application lies in classification within the ML domain. The SVM algorithm aims to create an optimal decision boundary, linear or non-linear, that effectively separates an n-dimensional space into specific classes, enabling precise classification of upcoming data points. This optimal boundary is often termed a hyperplane. SVM identifies key points or vectors crucial for forming this hyperplane. These significant instances are known as support vectors, hence the name SVM for the classification process. Figure 10.4 refers to the visual representation of classifying two distinct groups using a decision boundary or hyperplane.

A total of 15 features were taken and they were named the following:

Div_Count, Img_Count, Span_Count, Para_Count, List_Count, Script_Count, Iframe_Count, Btn_Count, Table_Count, Article_Count, Word_Count, Link_Count, Section_Count, Link, Article.

The dataset was prepared through web scraping of the articles. The content of different HTML tag elements was extracted and stored in the file. From initial analysis, it is found that most of the non-articles contain a large number of images, hyperlinks, and less word count. Most of the articles contain less number of image counts and high numbers of word counts.

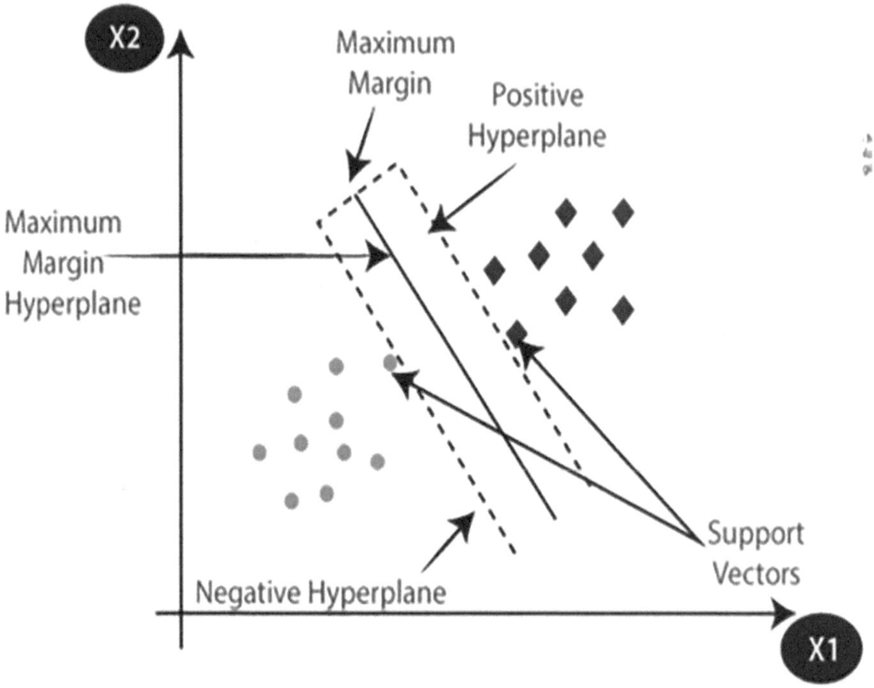

FIGURE 10.4
Explanation graph of support vector machine.

10.4 Result and Discussion

Results were obtained for three classification algorithms, RF, KNN, and SVM. The data was split into training and testing sets with a ratio of 70% for training and 30% for testing across all classifiers. The results from each classifier revealed that the RF model achieved an accuracy of 99%. This high accuracy suggests that the model is nearly perfect, indicating a potential for further testing, or it may be exhibiting signs of overfitting, as depicted in Figure 10.5.

The training data had an accuracy of 84.85%, while the test data showed an accuracy of 84.83%. Since the training score slightly surpassed the test score, it indicates that there was no overfitting. Therefore, the SVM produced satisfactory results, as illustrated in Figure 10.6.

From the plot below it can be observed that training data accuracy was more than that of testing data accuracy. The KNN model score was 93.78%, given in Figure 10.7.

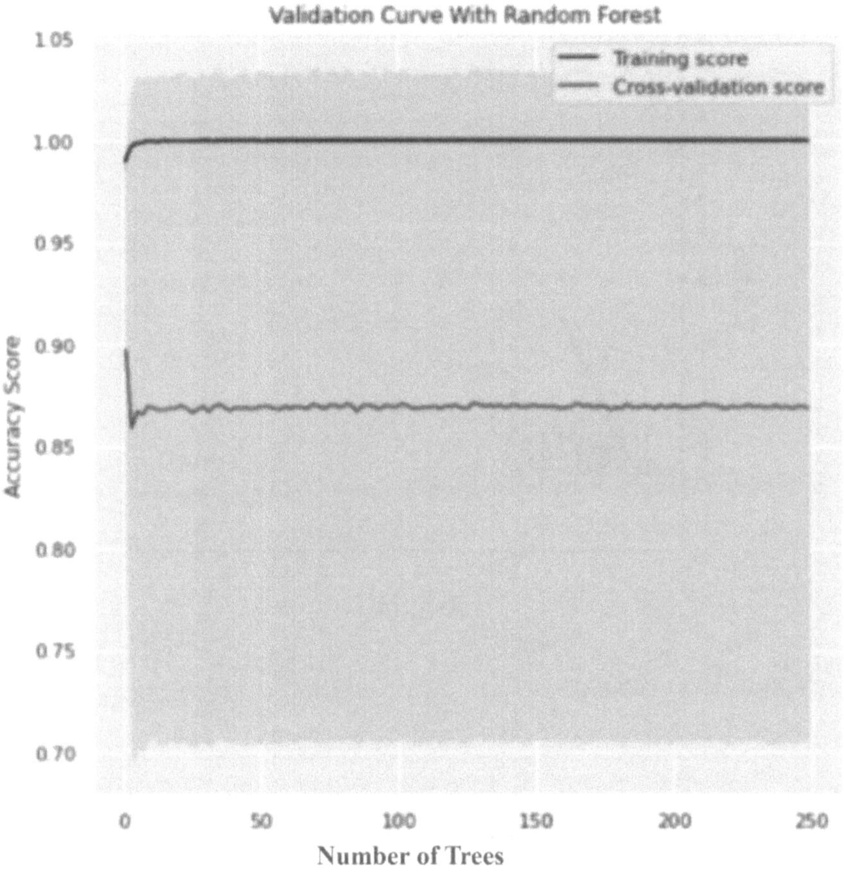

FIGURE 10.5
Validation curve of random forest algorithm.

Precision, recall, and accuracy of the model are given in Tables 10.1–10.3, which analyze the model accuracy with another browser.

10.5 Conclusion

This ML and fuzzy-wuzzy-based decision logic can play a key role in media analytic intelligence. From a research point of view, authors are trying the simplest deterministic strategy that can establish accuracy in media opinion. Based on a statistical analysis of the dataset, the mean was calculated,

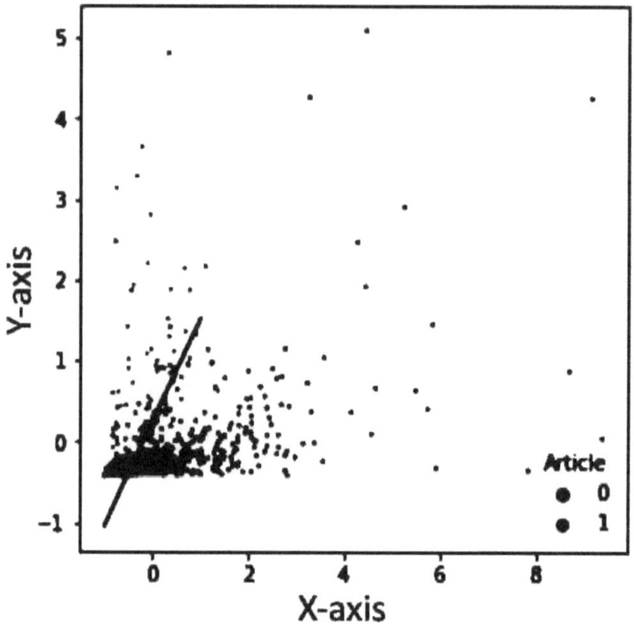

FIGURE 10.6
Accuracy of support vector machine.

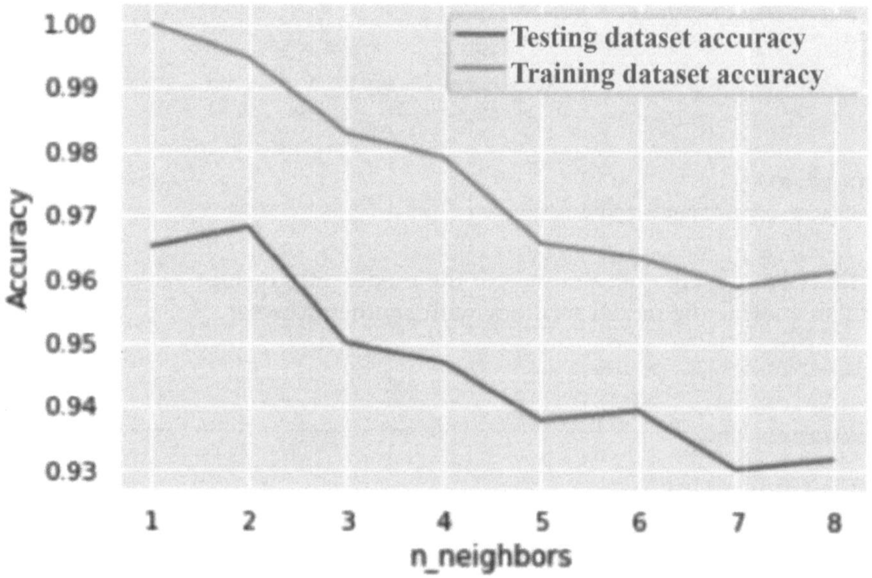

FIGURE 10.7
Accuracy of K-nearest neighbor algorithm.

TABLE 10.1

Model Selection Based on Precision, Recall, and Accuracy

Models Experimented	Precision	Recall	Accuracy
Random forest	0.99	0.99	0.99
Support vector machine	0.82	0.89	0.85
K-nearest neighbor	0.96	0.91	0.94

TABLE 10.2

Comparing Our Model to Other Browsers

Browser	ML Solution	Model	Precision	Recall
Samsung Internet	No	Readability	0.90	0.86
Mozilla Firefox	No	Readability	0.83	0.90
Google Chrome	Yes	AdaBoost	0.86	0.81
Proposed Solution 1	Yes	Random forest	0.99	0.99
Proposed Solution 2	Yes	Support vector machine	0.82	0.89
Proposed Solution 3	Yes	K-nearest neighbor	0.96	0.94

TABLE 10.3

Mean of Features of the Dataset

Data Set	Div	Span	Link	Para	List	Img	Article	Word	Section
Non-article	266.13	95.90	174.94	24.15	98.15	36.59	7.15	8,070.8	6.8
Article	210.27	246.17	395.20	31.36	191.57	54.73	0.19	11,842	1.38

and, through that, it was found that the articles have high span, link and word count compared to the non-articles. For the non-articles, div count was more than that of articles. Logistic regression is not holding well for the dataset. However, RF was seen as having some overfitting of data. KNN and SVM algorithms gave better results. To improve the accuracy, more data is required, i.e. websites of different languages and RTL (right-to-left) types. We would like to further improve the model score of the algorithms used and scrap more websites to add more data points. To address the issue of overfitting, the RF algorithm will be utilized in the future. In the future, we will have to pull important content from websites that have infinite scrolling capabilities and make any necessary assumptions based on information gleaned from nearby pages.

References

1. Choi, B. and Yao, Z. (2005). "Web page classification." In Chu, W., Young Lin, T. (eds) *Foundations and Advances in Data Mining*. Springer, Berlin, Heidelberg. pp. 221–274.
2. Qi, X. and Davison, B. D. (2009). "Web page classification: Features and algorithms," *ACM Computing Surveys (CSUR)*, vol. 41, no. 2, p. 12.
3. Safae, L., El Habib, B., and Abderrahim, T. (2018). "A review of machine learning algorithms for web page classification." In *2018 IEEE 5th International Congress on Information Science and Technology (CiSt)*. IEEE 5th International Congress on Information Science and Technology (CiSt), Marrakech, Morocco, pp. 220–226.
4. Vincent, J. B. (2017). "New evidence against chromium as an essential trace element," *The Journal of Nutrition*, vol. 147, no. 12, pp. 2212–2219.
5. Nayak, P., Thadichi, B., and Patidar, H. (2019, December). "Article detection on the web using DOM and text based features." In *2019 IEEE 16th India Council International Conference (INDICON)*, Rajkot, India. IEEE, pp. 1–4. https://doi.org/10.1109/INDICON47234.2019.9030324
6. Kohlschutter, C., Fankhauser, P., and Nejdl, W. (2010). "Boilerplate detection using shallow text features." In *Proceedings of the Third ACM International Conference on Web Search and Data Mining*. Association for Computing Machinery, New York, NY, USA, pp. 441–450.
7. Ou-Yang, L. (2013). Newspaper3k: Article Scraping & Curation. Newspaper 0.0.2 Documentation.
8. Nayak, P., Thadichi, B., and Patidar, H. (2019, December). "Article detection on the web using DOM and text based features." In *2019 IEEE 16th India Council International Conference (INDICON)*, Rajkot, India. IEEE,, pp. 1–4. https://doi.org/10.1109/INDICON47234.2019.9030324.
9. Mathur, S., Nikam, P., Patidar, H., Gaikwad, R. B., and Nayakand, P. N. (2021). "Machine-learning directed article detection on the web using DOM and text-based features." In *2021 IEEE 18th Annual Consumer Communications & Networking Conference (CCNC)*. https://doi.org/10.1109/CCNC49032.2021.9369599
10. Nayak, P., Thadichi, B., and Patidar, H. (2019, December). "Article detection on the web using DOM and text based features." In *2019 IEEE 16th India Council International Conference (INDICON)*, Rajkot, India. IEEE,, pp. 1–4. https://doi.org/10.1109/INDICON47234.2019.9030324
11. Nag, M. S. R. K., Srinivas, G., Rao, K. V., Vakkalanka, S., and Nagendram, S. (2021). "WITHDRAWN: Comparative and experimental study in identifying the similarity between languages for plagiarism detection and efficient language translation," *Materials Today: Proceedings*. https://doi.org/10.1016/j.matpr.2020.12.1206
12. Shahmohammadi, H., Dezfoulian, M. H., and Mansoorizadeh, M. (2021). "Paraphrase detection using LSTM networks and handcrafted features," *Multimedia Tools and Applications*, vol. 80, no. 4, pp. 6479–6492.
13. Salma, Saeed, M., ur Rahim, R., Gufran Khan, M., Zulfiqar, A., and Bhatti, M. T. (2021). "Research article: Development of ANPR framework for Pakistani vehicle number plates using object detection and OCR," *Complexity*. https://doi.org/10.1155/2021/5597337

14. Moss, T. (2022). "'Fuzzy Wuzzy'soldiers: Race and Papua New Guinean soldiers in the Australian Army, 1940–60," *War in History*, vol. 29, no. 2, pp. 467–485.
15. Luo, Y., Card, D., and Jurafsky, D. (2020). "Detecting Stance in Media on Global Warming." arXiv preprint arXiv:2010.15149.
16. M. Ahammad et al., "A Proficient Approach to Detect Osteosarcoma Through Deep Learning," 2022 10th International Conference on Emerging Trends in Engineering and Technology-Signal and Information Processing (ICETET-SIP-22), Nagpur, India, 2022, pp. 1–6, doi: 10.1109/ICETET-SIP-2254415.2022.9791502.
17. M. Shariful Islam, M. Abu Tareq Rony, M. Safran, S. Alfarhood and D. Che, "Elevating Driver Behavior Understanding With RKnD: A Novel Probabilistic Feature Engineering Approach," in IEEE Access, vol. 12, pp. 65780–65798, 2024, doi: 10.1109/ACCESS.2024.3397725.
18. Rao, G. A., Srinivas, G., Rao, K. V., and Prasad Reddy, P. V. G. D. (2018). "A partial ratio and ratio based fuzzy-wuzzy procedure for characteristic mining of mathematical formulas from documents," *IJSC-ICTACT Journal on Soft Computing*, vol. 8, no. 4, pp. 1728–1732.
19. Dev, K., Khowaja, S. A., Bist, A. K., Saini, V., and Bhatia, S. (2021). "Triage of potential COVID-19 Patients from chest X-ray images using hierarchical convolutional networks," *Neural Computing and Applications (NCAA)*, vol. 35, pp. 23861–23876.
20. Nag, A., Das, B., Sil, R., Hameed, A.A., Bhushan, B., Jamil, A. (2024). A Survey on Image-Based Cardiac Diagnosis Prediction Using Machine Learning and Deep Learning Techniques. In: García Márquez, F.P., Jamil, A., Ramirez, I.S., Eken, S., Hameed, A.A. (eds) Computing, Internet of Things and Data Analytics. ICCIDA 2023. Studies in Computational Intelligence, vol 1145. Springer, Cham. https://doi.org/10.1007/978-3-031-53717-2_44
21. M. M. Hassan et al., "Machine Learning-Based Rainfall Prediction: Unveiling Insights and Forecasting for Improved Preparedness," in IEEE Access, vol. 11, pp. 132196–132222, 2023, doi: 10.1109/ACCESS.2023.3333876.
22. Basheer, S., Nagawanshi, K., Bhatia, S., Dubey, S., and Sinha, G. R. (2021). "FESD: An approach for biometric human footprint matching using fuzzy ensemble learning," *IEEE Access*, https://doi.org/10.1109/ACCESS.2021.3057931

11

Evolution of Computational Intelligence for Vehicle Surveillance and Recognition in Industry 6.0: Recent Innovations, Practices, and Challenges

Vallidevi Krishnamurthy, Nivedita P R, Akshaya S, Nithyashree S, Prasanna S, and Sonali Agarwal

11.1 Introduction

Vehicle surveillance and recognition systems have become increasingly important tools for traffic monitoring, law enforcement, toll collection, and security applications. The ability to automatically detect, track, and identify vehicles from video data can provide valuable insights into traffic patterns, assist in investigating incidents, and enhance public safety measures. However, developing robust and accurate vehicle recognition systems presents significant challenges due to the complexity of real-world environments, varying lighting conditions, occlusions, and the vast diversity of vehicle makes, models, and appearances. Computational intelligence techniques, including machine learning, computer vision, and pattern recognition algorithms, have emerged as powerful solutions to address these challenges. These techniques leverage advanced data processing and analysis methods to extract meaningful information from vast amounts of raw data, enabling the development of intelligent systems capable of perceiving, learning, and adapting to dynamic environments. In recent years, there have been numerous innovations in computational intelligence methods specifically tailored for vehicle surveillance and recognition tasks. These innovations span various areas, such as object detection and tracking, feature extraction and representation, classification and identification, and sensor fusion techniques. For instance, deep learning architectures, like convolutional neural networks (CNNs) and recurrent neural networks (RNNs), have demonstrated remarkable performance in vehicle detection and classification tasks, outperforming traditional computer vision methods.

DOI: 10.1201/9781003503934-11

License plate recognition (LPR) systems, also known as automatic license plate recognition or automatic number plate recognition, have emerged as a critical technology for a wide range of applications in intelligent transportation systems (ITSs), law enforcement, and security domains [1]. These systems leverage advanced computer vision and machine learning techniques to automatically detect, localize, and recognize vehicle license plates from digital images or video footage. The ability to accurately and efficiently identify vehicles based on their license plates has numerous practical benefits. In traffic monitoring and management scenarios, LPR systems can be employed for electronic toll collection, parking enforcement, and tracking vehicles of interest. Law enforcement agencies utilize LPR technology for investigating crimes, locating stolen vehicles, and enforcing traffic regulations. Additionally, LPR systems play a crucial role in enhancing security measures at critical infrastructure sites, such as government facilities, airports, and border crossings.

LPR systems are at the forefront of modern surveillance and traffic management, but they grapple with formidable challenges in handling the diverse formats of license plates and the unpredictable outdoor lighting conditions during image capture. A typical LPR system is a symphony of technology, comprising four pivotal components: image acquisition, license plate localization and segmentation, character segmentation and standardization, and character recognition [2,3]. This chapter embarks on a nuanced exploration of these facets, shedding light on the pervasive conundrums, such as sensitivity to ambient brightness and protracted processing times. Emphasis is squarely placed on the foundational role of precise license plate localization, serving as the linchpin for the system's overarching efficiency.

11.2 The Challenges of LPR

Image Acquisition: The journey of LPR begins with image acquisition. However, this stage is fraught with challenges, primarily due to unpredictable outdoor lighting conditions. License plates can be obscured by shadows, reflections, or glare from the sun, making it challenging to capture clear and consistent images. Moreover, varying weather conditions, such as rain or snow, further exacerbate image quality issues.

License Plate Localization and Segmentation: Once an image is acquired, the next critical step is license plate localization and segmentation. This involves identifying the region of the image that contains the license plate and isolating it from the background. Precise localization is paramount, as errors at this stage can lead to the misinterpretation of characters on the license plate.

Character Segmentation and Standardization: With the license plate isolated, the system must segment individual characters or symbols for recognition. This process can be particularly challenging when dealing with license plates from different regions or countries, as they may have varying character formats and sizes. Standardizing characters to a consistent format is essential to ensure accurate recognition.

Character Recognition: Character recognition is the heart of any LPR system. It involves deciphering the segmented characters and converting them into a machine-readable format. Traditional methods often struggled with non-standard characters or symbols, especially in cases where license plates contained a mix of letters, numerals, and special characters.

Environmental Conditions: LPR systems operate in dynamic and often unpredictable environments, which can significantly impact their performance. Varying lighting conditions, such as direct sunlight, shadows, or low-light scenarios, can make license plate detection and character recognition more challenging. Adverse weather conditions, such as rain, snow, or fog, can also introduce distortions, obstructions, or noise in the captured images.

Occlusions and Obstructions: License plates can be partially or fully occluded by various objects, such as other vehicles, road signs, or debris. Dirt, scratches, or damages to the license plates themselves can also obscure or distort the characters, hindering accurate recognition.

Vehicle Motion and Orientation: Vehicles in motion can introduce blur or distortions in the captured images, making it difficult for LPR systems to accurately detect and recognize license plates. Additionally, the orientation of vehicles relative to the camera can lead to partial occlusions or skewed perspectives, further compounding the recognition task.

Addressing these challenges requires multifaceted approaches that combine advanced computer vision techniques, machine learning algorithms, robust data preprocessing, and efficient hardware implementations [4].

11.3 Innovations in LPR

LPR systems have witnessed significant advancements in recent years, driven by innovations in computer vision, machine learning, and computational intelligence techniques. These innovations have focused on enhancing the accuracy, robustness, and efficiency of LPR systems, enabling their broader adoption and applications in various domains [5]. Delving deeper into the LPR domain, this chapter sets its gaze on the diverse techniques underpinning license plate recognition. Many extant methods relegate themselves to processing a solitary line of characters and a restricted gamut of symbols, predominantly English letters and numerals. However, innovation is a driving force in this field, with researchers continuously pushing the boundaries to overcome these limitations.

Handling Uneven Lighting: Recognizing license plates in diverse lighting conditions, from glaring sunlight to dimly lit streets, necessitates advanced algorithms. Adaptive image enhancement techniques, such as histogram equalization and contrast stretching, help in improving image quality. Machine learning models, particularly CNNs, have shown remarkable capabilities in handling variations in lighting and contrast.

Character Recognition Advancements: The character recognition phase has witnessed significant advancements. Support vector machines (SVMs) have proven to be formidable tools in accurately identifying characters, even in scenarios with distorted or partially obscured symbols. These algorithms are trained on vast datasets containing diverse character formats and languages.

Multilingual and Multisymbol Recognition: License plates are not limited to English letters and numerals. Recognizing characters from different languages and accommodating various symbols is crucial for international applicability. Modern LPR systems are designed to recognize a wide range of characters, from Chinese characters to Kana characters, making them versatile for global use.

Shadow Eradication Mechanisms: Shadows can significantly impact the clarity of license plate images. To address this issue, innovative algorithms have been developed to detect and remove shadows from images, ensuring that characters are correctly identified, even in the presence of shadows.

Character Feature Extraction for Chinese Characters: Recognizing Chinese characters presents a unique set of challenges due to their complexity and stroke-based nature. Character feature extraction techniques tailored specifically to the nuances of Chinese characters have emerged, improving the accuracy of recognition.

11.4 The Landscape of Variable Factors

Vehicle LPR stands as a domain supplied with a myriad of dynamic variables, each contributing to the intricacies of the recognition process. Here we navigate through the diverse landscape of variability inherent in LPR applications, shedding light on the nuanced interplay of factors that shape recognition outcomes.

Illumination Dynamics: The fluctuating luminance conditions pose a significant challenge in LPR, influencing the clarity and contrast of captured license plate images. Variations in natural lighting, artificial illumination, and adverse weather conditions necessitate adaptive algorithms capable of robustly processing images under diverse lighting scenarios [6].

Research efforts focusing on illumination normalization techniques and adaptive thresholding algorithms are imperative to enhance the resilience of LPR systems against illumination dynamics.

Camera Viewpoint Variation: The perspective from which license plate images are captured can vary significantly depending on the placement and mobility of cameras. Fixed-camera setups may exhibit relatively stable viewpoints, whereas mobile or patrolling vehicles encounter dynamic shifts in camera angles and distances. Addressing viewpoint variation requires the development of algorithms resilient to changes in perspective, encompassing techniques such as perspective transformation and geometric normalization.

Distance and Scale Considerations: The distance between the camera and the license plate plays a crucial role in determining the scale and resolution of captured images. In fixed-camera environments, distance variations may be relatively limited, while patrolling vehicles encounter a broader range of distances leading to scale variations. Robust scale-invariant feature extraction techniques and adaptive scaling algorithms are indispensable to mitigate the effects of distance variations on recognition accuracy [7].

Application-Specific Variability: The scope of variability within each factor is inherently shaped by the specific requirements and constraints of the LPR application at hand. Urban surveillance systems may encounter distinct challenges compared to highway tollbooths or parking lot management applications. Tailoring recognition algorithms to accommodate the unique characteristics and operational contexts of diverse LPR applications is paramount to ensuring optimal performance.

Integration of Machine Learning: Leveraging machine learning approaches such as CNNs and RNNs can empower LPR systems to adaptively learn and generalize from diverse variability patterns [8]. Training models on diverse datasets encompassing a wide range of variability scenarios is essential to enhance the robustness and generalization capabilities of LPR systems.

11.5 Tailoring Solutions to Variable Scopes

Applications with wider variability scopes beckon more advanced computational prowess and processing acumen compared to their narrower-scoped counterparts. A transposition of methods between these applications emerges as a futile pursuit, given the incongruities in variable scopes. Each application necessitates bespoke solutions precisely tailored to the choreography of variability in its variables.

Fixed vs. Mobile Cameras: In environments where cameras remain stationary, such as toll booths or parking lots, variability in lighting conditions and camera-to-license-plate distance is relatively constrained.

Specialized algorithms can be developed and optimized for these conditions, capitalizing on the stability of variables to achieve heightened recognition accuracy [9]. Techniques like background subtraction and contrast enhancement may prove particularly effective in enhancing the clarity of license plate images captured by fixed cameras.

Patrolling Vehicles: LPR systems installed in patrolling vehicles navigate through a dynamic environment characterized by a wide range of conditions. The camera-to-license-plate distance fluctuates significantly, and lighting conditions undergo rapid changes as the vehicle traverses different environments. Recognizing the necessity for adaptability, these systems integrate real-time adjustments and robust algorithms capable of dynamically adapting to evolving scenarios. Employing techniques such as adaptive thresholding and feature fusion enables LPR systems in patrolling vehicles to maintain optimal performance amid variable conditions [10].

Real-Time Adaptability: The effectiveness of LPR systems in dynamic environments hinges on their ability to adapt in real time to shifting variables. Adaptive algorithms that dynamically adjust parameters based on contextual cues, such as lighting intensity and vehicle speed, are pivotal for ensuring robust performance. Integrating machine learning models trained on diverse datasets encompassing variability scenarios enhances the adaptability and generalization capabilities of LPR systems in patrolling vehicles [11].

Fusion of Sensor Data: Patrolling vehicles equipped with LPR systems can leverage data fusion techniques to enhance recognition accuracy and reliability. Integration of additional sensor modalities, such as GPS, inertial sensors, and environmental sensors, enables comprehensive contextual awareness and facilitates more informed decision-making. Multi-sensor fusion frameworks coupled with machine learning algorithms offer a synergistic approach to mitigating the effects of variable factors and improving overall recognition performance [40].

11.6 The Role of Benchmark Databases

11.6.1 Application-Oriented License Plate Benchmark Database

This chapter casts a spotlight on the heralded Application-Oriented License Plate benchmark database, an indispensable resource boasting 2,049 meticulously curated images. These images are meticulously classified into three distinct subsets, each meticulously mirroring the distinctive demands of one of the three primary LPR applications. Originating from diverse locales in Taiwan, these samples span a kaleidoscope of temporal dimensions, traffic scenarios, meteorological nuances, and geographic settings. This arsenal of data forms the crucible wherein a novel approach to LPR, characterized by three elemental modules—plate detection, character segmentation, and recognition—is rigorously forged. Each module ushers in its dimension of innovation, encompassing clustering for plate detection, maximally stable extreme region for character segmentation, and a bilayer classifier brimming with sophistication, crowned by the presence of a null class for character recognition [12,13]. The performance of this multifaceted solution is meticulously scrutinized, and the modules' merits unfurl under the meticulous scrutiny of comparative analyses [41].

11.7 Vehicle Make-and-Model Recognition

Turning the lens toward vehicle make-and-model recognition (MMR), the chapter underscores the seismic shift in the technological landscape. The growing demand for heightened security and the ubiquity of surveillance cameras coalesce, birthing an exigency for vehicle identification and classification technologies. In this evolving frontier, the orchestration of image-based vehicle recognition encounters a panoply of challenges. These entail a burgeoning landscape of vehicle models, perplexing similarities between certain models, and the variations accompanying the acquisition of vehicle images. Figure 11.1 shows the car image recognition with CNN applications. Car image recognition using CNNs is utilized across diverse

FIGURE 11.1
Car image recognition with convolutional neural network applications.

domains for its ability to accurately analyze and classify vehicle images. In automated driving systems, CNNs aid in real-time decision-making for navigation and collision avoidance. Traffic management systems benefit from CNN-based detection and classification of vehicles to optimize traffic flow and enforce regulations. Parking management systems use CNNs to identify available parking spaces and manage occupancy efficiently. In security and surveillance, CNNs detect and track vehicles of interest, enhancing law enforcement and border control efforts. Vehicle identification and authentication systems leverage CNN-based recognition for automated toll collection and access control. Market research and analytics benefit from CNNs analyzing vehicle images for consumer insights. Additionally, CNN-based car image recognition contributes to environmental impact assessments by analyzing vehicle types and emissions in urban areas. Overall, CNN-based car image recognition plays a crucial role in transportation, safety, security, urban planning, and environmental assessment [42].

11.8 Challenges in Vehicle MMR

Recognizing vehicle make-and-model presents a multifaceted challenge due to the diverse nature of vehicles themselves. Here we look into the intricate hurdles encountered in this domain, shedding light on the complexities that demand sophisticated solutions.

Expanding Vehicle Variety: With each passing year, automotive manufacturers unveil an array of new models and iterations, exponentially augmenting the spectrum of vehicles for recognition. This expanding diversity necessitates a dynamic database that is continuously updated to accommodate new additions [14,15]. Robust algorithms are indispensable to discern subtle design variances amid this vast pool of vehicles.

Addressing Model Similarities: Certain vehicle models exhibit striking resemblances in appearance, posing a formidable challenge for recognition systems to accurately differentiate between them. Overcoming this hurdle demands the employment of advanced image processing techniques and feature extraction methodologies. Exploring innovative approaches such as deep learning architectures may offer promising avenues for enhancing discrimination capabilities.

Managing Image Quality Variability: Images of vehicles captured in real-world scenarios manifest considerable variability in terms of lighting conditions, angles, and resolutions. The efficacy of an MMR system hinges on its ability to navigate through these diverse image qualities. Techniques for robust feature extraction and classification in the presence of image artifacts and variations are imperative to ensure reliable recognition outcomes.

Incorporating Contextual Awareness: Beyond the vehicle's physical appearance, contextual cues such as location, time of day, and surrounding environment can offer valuable insights for accurate recognition. Integrating contextual awareness into recognition algorithms presents an exciting frontier, enabling systems to adapt and refine their recognition capabilities based on situational contexts.

Ensuring Scalability and Efficiency: As the volume of vehicles on roads continues to escalate, scalability and efficiency become paramount concerns for MMR systems. Designing algorithms that strike a balance between computational complexity and recognition accuracy is essential to ensure practical applicability across diverse real-world scenarios. Leveraging cloud-based architectures and parallel processing techniques may offer scalable solutions to accommodate the ever-growing demands of real-time recognition applications [43].

11.9 Feature-Based Approaches in MMR

In the pursuit of fine-grained classification within the vehicle taxonomy, contemporary research in vehicle MMR is witnessing a paradigm shift toward the utilization of advanced machine learning paradigms. This shift is driven by the need for nuanced classification, necessitating the exploration of sophisticated feature-based approaches.

Feature-based approaches in MMR rely on a symphony of features extracted from vehicle images, encompassing a diverse array of attributes [16]. These attributes include edge-based characteristics, oriented-contour point models, and transform-based image features, each offering unique insights into the underlying structure and composition of vehicles. At the heart of feature-based approaches in MMR lie the Pyramid Histogram of Oriented Gradients (PHOG) edge descriptor and the versatile Gabor filter [17,18]. The PHOG edge descriptor excels in efficient feature encapsulation, unraveling gradient directions and magnitudes per pixel, thereby capturing intricate details essential for vehicle classification. The Gabor filter plays a pivotal role in multi-resolution and multi-orientation measurement of local spatial frequencies, providing a robust framework for feature extraction in MMR applications. The coherence in feature extraction emerges as a linchpin in diverse image processing quests, including MMR. By harmonizing various feature extraction techniques, researchers can enhance the discriminative power and robustness of MMR systems, enabling more accurate and reliable recognition outcomes. Despite the advancements in feature-based approaches, challenges such as computational complexity and robustness to variations in image quality persist.

11.10 Classification Paradigms in MMR

Within the realm of machine learning, the landscape of classification paradigms in vehicle MMR unfolds rich methodologies and techniques. Neural networks stand as foundational pillars in the domain of MMR, leveraging interconnected layers of artificial neurons to extract complex patterns and relationships from vehicle images. CNNs have emerged as particularly potent tools for image classification tasks, demonstrating remarkable performance in discerning intricate features inherent to vehicle MMR. SVM represents a powerful class of supervised learning algorithms that excel in binary classification tasks. SVMs are adept at delineating decision boundaries between different vehicle classes, leveraging kernel functions to map input data into high-dimensional feature spaces where classification becomes more straightforward. Random forests attach the collective wisdom of decision trees to achieve robust classification performance [19]. By training multiple decision trees on random subsets of the data and aggregating their predictions, random forests mitigate overfitting and enhance generalization capabilities, making them well-suited for MMR tasks. Ensemble classification methodologies, such as bagging, AdaBoost, and random subspace, orchestrate a symphony of diverse classifiers to enhance overall performance. The ensemble classifier represents an archetype of collaborative wisdom, leveraging the collective insights of multiple component classifiers to achieve superior classification accuracy and robustness. Rotation forests emerge as an innovative paradigm within the ensemble classification framework, imbuing classifier diversity while preserving accuracy. By employing principal component analysis to creatively divide object subsets, rotation forests offer a potent strategy for enhancing classification performance, particularly on benchmark datasets [20]. Discriminating reliable classifications amid uncertainty becomes paramount, highlighting the need for methodologies capable of ascertaining classification reliability for specific cases.

11.11 Vision-Based Vehicle Classification in Real-World Conditions

Within the dynamic realm of real-world traffic systems, the challenge of classifying vehicles stands as a typical endeavor with far-reaching implications for traffic optimization, law enforcement, automated navigation, and parking optimization. In this landscape, vision-based methods emerge as the precursor, harnessing external physical characteristics for vehicle identification and classification [21]. However, designing a robust and efficient automatic vehicle classification system for real-world deployment unfolds a web of mysteries. Here we discuss the multifaceted challenges encountered in vision-based vehicle classification in real-world conditions and explore strategies to overcome them.

Occlusion: Occlusion, caused by objects obstructing the view of vehicles, poses a significant challenge for vision-based classification systems. Partial occlusion can lead to inaccuracies in vehicle detection and classification, necessitating the development of robust algorithms capable of inferring vehicle characteristics even in partially obscured scenarios [22,23]. Techniques such as multi-view fusion and context-aware reasoning can mitigate the effects of occlusion by leveraging information from multiple camera viewpoints and contextual cues to enhance classification accuracy.

Object Tracking: Tracking vehicles across consecutive frames in real-world traffic scenes presents another formidable challenge. Rapid changes in vehicle position, scale, and orientation, compounded by occlusions and interactions with other objects, demand robust object-tracking algorithms capable of maintaining accurate trajectories [24]. Utilizing techniques such as Kalman filters, particle filters, and deep learning-based tracking algorithms can improve the robustness and reliability of object tracking in dynamic traffic environments.

Shadowing: Shadows cast by surrounding objects or changes in lighting conditions can obscure vehicle features, leading to misclassifications in vision-based classification systems.

Addressing shadowing requires the development of algorithms capable of distinguishing between shadow regions and actual vehicle features. Techniques such as shadow detection and removal, as well as illumination-invariant feature extraction, can help mitigate the effects of shadowing on classification accuracy.

Rotation and Viewpoint Variability: Vehicles in real-world traffic scenes exhibit variability in orientation and viewpoint due to factors such as camera placement, vehicle movement, and road geometry. Handling rotation and viewpoint variability necessitates the development of rotation-invariant and viewpoint-invariant feature extraction techniques [25]. Methods such as 3D modeling, geometric transformations, and viewpoint normalization can help mitigate the effects of rotation and viewpoint variability on vehicle classification accuracy.

Color Constancy: Changes in lighting conditions can lead to variations in color appearance, posing challenges for vision-based vehicle classification systems that rely on color features. Ensuring color constancy, or consistency in color perception across different lighting conditions, is essential for maintaining classification accuracy. Techniques such as color calibration, histogram equalization, and color space transformations can help mitigate the effects of color constancy on classification performance.

11.12 Feature-Based Methodologies for Vehicle Classification

Feature-based methodologies, notably the venerable Scale-Invariant Feature Transform, hold sway as stalwart choices for object classification, for they confer invariance and resilience in the face of surface fluctuations and local deformations. However, their effectiveness wanes when processing low-resolution images under real-world conditions. It is within this crucible that a beacon of hope emerges, a multi-class vehicle classification system underpinned by a hybrid dynamic Bayesian network. This system defies the convention of high-resolution dependence, heralding a new era of classification using low-level features such as height, width, and angle. Its computational parsimony and independence from proximity or image resolution make it a formidable contender in addressing the limitations of feature-based methods.

11.12.1 Optical Data Storage for Navigation

The narrative meanders into the labyrinth of optical data storage, an uncharted territory infused with promise for navigation applications. Optical data storage, endowed with inherent 2D parallel optical capabilities, beckons as an optimal ally. Rendering high storage capacity and expeditious parallel access, it affords a promising palette for vehicle navigation endeavors [26]. One pioneering experiment exemplifies the fusion of optical data storage with navigation, steering a vehicle along a predetermined path guided solely by visual input. The charge coupled devices (CCD) camera, serving as the sentinel eye, records the landscape in the vehicle's path as it traverses manually. Stills plucked from this recorded tapestry form "reference points" ensconced within a holographic database, constituting an enigmatic yet promising marriage of technology and navigation.

11.12.2 Anomaly Detection for Safer Roads

Anomaly detection and security draw the spotlight once more, this time in the context of optical correlation and vehicle navigation. This chapter orchestrates a vision of an automated vehicle search system amid urban surveillance videos, steeped in semantic attributes. This system empowers users with the capacity to search for vehicles predicated on attributes ranging from color and size to speed, the direction of travel, date/time, and location. Robust and adaptable, this system thrives in tempestuous urban landscapes, supported by massive data indexing capabilities. Unveiling a groundbreaking vehicle detection method, the chapter interweaves co-training, synthetic occlusion, and shape/aspect deformable sliding windows into the narrative, culminating in the encapsulation of occlusions and diverse vehicle types [27,28]. Once detected and tracked, attributes are meticulously extracted and immortalized within a database, illuminating a path toward vehicle retrieval grounded in semantic attributes.

11.12.2.1 Real-Time Anomaly Detection in ITSs

The specter of anomalies, lurking as a potential menace on the roadways, kindles the pursuit of anomaly detection in the realm of ITSs. Of these anomalies, vehicles veering in unintended directions pose a grave risk to road safety. Traditional methods, exemplified by road sensors, falter in congested lanes, while image processing emerges as the lodestar, fostering enhanced reliability and maintenance ease [29,30]. Here, a novel vehicle flow detection method surges into the fray, fusing light curtain intelligence, closed-circuit television (CCTV) cameras, and machine learning magic. The amalgamation of these forces bestows the capacity to ascertain traffic flow direction with precision, a potential boon for averting accidents [31].

11.12.2.2 Nighttime Vehicle Recognition

Nighttime, a realm cloaked in obscurity, presents vehicle recognition with a fresh set of challenges. The interplay of darkness, paucity of lighting, and vehicle headlights manifests as the crux of the nighttime challenge. Conventional vehicle detection mechanisms find their vigor severely hampered in this setting, for attributes such as color, shadow, edge, and motion information crumble in the face of obscurity [32]. In this unwavering pursuit of recognition under the cloak of night, headlights and taillights emerge as the defining characteristics of vehicles navigating the inky darkness. Yet, the realm of nocturnal traffic proves complex, teeming with potential light sources beyond vehicle headlights, from traffic signals to street lamps and headlight reflections.

Figure 11.2 shows the nighttime vehicle detection and tracking with occlusion handling by pairing headlights and taillights. Here, a novel algorithm

FIGURE 11.2
Nighttime vehicle detection and tracking with occlusion handling by pairing headlights and taillights.

steps into the spotlight, stitching together headlight segmentation, matching, and tracking. Independent of lane detection algorithms, this paradigm first extracts headlight pixels through a thresholding alchemy, before uniting them under the banner of grouping and labeling. The ensuing tracking choreography unearths the vehicles, unveiling a promising approach for nighttime vehicle recognition [36].

11.13 CCTV in Modern Policing Surveillance

The narrative meanders into the realm of modern policing, where traditional methods find themselves eclipsed by the omnipresence of technology. Traditional policing's limited impact on crime reduction beckons the embrace of contemporary strategies, typified by the pervasive presence of CCTV cameras. In this foray into the modern policing landscape, the chapter explores the nuanced geography of CCTV camera placement atop traffic lights in Gaborone, Botswana. Geographical considerations, steeped in the tenets of situational crime prevention theory, stand as the fulcrum of this evaluation. The juxtaposition of conventional cameras and internet protocol (IP) cameras lays bare the transformative influence of technology in modern policing.

11.13.1 The Future of Policing Surveillance

The crescendo of this discourse resounds with the clarion call for advanced video analytics. In the realm of security and surveillance, the chapter casts a discerning eye on real-time threat detection and forensic investigation. Automated video analytics take center stage, circumventing the fallibility of human attention and the cumbersome nature of manual video perusal. Multi-scale tracking emerges as the vanguard, bestowing real-time threat detection and forensic prowess upon surveillance endeavors. The tapestry of advanced video analytics unfurls its component technologies for smart surveillance systems, bridging the chasm between theory and practice [33]. Yet, the path forward teems with challenges, as the trifecta of detecting, tracking, and identifying objects grapples with the mercurial nature of video glitches.

11.13.2 Real-Time Image Processing for Enhanced Security

Security and video surveillance ascend once more as sentinels guarding against intrusions and unwanted phenomena. In this realm, real-time image processing stands as the quintessential arsenal. Object tracking, in particular, weaves its magic across video surveillance systems, a linchpin for real-time monitoring and the detection of potential security threats. Within this sphere of influence, the chapter charts a course through the landscape of edge detection, where the Canny Edge detection method rises to prominence. This method, endowed with the capacity to classify and detect objects within complex videos, harnesses the alchemy of computational modeling to detect moving objects in parking lots [34,35]. Human error is thus supplanted by the unyielding precision of computational vigilance.

11.13.3 Semantic Attributes for Urban Surveillance

Urban surveillance emerges as the crowning jewel, where the architecture of semantic attributes propels vehicle search into uncharted territories. The system proffers users the potent capability to scour urban landscapes for vehicles based on a myriad of attributes: color, size, length, width, height, speed, direction, and spatiotemporal characteristics. Robust and adaptable, this system weaves a comprehensive solution underpinned by innovative vehicle detection methods, leveraging co-training, synthetic occlusion, and shape-deformable sliding windows to navigate the labyrinth of occlusions and a cacophony of vehicle types. These detected vehicles surrender their attributes to a database, paving the way for future investigations.

11.14 Automated Traffic Systems

The complex world of automated traffic systems rekindles the discourse, surmounting challenges ranging from incident detection and traffic monitoring to real-time traffic updates and automated traffic signals. A symphony of tools—loop detectors, video cameras, and GPS—rallies to monitor traffic. Loop detectors and GPS, though stalwart choices, come encumbered with high installation and maintenance costs. In contrast, video cameras unveil themselves as cost-effective sentinels, bereft of exorbitant overheads and personal costs for commuters [37]. Figure 11.3 illustrates the automatic traffic control and monitoring system. These systems utilize various technologies such as sensors, cameras, and data analysis software to monitor traffic conditions, detect congestion or accidents, and adjust traffic signal timings accordingly. They facilitate efficient traffic management, reduce congestion, improve transportation efficiency, and enhance overall road safety.

11.14.1 Pioneering Approaches in Motion Detection

Delving into the realm of motion detection within traffic monitoring systems, the journey unfolds through three distinct categories: time difference, optical

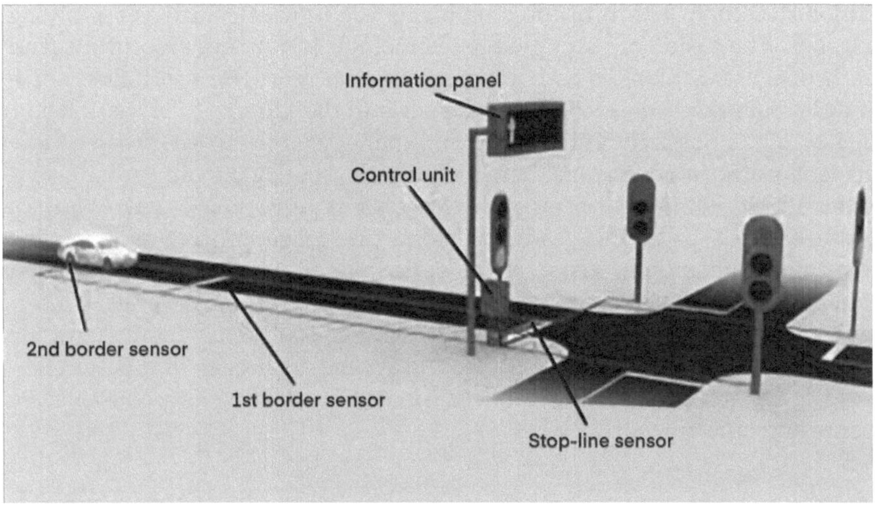

FIGURE 11.3
Automatic traffic control and monitoring system.

flow, and background subtraction. Time difference emerges as the adaptable sentinel, yet it falters in delineating the shapes of moving objects. Optical flow extends a hand, projecting motion into the image plane, but begets noise and computational load. In this cacophony, background subtraction strides forward as the reliable choice. Comparing current image pixel characteristics with a reference background model bestows accuracy sans excessive computational burdens. Yet, peculiar nuances emerge, such as the grey roofs of Bengaluru buses, mirroring the grey of the road. An enigmatic algorithm materializes, fostering a two-stream harmony, and ventures to illuminate a path for night and day traffic conditions.

11.14.2 Rising to the Challenge of Nighttime Vehicle Recognition

The cloak of night presents a fresh canvas for vehicle recognition, yet its nuances perplex the traditional recognition mechanisms. Here, the symphony of color, shadow, edge, and motion information crumbles, overshadowed by the spectral grace of headlights and taillights. Yet, the nocturnal landscape introduces complexity, replete with myriad light sources beyond headlights. Enter a novel algorithm, weaving together headlight segmentation, matching, and tracking, encapsulating the challenge of nighttime vehicle recognition.

The realms of LPR systems, vehicle MMR, vision-based vehicle classification, optical data storage, anomaly detection, real-time image processing, semantic attributes for urban surveillance, automated traffic systems, and motion detection within traffic monitoring are dynamic and ever-evolving domains. The fusion of cutting-edge technology, innovative algorithms, and real-world applications is reshaping the landscape of transportation, security, and surveillance. As we delve deeper into the intricacies of these fields, the horizon of possibilities continues to expand, promising safer roads, more efficient traffic management, and enhanced security in our urban environments [38,39]. In an era where technology is reshaping our world at an unprecedented pace, these advancements in image processing, machine learning, and computer vision stand as beacons of progress, illuminating a path toward a safer, more connected, and smarter future. Whether it's recognizing a license plate in challenging lighting conditions, identifying the make-and-model of a vehicle, or detecting anomalies in real time, these technologies are at the forefront of modern innovation, driving us toward a more secure and efficient world.

11.15 Conclusion

In conclusion, this chapter has delved into the intricate landscape of vehicle surveillance and recognition, particularly focusing on the innovations and challenges within LPR systems in Industry 6.0. We have explored the evolution of LPR technologies, discussing the advancements made in feature-based approaches, classification paradigms, and real-time anomaly detection. Additionally, the role of benchmark databases in driving advancements in LPR systems has been emphasized, highlighting the significance of application-oriented datasets in benchmarking and evaluating the performance of various recognition algorithms. Furthermore, we have examined the practical applications of LPR systems in modern policing surveillance, automated traffic systems, and enhancing security through real-time image processing. The discussion has also extended to address the challenges associated with nighttime vehicle recognition, emphasizing the importance of developing robust solutions capable of operating effectively in low-light conditions. Overall, the chapter underscores the critical role of computational intelligence in advancing the field of vehicle surveillance and recognition. By tailoring solutions to variable scopes and leveraging innovative methodologies such as optical data storage and anomaly detection, we can overcome existing challenges and pave the way for safer roads and more efficient transportation systems. As technology continues to evolve, it is imperative to stay abreast of emerging trends and developments, ensuring that our systems remain adaptive and resilient in addressing the ever-evolving landscape of vehicle surveillance and recognition.

References

1. Weihong, W. & Jiaoyang, T. Research on license plate recognition algorithms based on deep learning in complex environment. *IEEE Access*, 8, 91661–91675 (2020).
2. Kabir, A.T., Saha, P.K., Hasan, M.S., Pramanik, M., Ta-Sin, A.J., Johura, F.T., & Hossain, A.M. An IoT based intelligent parking system for the unutilized parking area with real-time monitoring using mobile and web application. In *2021 International Conference on Intelligent Technologies (CONIT)*, Hubli, IEEE, Karnataka, India, pp. 1–7 (2021).
3. Silpa, C. Implementing deep learning algorithms for automatic license plate recognition. *Turkish Journal of Computer and Mathematics Education*, 14, 2, 203–212 (2023).
4. Xu, H., Zhou, X.-D., Li, Z., Liu, L., Li, C., & Shi, Y. EILPR: Toward end-to-end irregular license plate recognition based on automatic perspective alignment. *IEEE Transactions on Intelligent Transportation Systems*, 23, 1–10 (2021).
5. Wang, Y., Bian, Z., Zhou, Y., & Chau, L.-P. Rethinking and designing a high-performing automatic license plate recognition approach. *IEEE Transactions on Intelligent Transportation Systems*, 23, 8868–8880 (2022).
6. Ali, M.Y., Haque, N., Suma, S.A., & Hadiuzzaman, M. Video sensor-based automatic license plate recognition of static and moving vehicles. *International Research Journal of Innovations in Engineering and Technology - IRJIET*, 6(8), 25–30, (2022).
7. Gao, L., & Zhang, W. Research on license plate detection and recognition based on deep learning. In *Proceedings of the 2021 International Conference on Computer Engineering and Artificial Intelligence (ICCEAI)*, Shanghai, China, 27–29 August 2021, pp. 410–415 (2021).
8. Bi, R., Xiong, J., Tian, Y., Li, Q., & Choo, K.K.R. Achieving lightweight and privacy-preserving object detection for connected autonomous vehicles. *IEEE Internet Things of Journal*, 10, 2314–2329 (2023).
9. Yan, C., Li, Z., Zhang, Y., Liu, Y., Ji, X., & Zhang, Y. Depth image denoising using nuclear norm and learning graph model. *ACM Transactions on Multimedia Computing, Communications and Applications*, 16(4), 122:1–122:17 (2020).
10. Shashirangana, J., Padmasiri, H., Meedeniya, D., & Perera, C. Automated license plate recognition: A survey on methods and techniques. *IEEE Access*, 9, 11203–11225 (2021).
11. Zhang, L., Wang, P., Li, H., Li, Z., Shen, C., & Zhang, Y. A robust attentional framework for license plate recognition in the wild. *IEEE Transactions on Intelligent Transportation Systems*, 22(11), 6967–6976 (2020).
12. Kong, X., Wang, K., Hou, M., Hao, X., Shen, G., Chen, X., & Xia, F. A federated learning-based license plate recognition scheme for 5G-enabled internet of vehicles. *IEEE Transactions on Industrial Informatics*, 17(12), 8523–8530 (2021).
13. Henry, C., Ahn, S.Y., & Lee, S.-W. Multinational license plate recognition using generalized character sequence detection. *IEEE Access*, 8, 35185–35199 (2020).
14. Vaiyapuri, T., NandanMohanty, S., Sivaram, M., Pustokhina, I.V., Pustokhin, D.A., & Shankar, K. Automatic vehicle license plate recognition using optimal deep learning model," *Computers, Materials & Continua*, 67(2), 1881–1897 (2021).

15. Boukerche, A., & Ma, X. A novel smart lightweight visual attention model for fine-grained vehicle recognition. *IEEE Transactions on Intelligent Transportation Systems*, 23, 13846–13862 (2021).
16. Touvron, H., Cord, M., Douze, M., Massa, F., Sablayrolles, A., & Jégou, H. Training data-efficient image transformers & distillation through attention. In *International Conference on Machine Learning*, Vienna, PMLR, Vienna, Austria, pp. 10347–10357 (2021).
17. Wu, H., Luo, X., & Zhou, M. Advancing non-negative latent factorization of tensors with diversified regularizations. *IEEE Transactions on Services Computing*, 15(3), 1334–1344 (2020).
18. Xia, Z., Peng, W., Khor, H.Q., Feng, X., & Zhao, G. Revealing the invisible with model and data shrinking for composite-database micro expression recognition. *IEEE Transactions on Image Processing*, 29, 8590–8605 (2020).
19. Ke, X., & Zhang, Y. Fine-grained vehicle type detection and recognition based on dense attention network. *Neurocomputing*, 399, 247–257 (2020).
20. Neupane, B., Horanont, T., & Aryal, J. Real-time vehicle classification and tracking using a transfer learning-improved deep learning network. *Sensors*, 22, 3813 (2022).
21. Abdelhalim, A., & Abbas, M. Towards real-time traffic movement count and trajectory reconstruction using virtual traffic lanes. In 2020 *IEEE/CVF Conference on Computer Vision and Pattern Recognition Workshops (CVPRW)*, Seattle, WA, pp. 2527–2533 (2020).
22. Abishek, T.S., Jose, A., Davis, A., Thomas, A., & Davies, J. Accident detection and warning system. *International Journal of Innovative Science and Research Technology*, 6, 1532–1535 (2021).
23. Nancy, P., Dilli Rao, D., Babuaravind, G., & Bhanushree, S. Highway accident detection and notification using machine learning. *International Journal of Computer Science and Mobile Computing* 9, 168–176 (2020).
24. Pillai, M.S., Chaudhary, G., Khari, M., & Crespo, R.G. Real-time image enhancement for an automatic automobile accident detection through CCTV using deep learning. *Soft Computing*, 25, 11929–11940 (2021).
25. Lee, C., Kim, H., Oh, S., & Doo, I. A study on building a "real-time vehicle accident and road obstacle notification model" using AI CCTV. *Applied Sciences*, 11, 8210 (2021).
26. Sergiyenko, O.Y., & Tyrsa, V.V. 3D optical machine vision sensors with intelligent data management for robotic swarm navigation improvement. *IEEE Sensors Journal*, 21, 11262–11274 (2021).
27. Arulananth, T., Baskar, M., Thrishma, K., Srilekha, N., Supraja, S., & Ravalika, C. Pothole detection using arduino and ultrasonic sensors. In *Advanced Techniques for IoT Applications: Proceedings of EAIT 2020*, pp. 435–442. Springer: Berlin/Heidelberg, Germany (2022).
28. Gupta, S., Sharma, P., Sharma, D., Gupta, V., & Sambyal, N. Detection and localization of potholes in thermal images using deep neural networks. *Multimedia Tools and Applications*, 79, 26265–26284 (2020).
29. Yik, Y.K., Alias, N.E., Yusof, Y., & Isaak, S. A real-time pothole detection based on deep learning approach. *Journal of Physics: Conference Series*, 1828, 012001 (2021).
30. Ucar, S., Patnayak, C., Oza, P., Hoh, B., & Oguchi, K. Management of anomalous driving behavior. In *2019 IEEE Vehicular Networking Conference (VNC)*, Los Angeles, CA, USA, pp. 1–4 (2019).

31. Ryan, C., Murphy, F., & Mullins, M. End-to-end autonomous driving risk analysis: A behavioural anomaly detection approach. *IEEE Transactions on Intelligent Transportation Systems*, 22, 1650–1662 (2021).
32. Yu, Y., Chen, W., Chen, F., Jia, W., & Lu, Q. Night-time vehicle model recognition based on domain adaptation. *Multimedia Tools and Applications*, 83, 9577–9596 (2023).
33. Rameau, F., Choe, J., Pan, F., Lee, S., & Kweon, I. CCTV-Calib: A toolbox to calibrate surveillance cameras around the globe. *Machine Vision and Applications*, 34, 1–21 (2023).
34. Tahir, M., Qiao, Y., Kanwal, N., Lee, B., & Asghar, M.N. Real-time event-driven road traffic monitoring system using CCTV video analytics. *IEEE Access*, 11, 139097–139111 (2023).
35. Shin, H., Na, K., Chang, J., & Uhm, T. Multimodal layer surveillance map based on anomaly detection using multi-agents for smart city security. *ETRI Journal*, 44, 183–193 (2022).
36. Ammar, A., Koubâa, A., Boulila, W., Benjdira, B., & AlHabashi, Y. A multi-stage deep-learning-based vehicle and license plate recognition system with real-time edge inference. *Sensors*, 23(4), 2120 (2023).
37. Aribilola, I., Asghar, M.N., Kanwal, N., Fleury, M., & Lee, B. SecureCam: Selective detection and encryption enabled application for dynamic camera surveillance videos. *IEEE Transactions on Consumer Electronics*, 69, 156–169, (2022).
38. Li, T., Han, X., & Ma, J. Cooperative perception for estimating and predicting microscopic traffic states to manage connected and automated traffic. *IEEE Transactions on Intelligent Transportation Systems*, 23, 13694–13707 (2022).
39. An, L., Yang, X., & Hu, J. Modeling system dynamics of mixed traffic with partial connected and automated vehicles. *IEEE Transactions on Intelligent Transportation Systems*, 23, 15755–15764 (2022).
40. Bhatia, S., & Alojail, M. A novel technique for behavioral analytics using ensemble learning algorithms in E-commerce. *IEEE Access*, 1–9, ISSN: 2169–3536 (2020).
41. Reddy, C.K.K., Anisha, P.R., Hanafah, M.M., Doss, S., & Lipert, K.J. *Intelligent Systems and Industrial Internet of Things for Sustainable Development*, CRC Press, Taylor & Francis, New York. (2024).
42. Reddy, C.K.K., Anisha, P.R., Khan, S., Hanafiah, M.M., Pamulaparty, L., & Mohana, R.M. *Sustainability in Industry 5.0: Theory and Applications*, CRC Press, Taylor & Francis, New York. (2024).
43. Chakraborty, K., Bhatia, S., Bhattacharyya, S., Platos, J., Bag, R., & Hassanien, A.E. Sentiment analysis of COVID-19 tweets by deep learning classifiers—A study to show how popularity is affecting accuracy in social media. *Applied Soft Computing*, 97, 106754 (2020).

12

Piecewise Mixed-Integer Non-Linear Optimization for Supply Chain Network Design (SCND) Model: Confirmation through Cluster Analysis

Monalisha Pattnaik, Alipsa Pattnaik, Ashirbad Mishra, and Agnibrata Nayak

12.1 Introduction

Supply chain network design (SCND) is one of the most decisive planning problems in the supply chain management (SCM) system. In the current business landscape, design decisions must be practical enough to thrive in dynamic, complex, uncertain, unpredictable, and robust environments over several decades. Given the intricate nature of SCN today, it is crucial to consider various planning decisions alongside location science choices to create an integrated system. At the tactical level, dynamic supply chain (SC) decisions include determining the facility locations allocation, number of facility locations, influence areas, and facility capacities. This chapter is related to a real-life-based case study of an SCND of a single milk product, i.e., premium milk of the leading milk-producing federation of Odisha. Given the product's high perishability, SCND involves strategic decision-making. Network design serves as the foundation for efficient supply chain operations and is one of the most critical challenges faced by supply chain managers. This chapter conducts modeling using empirical data to address the slit in the area of SCND. This chapter delves into the design of milk SCN, introducing a universal model that integrates both economic and social goals, implementing advanced preservation methods. Additionally, we've tailored an optimization SC mathematical model specifically for a prominent milk federation in Odisha. These models are constructed and resolved through the application of piecewise mixed-integer non-linear optimization techniques.

Lee and Wen (2007) examined the challenges associated with transporting a single product. Their goal was to reduce holding inventory and the combined expenses of transportation and ordering, over an unlimited time period, employing a method of heuristic to do so. Li and Ouyang (2010)

 DOI: 10.1201/9781003503934-12

explored the dependable location problem with a fixed charge, considering facilities vulnerable to disruptions that are spatially correlated and have varying probabilities based on location. They analyzed the scenario where, upon a facility's failure, its clientele is rerouted to an alternate facility, resulting in increased transportation costs. Murat et al. (2010) formulated a discrete model that incorporates uncertainties regarding costs, extensive demand data, and potential site locations. Their objective was to minimize the overall cost by optimally assigning customers and situating facilities to them. Paksoy and Chang (2010) examined an SCND problem that involved popup stores, aiming to minimize the total expenses, which included transportation costs at all stages, costs of holding inventory, backordering costs, and setup costs for popup stores. Wang et al.'s (2010) research centered on the company's worldwide SC model with the primary performance metrics being the overall inventory expense, the rate of inventory turnover, and the magnitude of the bullwhip effect for producing thin film liquid crystal display products. Imran and Ramzan (2012) highlighted the optimization of a multi-objective and multi-period pharmaceutical SC model tailored for a cohesive healthcare system, utilizing fuzzy-based optimization methods for the analysis. Alejandro et al. (2013) introduced a lot-sizing problem of multi-product for developing an integer linear programming model. This model takes into account various factors such as demand, costs associated with backorders, inventory strategies, and the pursuit of resource utilization efficiency. Eskandarpour et al. (2015) examined the mixed-integer mathematical model that incorporates considerations of environmental, economic, and social factors. Pal et al. (2015) and Pattnaik (2018) crafted a model aimed at optimizing the promotional activities and retail selling price of the retailer, whereas the manufacturer was responsible for ensuring product quality and setting the optimal wholesale price. Taxakis and Papadopoulos (2016) introduced two distinct models; the first combines both linear and non-linear programming methods with mixed integers to address the design of SCN, with the objective of reducing the overall costs of these networks. Ackerman et al. (2017) formulated a mixed-integer model for a green SC, aiming to optimize the net present value, a key metric reflecting a project's economic viability through the sum of money flows over a time period. Govindan et al. (2017) conducted an extensive survey of research in SCN and logistics network design under conditions of uncertainty. Their chapter systematically covers planning decisions, paradigms, and aspects pertinent to SCND, network structure, applying robust optimization, stochastic programming, risk-averse stochastic programming, and fuzzy optimization techniques. Varesi et al. (2017) integrated the environmental, social, and economic aspects of sustainability into the design of SCN within the wine industry. Their approach aimed to reduce SC expenses while enhancing the social influence of the network, culminating in the development of a mixed-integer programming multi-objective model. Pham and Yendradee (2017) developed a model to optimize the total profit as the revenue is very much affected by the marketing endorsement policy, which, being short term, was not incorporated in the strategic nature of SCND. Ramezaanian et al. (2017) aimed to encourage blood donations and enhance the utility, while also reducing the logistical expenses from facilities to blood centers and the costs associated with relocating temporary facilities. Zhong et al. (2017) devised an integrated SCND model that reduces costs related to system location, transportation, financing, and multi-level inventory. This model concurrently establishes warehouse locations, allocates retailers to these warehouses, manages warehouse-retailer inventory policies, and sets the deferred payment periods that warehouses permit for each retailer. Jiang et al. (2019) incorporated environmental factors into conventional SCM, focusing on a sustainable supply chain that accounts for carbon emissions. He utilized a mixed-integer linear programming model to facilitate optimal decision-making regarding the selection of partners, transportation mode, and technologies, in addition to product distribution, processes of material procurement, and recovery operations. Saravanan and Anubama (2017) examined the process of selecting and prioritizing the providers of cold chain logistics within the pharmaceutical sector, focusing on the distribution of diabetic medications such as insulin. In this study, for the cold chain, diabetic drugs were regularly used in supplying insulin to various vendors. A novel algorithm for a customer-to-customer SCM system is developed by Anand and Pandian (2019). Concurrently, the study takes into account cost savings through quantity rebates for both inbound and outbound logistics transportation. Asgharizadeh et al. (2019) conducted a comprehensive literature review on sustainable SCND, covering the period from 1990 to 2016 and analyzing 261 papers. This study employed a systematic qualitative approach for review. To enhance the precision of the literature review, a content analysis method was utilized, which involved the collection of data, coding, statistical and bibliometric analysis, and interpretation of data. The chapter's primary contribution lies in its focus on sustainable SCND, a critical strategic issue that intersects with management, operations research, strategy, and logistics. Furthermore, this research significantly aids in the comprehension of sustainability and pinpoints both research gaps and the distinctive features of scientific publications on this subject. Cortinhal et al. (2019) introduced a model of linear mixed-integer programming, and it is tailored for a new multi-stage SCND challenge. His model harmonizes the selection of plant and warehouse locations and capacities with the transportation methods, choice of suppliers, as well as the distribution of various products across the network. The aim is to pinpoint the network setup that minimizes the total cost while adhering to constraints on technology, resources, and customer service standards. The model also considers the balance between manufacturing products in-house and procuring finished goods from third parties for consolidation in warehouses, determining the optimal combination of outsourcing and production. Javid and Kannan (2020) crafted a mathematical programming model with a multi-objective to streamline costs, traffic congestion, and energy uses in SC operations. This model distinctly accounts for the uncertainty of product lifetimes through a Weibull distribution and considers the impact of refrigeration in vehicles on food spoilage as a variable. Additionally, multiple product types and multiple vehicle types are considered. It also takes into account various types of vehicles and products. The chapter delves into a dairy supply chain scenario, examining how the three pillars of sustainability, also known as the three triple bottom lines of sustainability, are studied. Findings indicate that decision-makers can notably mitigate the social and environmental impacts of the SC without significantly sacrificing economic benefits. Khalaf et al. (2020) focused on enhancing the dependability and societal accountability of the hydrogen SC, in addition to its ecological and economic dimensions. A fresh objective function was introduced to bolster the network's capacity for prompt product delivery. Furthermore, an innovative reliability strategy was crafted to shield the network from interruptions. As a new sustainability pointer, social considerations were also integrated as a new measure of sustainability in the hydrogen SC's framework. Lastly, a hybrid possibilistic flexible programming technique was suggested to guarantee the dependability of the outputs. Biuki et al. (2020) presented an integrated model addressing the trifocal of routing, location, and inventory problems, which are crucial for refining logistics systems. In light of the specific decision-making context in these industries, a dual-phase method was developed to weave the three facets of sustainability into SC operations. Following the identification of suppliers with a stronger commitment to sustainability, the issue was structured as a mixed-integer programming multi-objective model to facilitate the design of a sustainable SC. Given the nondeterministic polynomial time (NP)-hard nature of achieving the optimal solution, two meta-heuristic hybrid strategies such as sequential and parallel combinations of particle swarm optimization and genetic algorithm were employed to solve the problem. Testing on various instances revealed the parallel approach's dominance over the sequential. Additionally, it was noted that, as sustainability becomes a focal point, the manufacturing industry is inclined toward local production and decentralization in moderately and less developed areas, which tends to yield the most sustainable outcomes. Over the past several decades, scholars such as Dave and Patel (1981), Kang and Kim (1983), Wee (1997), Lodree and Uzochukwu (2008), Bhunia et al. (2009), Chang et al. (2010), and Pattnaik (2018) have conducted research on deteriorating inventory. Their studies have primarily concentrated on aspects like quantity discount, constant or functional deterioration rate, and supply chain co-ordination. Despite this, the investment in preservation technologies (PTs) to diminish the rate of inventory deterioration has not been extensively explored in recent years. The application of PT could shrink the deterioration rate significantly. Implementing PT could significantly lower the deterioration rate. Furthermore, the figures for sales, order quantities, and stock levels, are highly influenced by the deterioration rate, particularly in the case of products that deteriorate faster. Yang and Wee (2006) observed that an increased deterioration rate leads to higher overall annual costs and reduced demand. Ouyang et al. (2006) and Pattnaik (2018) discovered that, by enhancing storage facilities to effectively lower the deterioration rate, retailers could cut down on the total yearly inventory costs. Wang et al. (2007a,b) concentrated on selecting a resource portfolio and distributing these resources across different orders of each replenishment cycle. Many manufacturers invest in equipment that slows down deterioration rates and prolongs product shelf life. For example, supermarkets use refrigeration equipment to decrease the spoilage rate of perishables like fruits, flowers, and seafood. Murr and Morris (1975) demonstrated that lower temperatures can extend storage duration and reduce spoilage. Additionally, vacuum technology has been employed to slow the spoilage of medicines and food items. Zauberman et al. (1990) innovated a technique to preserve the color of litchi fruits using SO_2 fumigation. This chapter primarily examines the balance between the higher costs of this investment and the potential profit gains from a reduced rate of fruit deterioration. Previous research on perishable goods has treated the deterioration rate as an exogenous parameter, which could not be controlled by the company. However, in reality, companies can employ specific equipment or methods, like freezers and cold storage, to manage deterioration. As noted by Dye and Hsieh (2012), numerous companies have implemented different technologies of preservation to mitigate the effects of deterioration for enhancing profitability. Hsu et al. (2010) established policies for restocking and investing in PT, allowing for partial backlogging and assuming a steady rate of deterioration. Dye and Hsieh (2012) and Dye (2013) examined the costs of PT, which vary with the duration of the restocking cycle. Dye and Hsieh (2013) also analyzed how the costs of investing in PT impact the equipment used to slow deterioration, within the context of two-level forward financing. The present study takes into account both the costs of PT and how the rate of deterioration is affected by the efforts of preservation.

This chapter introduces an SCND model that determines the scope of influence, the specific number of facility locations, such as distribution centers (DCs), the timing of joint replenishment cycles, and the flows of materials and finances, all within the context of investing in PT. This pioneering real-world case study simultaneously refines the SCND and the multi-tiered inventory restocking for a perishable item, namely milk product. For this real-life case study, a specialized mathematical model has been crafted to address the comprehensive SCND challenge through a piecewise mixed-integer non-linear optimization method. Through real-life experiments, this novel model is evaluated against conventional SCND that do not utilize PT, focusing on how such technology influences key decisions like the ideal number of facility locations (proposed DC) and annual SCND profit per time unit. The findings suggest that incorporating financial factors can lead to substantial cost reductions. From the machine learning technique, cluster analysis is employed to corroborate the optimal DC outcomes for the milk SCND model.

The structure of this chapter is outlined below: Section 12.2 details the assumptions and notations employed. Section 12.3 develops the mathematical model, along with an explanation of the optimization methods and solution approaches. Section 12.4 delves into the case study specifics, the tailored mathematical model, and the problem definition. Section 12.5 is dedicated to the presentation and discussion of the results and analysis. The study culminates in Section 12.6 with a summary and concluding remarks.

12.2 Assumptions and Notations

The SCND model of milk products is derived using the following assumptions and notations:

1. In irregular service regions, different shapes are compared with circles, hexagons, and squares. This irregular service region is verified to have little control over the optimum solution of an SCND problem (Dasci and Verter, 2001).
2. In the service region, the facility location of each DC is centrally located.
3. Annual SCND profit per unit time for two cases are:

$$\pi_j(A_i,T,\tau)=\begin{cases}\pi_I(A_i,T,\tau), & \text{when } C_i \geq A_i \text{ or } C_i < A_i \\ \pi_{II}(A_i,T,\tau), & \text{when } C_i \geq A_i\end{cases}$$

4. In ith cluster number of existing DC is 1. So no. of DC=no. of cluster=1.
5. A particular DC is assigned and served for each retailer. The location of retailers should be located within the influence area of that DC.
6. After implementing the PT to the deteriorated product, the quantity of deterioration of the products declines. In this chapter, as the milk product is taken for study, the deterioration rate is indirectly related to the PT efforts.
7. Preservation effort cost is directly related to preservation efforts.

TABLE 12.1

Notations Used in Mathematical Model

T	Common cycle time of replenishment (decision variable)
N	Total number of clusters
i	Index of each distribution center (DC), $i = 1, 2, \ldots N$
m_i	Proposed DCs, $m_i = \frac{i\text{th service area}}{i\text{th influence area}} = \frac{C_i}{A_i}$ (decision variable); integer
j	Case index, where $j \in [\text{I}, \text{II}]$, for case-I: $C_i \geq A_i$ or $C_i < A_i$ and for case-II: $C_i \geq A_i$
A_i	Cluster i influence area for each DC & $A_i > 0$ (decision variable)
τ	Effort factor for preservation (decision variable)
Q_i	Cluster i ordering quantity for each DC (decision variable)
$I_i(t)$	ith cluster inventory level for time t
θ	Deterioration rate of the product, $\theta = \alpha - \beta\tau$, $\forall$ α and β are positive constants
F	Facility cost
δ_i	Cluster i store density
λ_i	ith cluster annual demand rate of a retailer (deterministic)
C_f	Per inbound shipping fixed cost
C_v	Per item variable cost for each inbound shipping
C_r	Per unit distance outbound transportation cost per item
f_r	Constant; depends on the distance metric and it determines the shape of the DC of service region
$K(\tau)$	Per unit item preservation effort cost, $K(\tau) = b\tau^2$, $\forall$ b is a positive constant
R	Per order ordering cost
h	Per unit holding cost per unit of time
C_i	ith cluster service area
p	Per unit selling price
c	Per unit purchasing cost
π_j	Per unit time annual SCND profit for case j

Table 12.1 presents the notations used in the mathematical models:

In this chapter π_j is evaluated for two cases. Hence, the objective is to determine A_i, T, τ for both the cases, m_i for Case II only, Q_i for the case j, $j \in [I, II]$.

12.3 Mathematical Model

In this section, mathematical model is formulated for SCND under PT.

$$\frac{dI_i(t)}{dt} = -\theta I_i(t) - \lambda_i \delta_i C_i, \; i = 1, 2, \ldots N \; \& \; 0 \leq t \leq T \tag{12.1}$$

$$\frac{dI_i(t)}{dt} + \theta I_i(t) = -\lambda_i \delta_i C_i$$

After multiplying the integrating factor (IF) $e^{\int \theta dt} = e^{t\theta}$with Equation (12.1)

$$\frac{dI_i(t)e^{t\theta}}{dt} = -\lambda_i\delta_i C_i e^{t\theta} \Rightarrow \int dI_i(t)e^{t\theta} = -\int \lambda_i\delta_i C_i e^{t\theta}dt \Rightarrow I_i(t)e^{t\theta}$$

$$= \frac{-\lambda_i\delta_i C_i e^{t\theta}}{\theta} + c_a \Rightarrow I_i(t) = \frac{-\lambda_i\delta_i C_i}{\theta} + \frac{c_a}{e^{t\theta}},\ c_a \text{ is an arbitrary constant} \tag{12.2}$$

It is known that $I_i(T) = 0 \Rightarrow$ for $t = T,\ I_i(t) = 0$.

So

$\Rightarrow 0 = \frac{-\lambda_i\delta_i C_i}{\theta} + \frac{c_a}{e^{T\theta}} \Rightarrow c_a = \frac{\lambda_i\delta_i C_i e^{T\theta}}{\theta}$, after putting the value of c_a in Equation (12.2).

$$I_i(t) = \frac{-\lambda_i\delta_i C_i}{\theta} + \frac{\lambda_i\delta_i C_i e^{T\theta}}{\theta e^{t\theta}} \Rightarrow I_i(t) = \frac{\lambda_i\delta_i C_i}{\theta}\left[e^{\theta(T-t)} - 1\right], \text{for } 0 \le t \le T \tag{12.3}$$

$I_i(0)$ is the initial level of inventory of ith cluster. It can be known as the order quantity of the product.

$$Q_i = I_i(0) = \frac{\lambda_i\delta_i C_i}{(\alpha - \beta\tau)}\left[e^{(\alpha-\beta\tau)T} - 1\right] \tag{12.4}$$

12.3.1 Components of Total SCND Profit per unit Time

- Total sales revenue (TSR): $TSR = \sum_{i=1}^{N} p\lambda_i\delta_i C_i$
- Total purchasing cost (TPC): $TPC = \sum_{i=1}^{N} c\frac{Q_i}{T} = \sum_{i=1}^{N} \frac{c\lambda_i\delta_i C_i\left[e^{(\alpha-\beta\tau)T} - 1\right]}{(\alpha-\beta\tau)T}$
- Total inventory holding cost is (THC):

$$THC = \sum_{i=1}^{N} h \times \int_0^T I_i(t)dt = \sum_{i=1}^{N} h \times \int_0^T \frac{\lambda_i\delta_i C_i}{\theta}\left[e^{\theta(T-t)} - 1\right]dt$$

$$= \sum_{i=1}^{N} \frac{h\lambda_i\delta_i C_i}{(\alpha-\beta\tau)^2 T}\left[e^{(\alpha-\beta\tau)T} - (\alpha-\beta\tau)T - 1\right]$$

Total facility cost (TFC): $TFC = \sum_{i=1}^{N} F\frac{C_i}{A_i}$

- Total inbound transportation cost (TITC):
$$\text{TITC}=\sum_{i=1}^{N}\frac{[C_f+[C_v\lambda_i\delta_iC_i\left[e^{(\alpha-\beta\tau)T}-1\right]]/(\alpha-\beta\tau)]}{T}$$
- Total outbound transportation cost (TOTC): $\text{TOTC}=\sum_{i=1}^{N}C_rf_r\sqrt{A_i}\lambda_i\delta_iC_i$
- Total preservation effort cost (TPEC):$\text{TPEC}=\sum_{i=1}^{N}b\tau^2\lambda_i\delta_iC_i$
- Total ordering cost (TOC): $\text{TOC}=\sum_{i=1}^{N}\frac{R}{T}\frac{C_i}{A_i}$

Per unit time annual SCND profit for case j, where $j\in[I,II]$; π_j is given by:

$$\pi_j=\text{TSR}-\text{TPC}-\text{THC}-\text{TFC}-\text{TITC}-\text{TOTC}-\text{TPEC}-\text{TOC}$$

$$\pi_j=\sum_{i=1}^{N}\left[\begin{array}{l} p\lambda_i\delta_iC_i-\dfrac{c\lambda_i\delta_iC_i\left[e^{(\alpha-\beta\tau)T}-1\right]}{(\alpha-\beta\tau)T}-\dfrac{h\lambda_i\delta_iC_i}{(\alpha-\beta\tau)^2T}\left[e^{(\alpha-\beta\tau)T}-(\alpha-\beta\tau)T-1\right]\\ -F\dfrac{C_i}{A_i}-\dfrac{[C_f+[C_v\lambda_i\delta_iC_i\left[e^{(\alpha-\beta\tau)T}-1\right]]/(\alpha-\beta\tau)]}{T}\\ -C_rf_r\sqrt{A_i}\lambda_i\delta_iC_i-b\tau^2\lambda_i\delta_iC_i-\dfrac{R}{T}\dfrac{C_i}{A_i} \end{array}\right] \tag{12.5}$$

12.3.2 Optimization

The objective is to determine A_i^* (optimum influence area) in each DC of ith cluster, τ^*(optimum preservation effort factor), T^* (optimum cycle time of replenishment), to maximize the annual SCND profit $\pi_j(A_1, A_2, A_{3,\ldots,} A_N, T, \tau)$, $j\in[\text{I,II}]$. It is a two-branch function with $N+2$ decision variables. The first order derivative of $\pi_j(A_1, A_2, A_{3,\ldots,} A_N|T, \tau)$ with respect to A_i for given T and τ is:

$$\frac{\partial\pi_j(A_1, A_2, A_3, \ldots, A_N|T,\tau)}{\partial A_i}=\left(\frac{FC_i}{A_i^2}\right)+\left(\frac{RC_i}{TA_i^2}\right)-\left(\frac{C_rf_r\lambda_i\delta_iC_i}{2A_i^{1/2}}\right),\ i=1,2,\ldots N$$

$$\frac{\partial^2\pi_j(A_1, A_2, A_3, \ldots, A_N|T,\tau)}{\partial A_i^2}=-\left(\frac{2FC_i}{A_i^3}\right)-\left(\frac{2RC_i}{TA_i^3}\right)+\left(\frac{C_rf_r\lambda_i\delta_iC_i}{4A_i^{3/2}}\right),\ i=1,2,\ldots,N$$

From $\dfrac{\partial \pi_j(A_1, A_2, A_3, \ldots, A_N|T, \tau)}{\partial A_i} = 0,$ it can get the threshold of $F = \left\{\dfrac{C_r f_r \lambda_i \delta_i A_i^{3/2}}{2} - \dfrac{R}{T}\right\}.$

This means $\dfrac{\partial^2 \pi_j(A_1, A_2, A_3, \ldots, A_N|T, \tau)}{\partial A_i^2} < 0$ when, $F > \left\{\dfrac{C_r f_r \lambda_i \delta_i A_i^{3/2}}{2} - \dfrac{R}{T}\right\}.$

For the large value of facility opening cost F, $\dfrac{\partial^2 \pi_j(A_1, A_2, A_3, \ldots, A_N|T, \tau)}{\partial A_i^2} < 0$ is satisfied in general case. A_i is independent to A_k, where $i = 1, 2 \ldots N; k = 1, 2 \ldots N$ but $i \neq k$. From the maximum theorem, $\pi_j(A_1, A_2, A_3, \ldots, A_N|T, \tau)$ is satisfying the property of concavity with a function of $(A_1, A_2, \ldots, A_N)$. Hence, optimum $A_i(T)$, $i = 1, 2, \ldots N$, can be derived by solving $\dfrac{\partial \pi_j(A_1, A_2, A_3, \ldots, A_N|T, \tau)}{\partial A_i} = 0.$

$$A_i(T) = \left[\frac{2\left(F + \frac{R}{T}\right)}{C_r f_r \lambda_i \delta_i}\right]^{2/3}, \quad i = 1, 2 \ldots N. \tag{12.6}$$

12.3.3 Solution Procedure

Substituting $A_i(T)$, $i = 1, 2, \ldots N$ of Equation (12.6) with $\pi_j(A_1, A_2, A_3, \ldots, A_N|T, \tau)$ reduces the model with a two-branch function under T and τ (two decision variables):

$$\pi_j(T, \tau) = \begin{cases} \pi_{\text{I}}(T, \tau), \text{ when } C_i \geq A_i \text{ or } C_i < A_i \\ \pi_{\text{II}}(T, \tau), \text{ when } C_i \geq A_i \end{cases}$$

$$\pi_j(T, \tau) = \sum_{i=1}^{N} \left[\begin{array}{l} p\lambda_i\delta_i C_i - \dfrac{c\lambda_i\delta_i C_i\left[e^{(\alpha-\beta\tau)T} - 1\right]}{(\alpha-\beta\tau)T} - \dfrac{h\lambda_i\delta_i C_i}{(\alpha-\beta\tau)^2 T}\left[e^{(\alpha-\beta\tau)T} - (\alpha-\beta\tau)T - 1\right] \\ -\left(F + \dfrac{R}{T}\right)\dfrac{C_i}{\left[\dfrac{2\left(F + \frac{R}{T}\right)}{C_r f_r \lambda_i \delta_i}\right]^{2/3}} - \dfrac{[C_f + [C_v\lambda_i\delta_i C_i\left[e^{(\alpha-\beta\tau)T} - 1\right]]/(\alpha-\beta\tau)]}{T} - \\ C_r f_r \lambda_i \delta_i C_i \left[\dfrac{2\left(F + \frac{R}{T}\right)}{C_r f_r \lambda_i \delta_i}\right]^{1/3} - b\tau^2\lambda_i\delta_i C_i \end{array}\right] \tag{12.7}$$

The solution procedures of cases like I and II to obtain the optimal values of T^* and τ^* from $\pi_j(T,\tau)$ are the same. So, for obtaining T^* and τ^* to maximize $\pi_j(T,\tau)$ must satisfy:

$$\frac{\partial \pi_j(T|\tau)}{\partial T} = \sum_{i=1}^{N} \left[\begin{array}{l} \left[\frac{c\lambda_i\delta_i C_i}{(\alpha-\beta\tau)T}\left[-(\alpha-\beta\tau)e^{(\alpha-\beta\tau)T} + \frac{e^{(\alpha-\beta\tau)T}-1}{T} \right]\right] \\ -\left[\frac{h\lambda_i\delta_i C_i}{T(\alpha-\beta\tau)^2}\left[e^{(\alpha-\beta\tau)T}(\alpha-\beta\tau)-(\alpha-\beta\tau)-\left[e^{(\alpha-\beta\tau)T}-(\alpha-\beta\tau)T-1\right]\frac{1}{T}\right]\right] \\ +\frac{1}{T^2}\left[C_f + \left[\frac{\left[C_v\lambda_i\delta_i C_i\left\{e^{(\alpha-\beta\tau)T}-1\right\}\right]}{(\alpha-\beta\tau)}\right]\right] \\ -\frac{1}{T}\left[C_v\lambda_i\delta_i C_i e^{(\alpha-\beta\tau)T}\right] + \frac{C_iR}{T^2}\left[\frac{2\left(F+\frac{R}{T}\right)}{C_r f_r\lambda_i\delta_i}\right]^{-2/3} \\ -\frac{2R}{C_r f_r\lambda_i\delta_i T^2}\left[\frac{2}{3}C_i\left(F+\frac{R}{T}\right)\left[\frac{2\left(F+\frac{R}{T}\right)}{C_r f_r\lambda_i\delta_i}\right]^{-5/3}\right] \end{array}\right]$$

$$-\frac{2R}{C_r f_r\lambda_i\delta_i T^2}\left[\frac{2}{3}C_i\left(F+\frac{R}{T}\right)\left[\frac{2\left(F+\frac{R}{T}\right)}{C_r f_r\lambda_i\delta_i}\right]^{-5/3}\right] - C_r f_r\lambda_i\delta_i C_i\left[\frac{2\left(F+\frac{R}{T}\right)}{C_r f_r\lambda_i\delta_i 0}\right]^{-2/3} = 0 \tag{12.8}$$

and

$$\frac{\partial \pi_j(\tau|T)}{\partial \tau} = \sum_{i=1}^{N} \left[\begin{array}{l} \frac{c\lambda_i\delta_i C_i\beta}{T}\left[\frac{Te^{(\alpha-\beta\tau)T}}{(\alpha-\beta\tau)} - \left[e^{(\alpha-\beta\tau)T}-1\right]\right] - \frac{h\lambda_i\delta_i C_i}{T(\alpha-\beta\tau)^2} \\ \left[-e^{(\alpha-\beta\tau)T}\beta T + \beta T + \frac{2\beta}{(\alpha-\beta\tau)}\left[e^{(\alpha-\beta\tau)T}-(\alpha-\beta\tau)T-1\right]\right] - \\ \frac{1}{T}\left[\frac{\beta C_v\lambda_i\delta_i C_{i\left\{e^{(\alpha-\beta\tau)T}-1\right\}}}{(\alpha-\beta\tau)^2} + \frac{T\beta C_v\lambda_i\delta_i C_{i\left\{e^{(\alpha-\beta\tau)T}\right\}}}{(\alpha-\beta\tau)}\right] - 2b\tau\lambda_i\delta_i C_i \end{array}\right] = 0 \tag{12.9}$$

As the closed-form solution could not be obtained directly from Equations (12.8) and (12.9), for the optimum solution of the SCND model, an iterative method is applied.

12.4 Case Study

Odisha's premier milk production entity is taken as a real-life-based case study in this chapter. The organization's core functions include the acquisition, processing, and distribution of milk and dairy products to economically uplift rural communities. It was founded in the year 1978, registered with the National Dairy Development Board in 1980, and operational from the year 1981. The Milk Union of Odisha serves the state where the majority of its 4.6 crore population relies on agriculture and animal husbandry, despite being economically disadvantaged with slow industrial growth, inadequate irrigation, and frequent natural disasters. The federation has been a beacon of hope providing livelihood to the rural inhabitants along with processed and safe hygienic milk to the urban inhabitants. Originating from the innovative concept of collective farming and marketing initiatives with local milk producers, the federation has bridged rural and urban development with a safe, hygienic, and secured supply of milk. Its vision is to become a globally efficient leader in milk production, with a broad and content customer base, enhancing stakeholder wealth and contributing to the state's economy. The federation's primary goal is to promote dairy farming, foster community engagement and education, and enhance reciprocated contribution. It strives to enhance per capita consumption and productivity, advocate for the production and distribution of clean milk using advanced technology, and ensure customer satisfaction through consistent service and high-quality products. The organization encourages a culture of performance and innovation, a work environment that motivates employee involvement and organizational growth, continuous skill enhancement, career progression for its workforce, and a commitment to improving the quality of life while maintaining ecological harmony.

12.4.1 Milk-Producing Federation Supply Chain of Odisha

The essential framework of Odisha's dairy federation is its SC, crucial for efficiently sourcing raw materials and distributing them to the final consumers. This system operates on a three-tier structure for the procurement of raw milk. These comprise village cooperative society (VCC), district milk union (DMU), and overarching milk federation (MF). VCC represents the foundational level of the entire SC system, functioning as a collaborative group of milk producers who collectively market their milk. This society is governed

TABLE 12.2

Information of DC and Marketing Center of Milk Product

Cluster (i)	DC	m_i (Existing Number of DC)	Marketing Center
1	Balasore	1	Balasore, Bhadrak, and Remuna
2	Tirtol	1	Tirtol, Kendrapada, and Jagatsinghpur
3	Berhampur	1	Ganjam, Berhampur, Malkanagiri, and Gajapati
4	Khurda	1	Khurda, Cuttack, Bhubaneswar, Jatni, and Puri

by a committee and a chairman, elected by the members to ensure effective management. Additional roles such as milk testers and a secretary are appointed by the committee. Producers deliver their milk to the society twice daily, and each batch is quality-checked using a lactometer. Subsequently, the society supplies the milk to the closest DMU. The Milk Union transports the collected milk to its processing facility, where it is cooled to 5°C under the supervision of the district collector, who serves as the chairman. Additionally, DMU offers technical assistance to the milk union, including services such as embryo transfer technology, artificial insemination, and feed and fodder programs. The DMU dispatches the cooled milk for processing and distribution of milk and its products to the DC or the MF. A key aspect of this SC is the financial returns from sales, which are distributed back to the producers every 10 days through the network. Authorized dealers of the premier milk-producing entity market the milk and dairy products across the state of Odisha. The urban consumer represents the final destination in the organization's supply chain. The milk is delivered in standardized hygienic polypacks using PT to certified retailers every morning via small trucks leased by the organization. These vehicles collect milk from the closest federation dairy and serve designated retail locations within their assigned cluster areas. Table 12.2 details the distribution and marketing centers for milk products of this prominent MF in the coastal region of Odisha. Figures 12.1–12.4 illustrate Odisha's location in India, the western and coastal regions of Odisha, and the DCs for milk products in Odisha and its coastal region, respectively.

12.4.2 Case Study Settings

The area of the state of Odisha is 155,707 km^2, 4.88% of the total area of the country, India, with a coastline of around 450 km. Odisha is divided into two geographical regions, namely western and coastal. Western and coastal regions have an area of 130,569.21 and 5,137.79 km^2, respectively. There are nine and four clusters in these two regions, respectively. This chapter studies

FIGURE 12.1
State Odisha in India.

four clusters of the coastal region, namely Balasore, Tirtol, Berhampur, and Khurda, and only one DC is available for distributing the milk product to the retailer. As the milk product's, i.e. premium milk, demand rate is very high, it is required to obtain the optimal A_i, T, τ, m_i, Q_i, and π_j for $i = 1,2,\ldots 4$ and j = I & II. In Case I, m_i is not possible to be found out. The condition of Case I is practically not possible since the influence area can never be greater than the service area. Theoretically and practically, the service area must be greater than or equal to the influence area, so, with practical and scientific validation, Case II model is acceptable.

12.4.3 Customized Model

π_j, $j \in [\text{I,II}]$ under PT is given by:

$$\pi_j(A_i, T, \tau) = \text{TSR} - \text{TPC} - \text{THC} - \text{TFC} - \text{TITC} - \text{TOTC} - \text{TPEC} - \text{TOC}$$

FIGURE 12.2
Western and coastal region of Odisha.

$$\pi_j = \sum_{i=1}^{4} \left[\begin{array}{l} p\lambda_i\delta_i c_i - \frac{c\lambda_i\delta_i c_i\left[e^{(\alpha-\beta\tau)T}-1\right]}{T(\alpha-\beta\tau)} - \frac{h\lambda_i\delta_i c_i}{T(\alpha-\beta\tau)^2}\left[e^{(\alpha-\beta\tau)T}-T(\alpha-\beta\tau)-1\right] \\ -F\frac{C_i}{A_i} - \frac{[C_f+[C_v\lambda_i\delta_i c_i\left[e^{(\alpha-\beta\tau)T}-1\right]]/(\alpha-\beta\tau)]}{T} - \\ C_r f_r\sqrt{A_i}\lambda_i\delta_i c_i - b\tau^2\lambda_i\delta_i c_i - \frac{R}{T}\frac{C_i}{A_i} \end{array} \right]$$

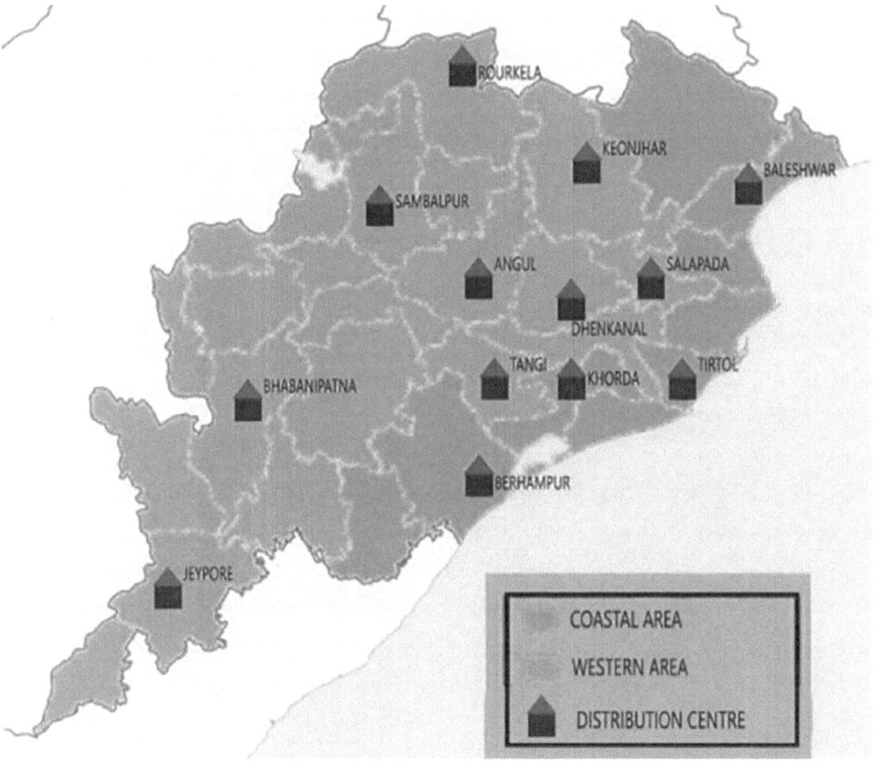

FIGURE 12.3
Distribution centers of milk products in Odisha.

12.5 Results and Analysis

Using the given datasets, the SCND profit model is justified under different cases with and without preservation efforts and is solved in a personal computer applying the software package of Lingo 13.0. The empirical data set (confidentiality) of Tables 12.3 and 12.4 is used for the present model. Tables 12.5 and 12.6 show the optimum values of the present model of Case I and Case II, respectively.

From Table 12.5, it is investigated that A_i^* of the three clusters except Tirtol is with a reduction compared to that of the service area, while A_i^* of Tirtol is more than that of the service area. Q_i^* of premium milk is the highest for Khurda (1,129,189 L) and low for Tirtol (984.9192 L). T^* and τ^* are 0.9976 and 0.4114, respectively. π_I^* is Rs. 8,605,204 with 295 iterations.

Similarly, from Table 12.6, it is investigated that A_i^* of the three clusters except Tirtol is with a reduction compared to that of the service area, while

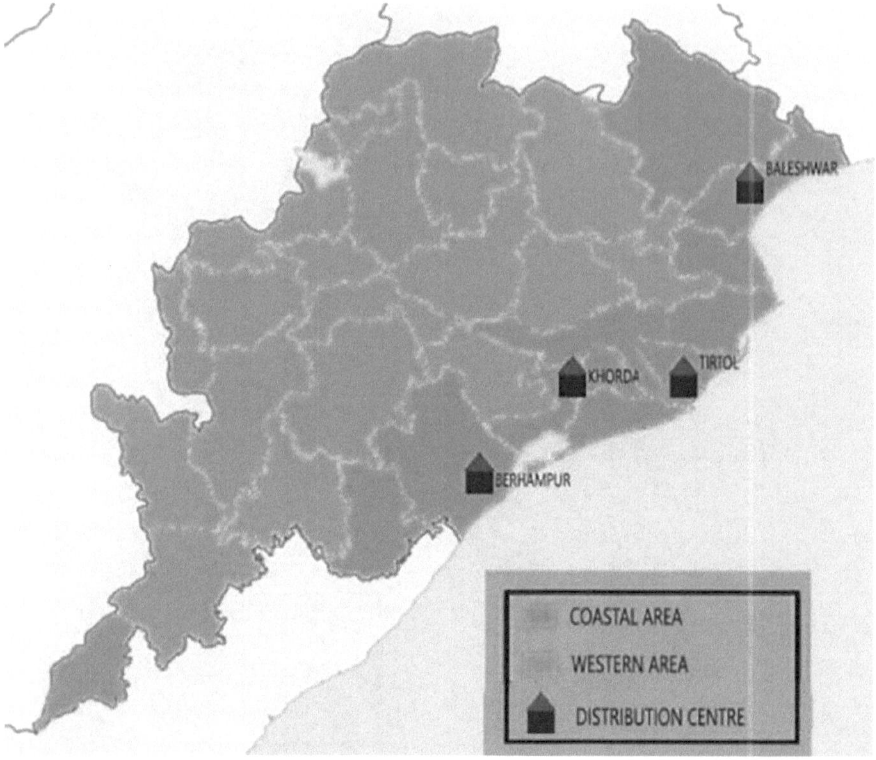

FIGURE 12.4
Distribution centers of milk products in the coastal region.

A_i^* of Tirtol is the same as its service area. So m_i^* in Balasore, Berhampur, and Khurda is 5, 3, and 23, respectively, and in Tirtol, there is only 1 DC. By adding 4 numbers of existing DCs with 28 numbers of proposed DCs, the total DCs in the coastal region should be 4+28=32. Q_i^* is the highest for Khurda (1,129,189 L) and lowest for Tirtol (684.9192 L). T^* and τ^* are 0.997645 and 0.4114745, respectively. π_{II}^* is Rs. 8,603,151 with 219 iterations.

12.5.1 Comparative Analysis

For four different models for both the cases of milk products with and without implications of PT, the comparative analysis is shown in Table 12.7.

Table 12.7 shows that there are two different SCND models under Case I, namely those with and without PT effort. It is investigated that π_{I}^* for the model with PT effort is highest at Rs. 8,605,204 and that without PT effort model is lowest at Rs. 8,584,982. Hence, the preservation effort is the catalyst for selling the premium milk product to raise SCND profit.

TABLE 12.3
Premium Milk Annual Demand Rate

Cluster (i)	DCs	C_i (km²)	Marketing Center	Daily Demand (L)	λ_i (Annual Demand Rate)
1	Balasore	16,557	Balasore	50,000	136.99
			Remuna		
			Bhadrak		
2	Tirtol	4,403	Tirtol	5,000	13.70
			Jagatsinghpur		
			Kendrapada		
3	Berhampur	17,712	Ganjam	30,000	106.85
			Berhampur		
			Malkanagiri		
			Gajapati		
4	Khurda	6,837	Khurda	200,000	547.95
			Puri		
			Cuttack		
			Bhubaneswar		
			Jatni		

Note: The data used here have been scaled to protect confidentiality.

TABLE 12.4
Data Set of Premium Milk in Coastal Region of Odisha

Parameters	Balasore	Tirtol	Berhampur	Khurda
C_i	16,557	4,403	17,712	6,837
λ_i	136.99	13.70	106.85	547.95
δ_i	0.03	0.011	0.016	0.292
F (Rs.)	3,000	3,000	3,000	3,000
P (Rs.)	38	38	38	38
c (Rs.)	28.55	28.55	28.55	28.55
h (Rs.)	0.47	0.47	0.47	0.47
R	30	30	30	30
C_r	0.75	0.75	0.75	0.75
f_r	0.01	0.01	0.01	0.01
C_f	17,531	17,531	17,531	17,531
C_v	0.83	0.83	0.83	0.83
α	0.07	0.07	0.07	0.07
β	0.005	0.005	0.005	0.005
b	0.10.1	0.1	0.1	0.1

Note: The data used here have been scaled to protect confidentiality.

TABLE 12.5

Results of Premium Milk under Case I

Cluster (i)	Distribution Center	Iteration	C_i	A_i	m_i (Existing DC)	Q_i (L)	T	τ	Cost (Rs.)	π_I (Rs.)
1	Balasore	295	16,557	3,381.226	1	70,237.52	0.9976443	0.4114745	Purchasing=35,377,240	860,5204
2	Tirtol		4,403	30,634.13	1	684.9192			Facility=93,560.90	
3	Berhampur		1,7712	6,067.615	1	31,256.44			IT=1,094,738	
4	Khurda		6,837	294.3508	1	1,129,189			OT=188,997.4; SR=45,330,880; PEC= 20,197.44;OC=937.8182; HC=0.00	

TABLE 12.6

Results of Premium Milk under Case II

Cluster (i)	Distribution Center	Iteration	C_i	A_i	m_i (Proposed DC)	Q_i (L)	T	τ	Cost (Rs.)	π_{II} (Rs.)
1	Balasore	219	16,557	3,381.226	5	70,237.52	0.997645	0.4114745	Purchasing=35,327,240	8,603,151
2	Tirtol		4,403	4,403.0	1	684.9192			Facility=96,129.72	
3	Berhampur		17,712	6,067.615	3	31,256.44			Inbound transportation=1,094,738	
4	Khurda		6,837	294.3508	23	1,129,189			Outbound transportation=188,456.6; SR=45,330,880; PEC=20,197.45; OC=963.567; HC=0.00	

TABLE 12.7

Comparative Analysis

Cluster (i)	Distribution Center	Iteration	C_i	A_i	m_i	Q_i (L)	T	τ	Total Cost (Rs.)	π_1 (Rs.)
Case I										
1	Balasore	295	16,557	3,381.226	1	70,237.52	0.9976478	0.4114747	Purchasing=35,377,240	88,605,204
2	Tirtol		4,403	30,634.13	1	684.9192			Facility=93,560.90	
3	Berhampur		17,712	6,067.615	1	31,256.44			TIT=1,094,738	
4	Khurda		6,837	294.3508	1	1,129,189			TOTC=188,997.4, TPEC=20,197.44, TOC=937.818, HC=0.00, TSR=45,330,880	
1	Balasore	192	16,557	3,381.23	1	70,299.8	0.9974981	0	Purchasing=35,366,590	8,584,982
2	Tirtol		4,403	30,634.16	1	685.5265			Facility=93,560.81	
3	Berhampur		17,712	6,067.621	1	31,284.15			TIT=1,095,807	
4	Khurda		6,837	294.3510	1	1,130,191			TOT=188,997.5, PEC=0, TOC=937.9547, HC=0.00, TSR=45,330,880	
Case II										
1	Balasore	219	16,557	3,381.226	5	70,237.52	0.997645	0.4114745	Purchasing=35,327,240	88,603,151
2	Tirtol		4,403	4,403	1	684.9192			Facility=96,129.72	
3	Berhampur		17,712	6,067.615	3	31,256.44			TIT=1,094,738	
4	Khurda		6,837	294.3508	23	1,129,189			TOT=188,456.6; TSR=45,330,880; TPEC=20,197.45; TOC=963.5637; THC=0.00	
1	Balasore	126	16,557	3,381.23	5	70,299.8	0.9974981	0	Purchasing=35,366,590	8,582,928
2	Tirtol		4,403	4,403	1	685.5265			Facility=96,129.63	
3	Berhampur		17,712	6,067.621	3	31,284.15			TIT=1,095,807	
4	Khurda		6,837	294.3510	23	1,130,191			TOT=188456.7, TPEC=0, TOC=963.7073, THC=0.00, TSR=45,330,880	

Similarly, there are two different SCND models under Case II, namely those with and without PT effort. It is seen that π_{II}^* for the model with PT effort is highest at Rs. 8,603,151 and that for the model without PT effort is lowest at Rs. 8,582,928. It is the appropriate model as the number of DCs is also increased by 1. It is also seen that A_i^* and service area of Tirtol is the same; however, A_i^* of Balasore, Berhampur, and Khurda is less than that of the service area. So m_i^* in Balasore, Berhampur, and Khurda is 5, 3, and 23, respectively, and there is only 1 DC in Tirtol. By adding 4 numbers of existing DCs with 28 numbers of proposed DCs, the total DCs in the coastal region should be 4+28=32. Q_i^* is the maximum in Khurda and minimum in Tirtol. T^* and τ^* are 0.997645 and 0.4114745, respectively. π_{II}^* of premium milk is Rs. 8,603,151 with 219 iterations. It indicates that the PT effort could be a mechanism for generating profits of SCND of milk products.

For the first model, the SCND profit is greater than that of the rest of the three models; however, practically, the last two models are adequate. So, from these last two models, it is seen that π_{II}^* of the model under PT effort is Rs. 8,603,151, which is the highest, while π_{II}^* of the model without PT effort is Rs. 8,582,928, which is the lowest. For the Case II model with PT, m_i^* is 5, 1, 3, and 23 for four clusters.

Figure 12.5 shows τ and TPEC, which indicates that it is directly related to the PT effort factor. Figures 12.6–12.9 show the mesh plot of influence areas Balasore, Tirtol, Berhampur, and Khurda, with respect to A_i, T, and π_j, respectively.

Figures 12.10–12.13 show the service area of four clusters, namely Balasore, Tirtol, Berhampur, and Khurda, respectively, of the coastal region. Figures 12.14–12.17 depict the influence area of the above four numbers of clusters under Case I, respectively, and similarly Figures 12.18–12.21 depict the influence area of the same four clusters under Case II, respectively. These figures show that the influence area of cluster Tirtol is more than that of the

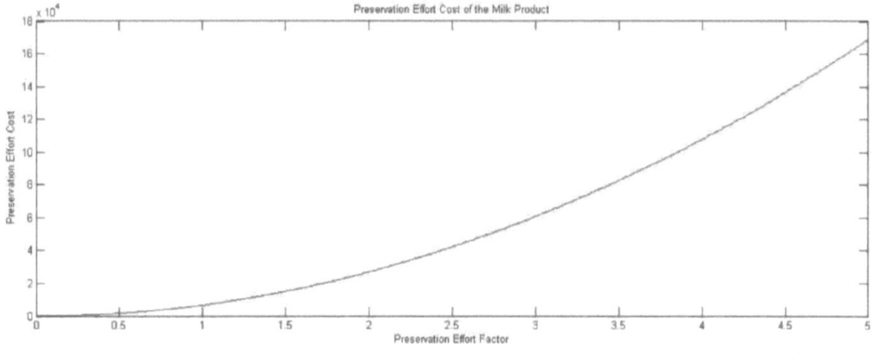

FIGURE 12.5
Preservation effort cost with respect to its effort factor.

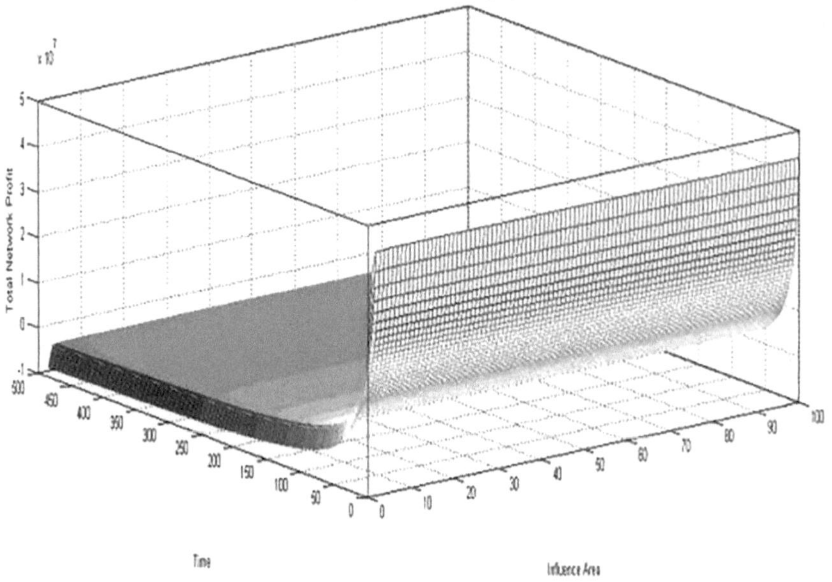

FIGURE 12.6
Mesh plot of annual SCND profit of Balasore with respect to influence area and time under Case II.

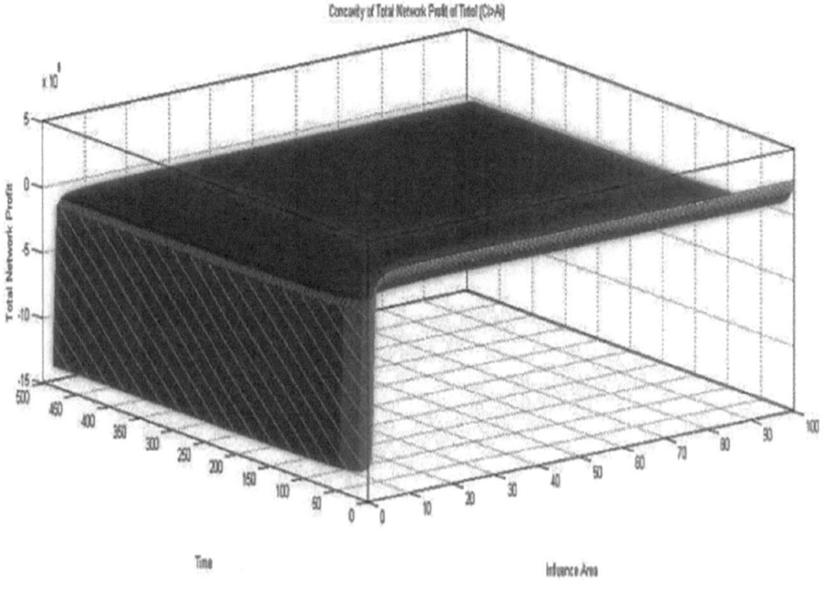

FIGURE 12.7
Mesh plot of annual SCND profit of Tirtol with respect to influence area and time under Case II.

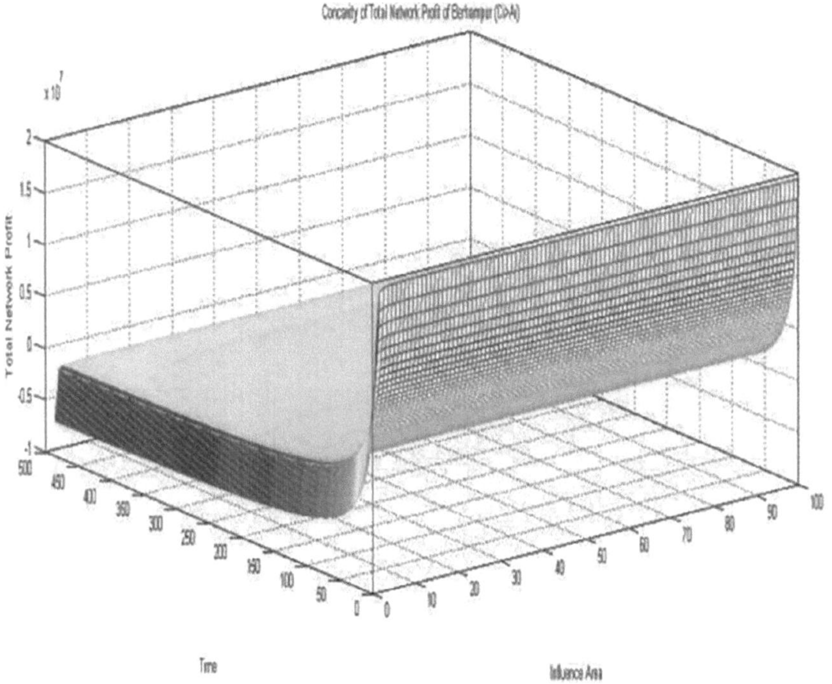

FIGURE 12.8
Mesh plot of annual SCND profit of Berhampur with respect to influence area and time under Case II.

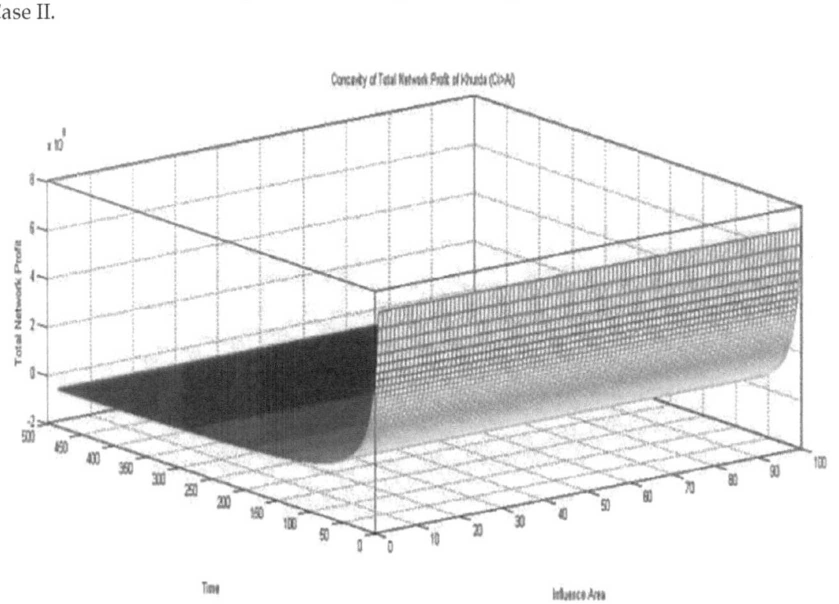

FIGURE 12.9
Mesh plot of annual SCND profit of Khurda with respect to influence area and time under Case II.

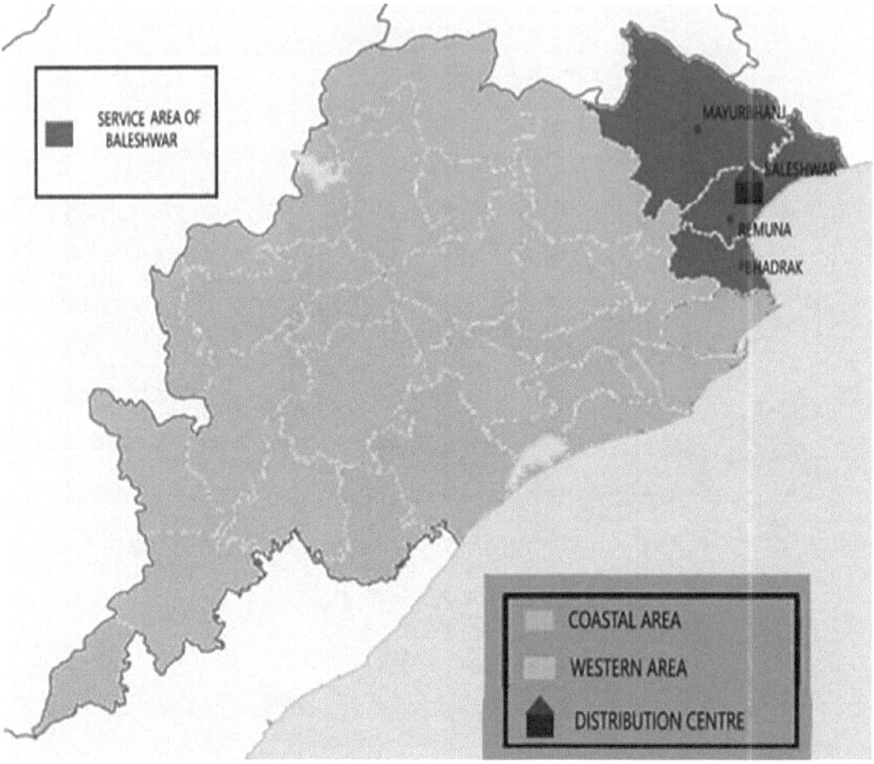

FIGURE 12.10
Service area of Balasore.

service area under Case I as it is not feasible basically. However, under Case II, the service area is more than that of the influence area for all four numbers of clusters, and it is shown in Figures 12.18–12.21. Figures 12.22–12.25 show the DCs of four clusters under Case II, respectively. Figure 12.26 shows four existing DCs, and Figure 12.27 shows 28 existing and proposed DCs like Remuna, Bhadrak, Soro, Mayurbhanj, Malkanagiri, Gajapati, Khandagiri, Bhubaneswar, Cuttack, Puri, Pipili, Kakatpur, Brahmagiri, Astaranga, Mancheswar, Patia, Tankapani Road, Mahanadi Vihar, Chauliaganj, Sikharpur, Phulnakhara, Tulasipur, CDA-6, Nua Bazar, Mangalabag, Jharpada, Bidanasi, and Jatni. Figure 12.28 shows the influence area of 28 proposed DCs like Remuna, Bhadrak, Soro, and Mayurbhanj. Figure 12.29 shows the influence area of the proposed 28 DCs like Malkanagiri and Gajapati. Figure 12.30 shows the influence area of 28 proposed DCs like Khandagiri, Bhubaneswar, Cuttack, Puri, Pipili, Kakatpur, Brahmagiri, Astaranga, Mancheswar, Patia, Tankapani Road, Mahanadi Vihar, Chauliaganj, Sikharpur, Phulnakhara, Tulasipur, CDA-6, Nua Bazar, Mangalabag, Jharpada, Bidanasi, and Jatni of coastal Odisha. Figure 12.31 shows the visualization of the distance matrices

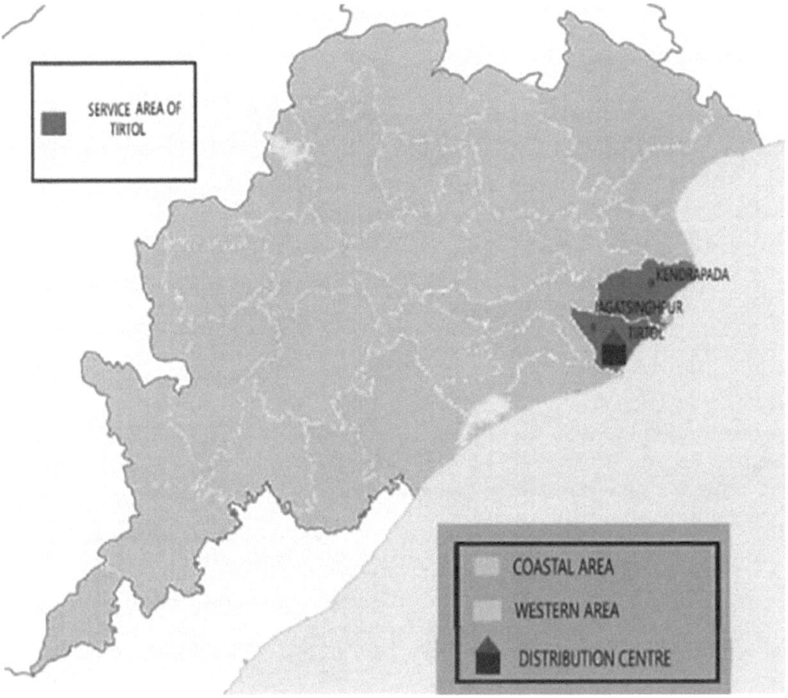

FIGURE 12.11
Service area of Tirtol.

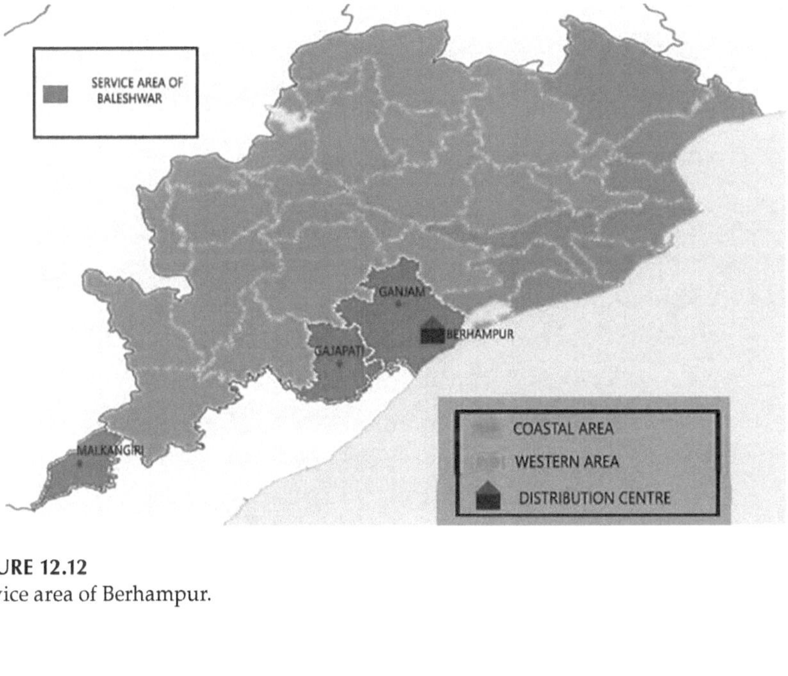

FIGURE 12.12
Service area of Berhampur.

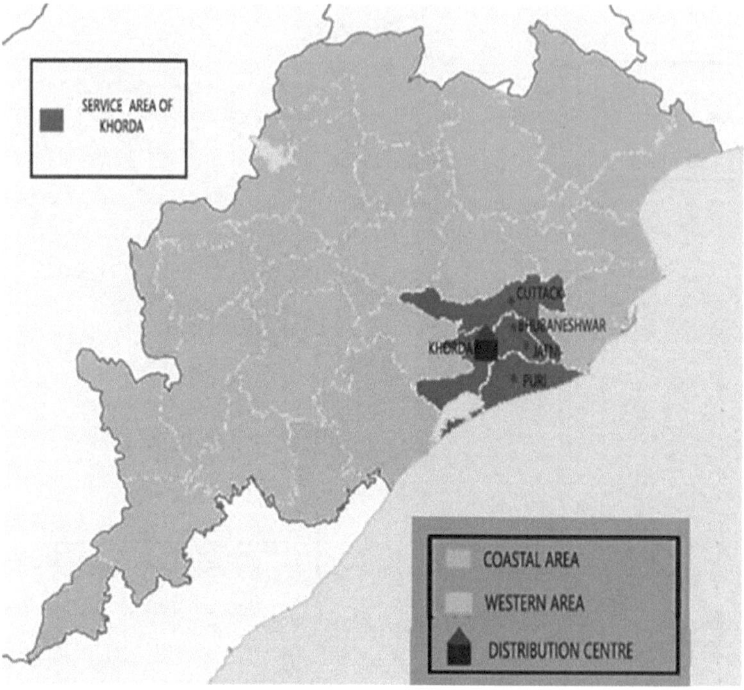

FIGURE 12.13
Service area of Khurda.

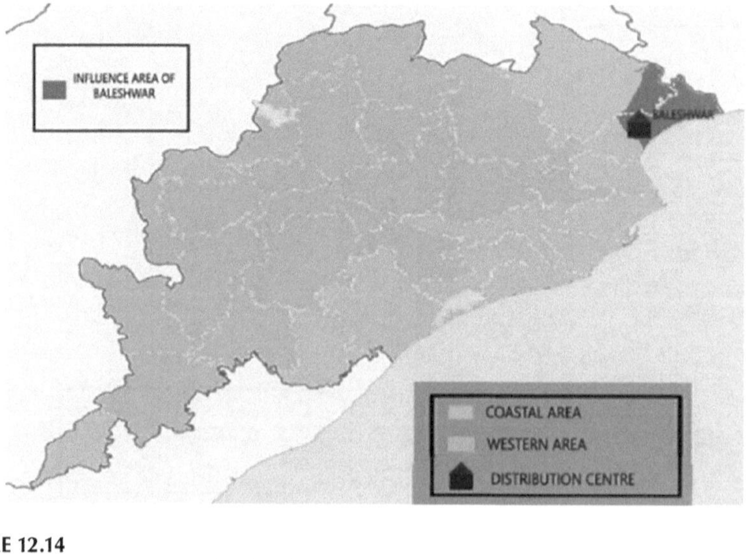

FIGURE 12.14
Influence area of Balasore under Case I.

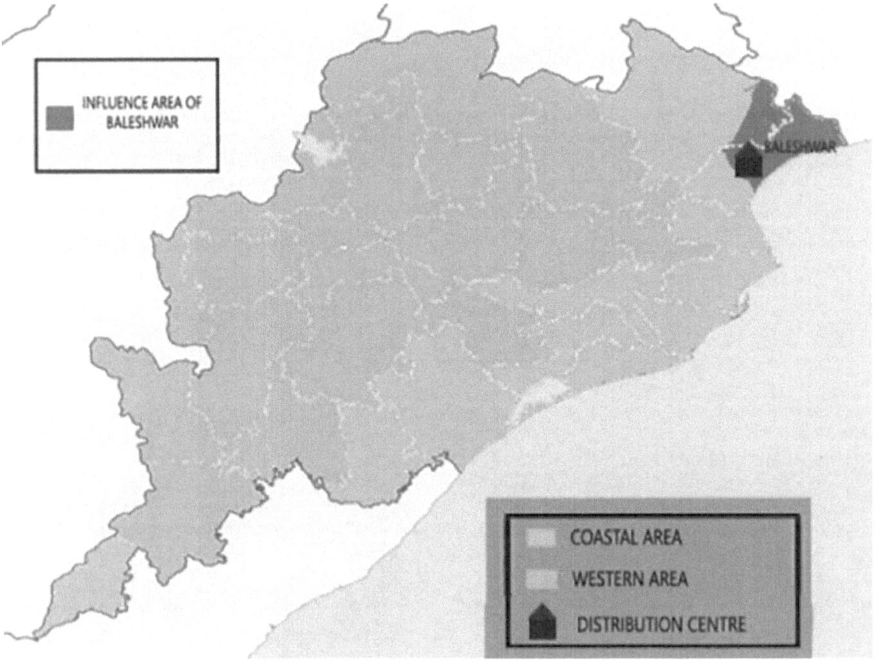

FIGURE 12.15
Influence area of Balasore under Case II.

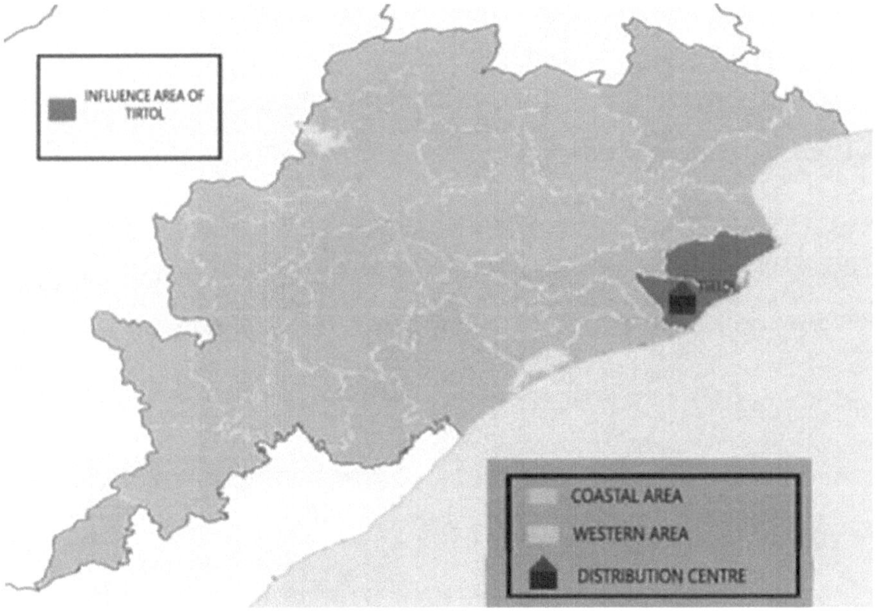

FIGURE 12.16
Influence area of Tirtol under Case I.

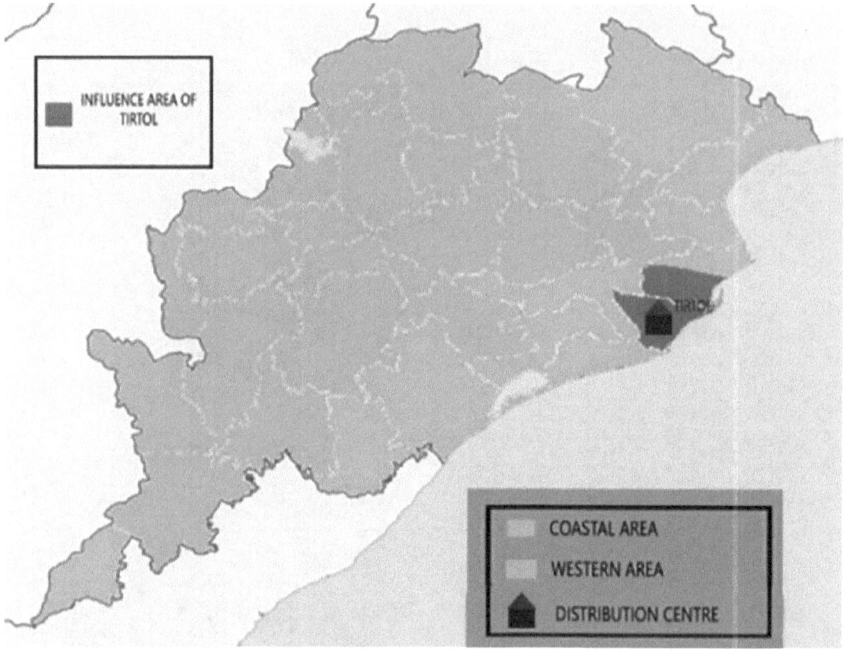

FIGURE 12.17
Influence area of Tirtol under Case II.

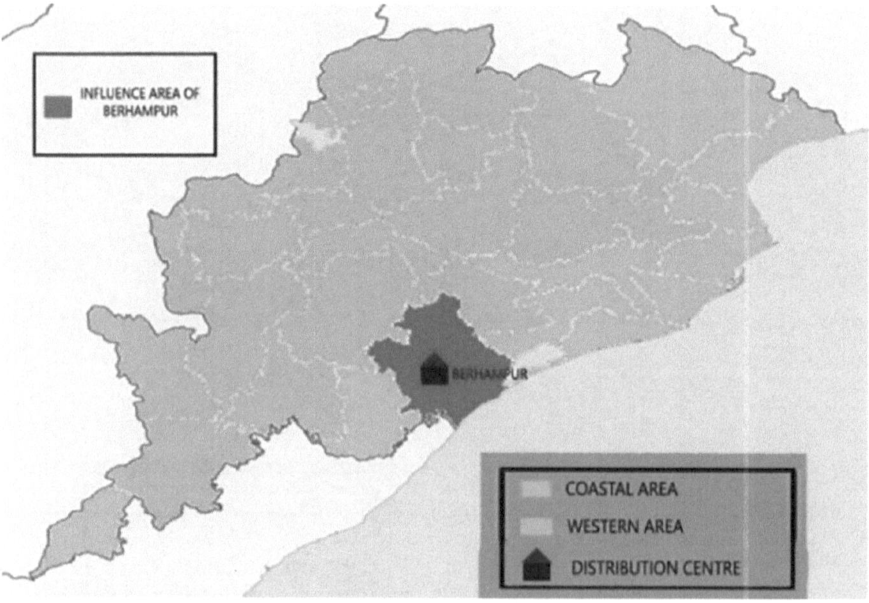

FIGURE 12.18
Influence area of Berhampur under Case I.

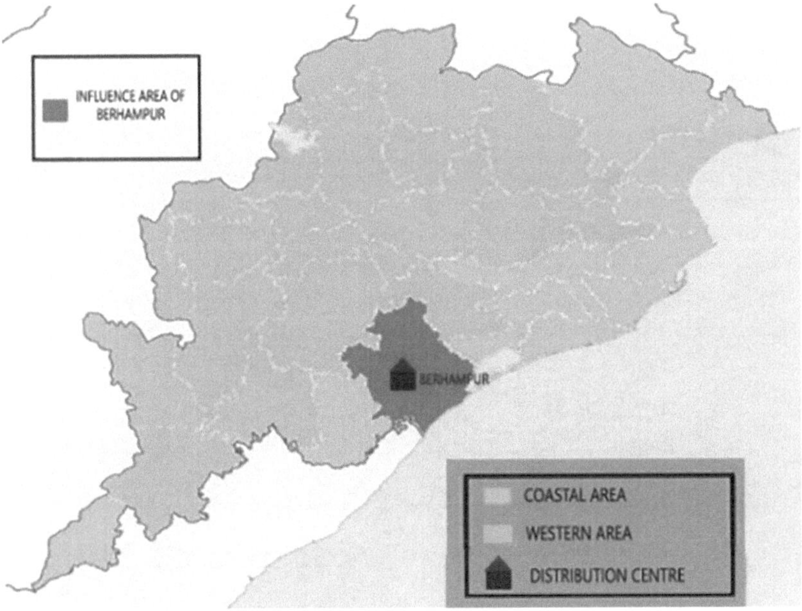

FIGURE 12.19
Influence area of Berhampur under Case II.

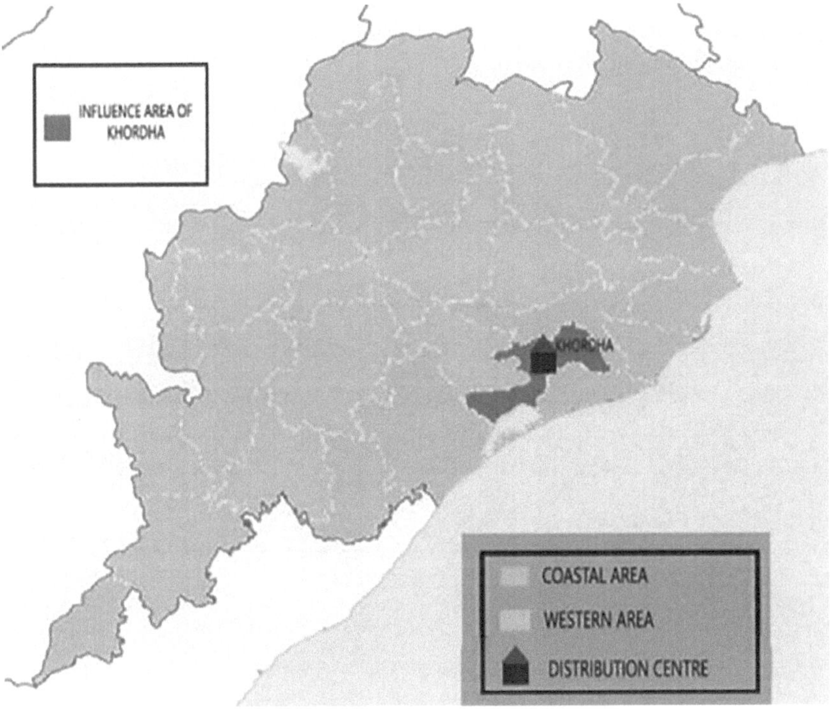

FIGURE 12.20
Influence area of Khurda under Case I.

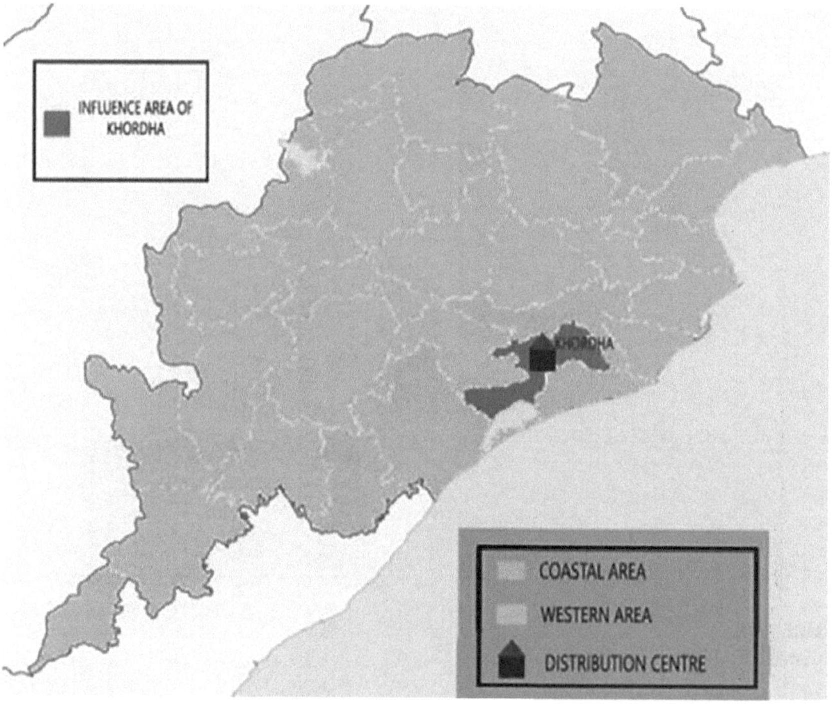

FIGURE 12.21
Influence area of Khurda under Case II.

FIGURE 12.22
DCs of Balasore under Case II.

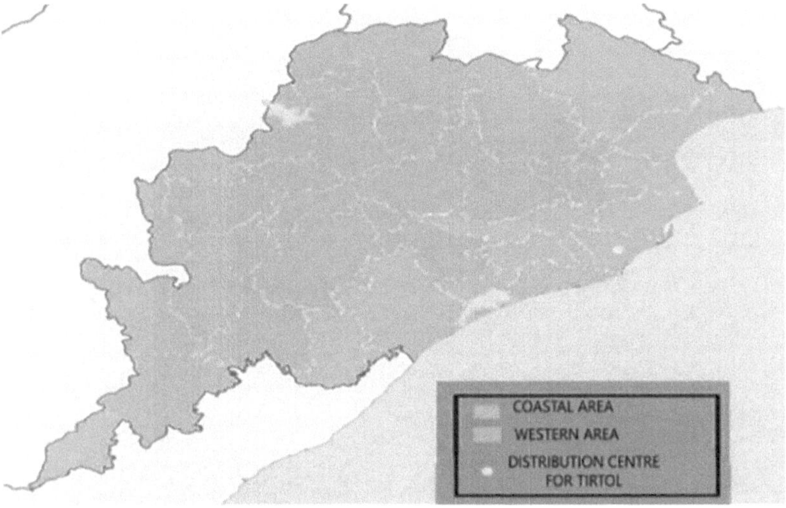

FIGURE 12.23
DCs of Tirtol under Case II.

FIGURE 12.24
DCs of Berhampur under Case II.

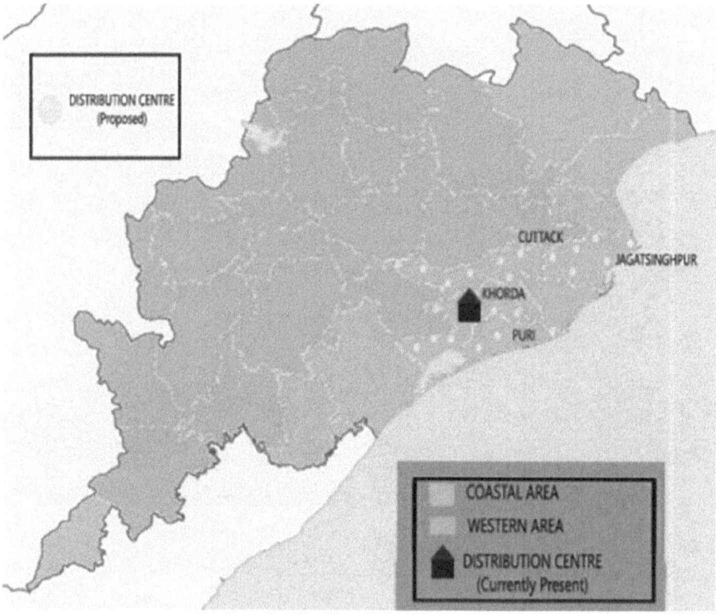

FIGURE 12.25
DCs of Khurda under Case II.

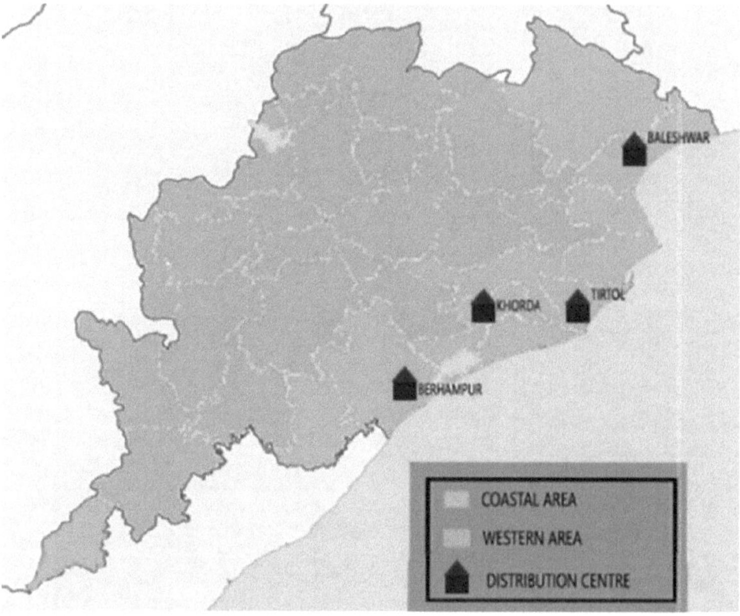

FIGURE 12.26
Existing distribution centers (four) of the coastal region of Odisha: Balasore, Tirtol, Berhampur, and Khurda.

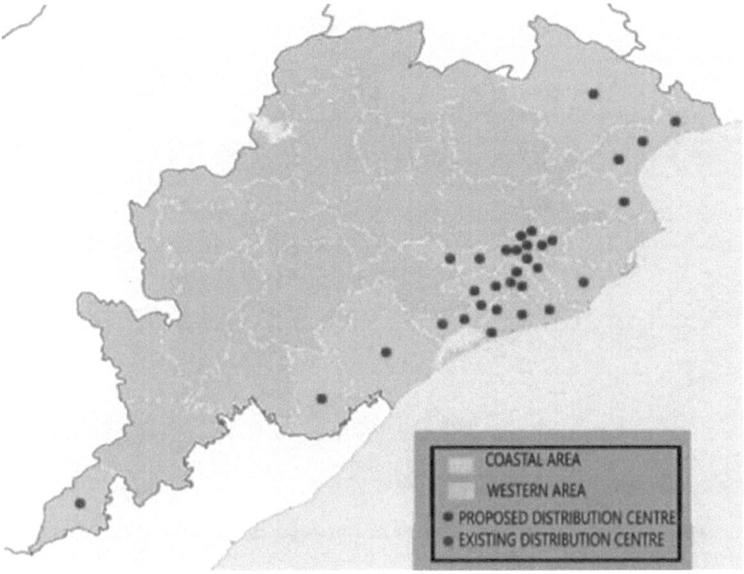

FIGURE 12.27
Existing and proposed distribution centers (4+28=32) of coastal Odisha: (existing) Balasore, Tirtol, Berhampur, and Khurda and (proposed) Remuna, Bhadrak, Soro, Mayurbhanj, Malkanagiri, Gajapati, Khandagiri, Bhubaneswar, Cuttack, Puri, Pipili, Kakatpur, Brahmagiri, Astaranga, Mancheswar, Patia, Tankapani Road, Mahanadi Vihar, Chauliaganj, Sikharpur, Phulnakhara, Tulasipur, CDA-6, Nua Bazar, Mangalabag, Jharpada, Bidanasi, and Jatni.

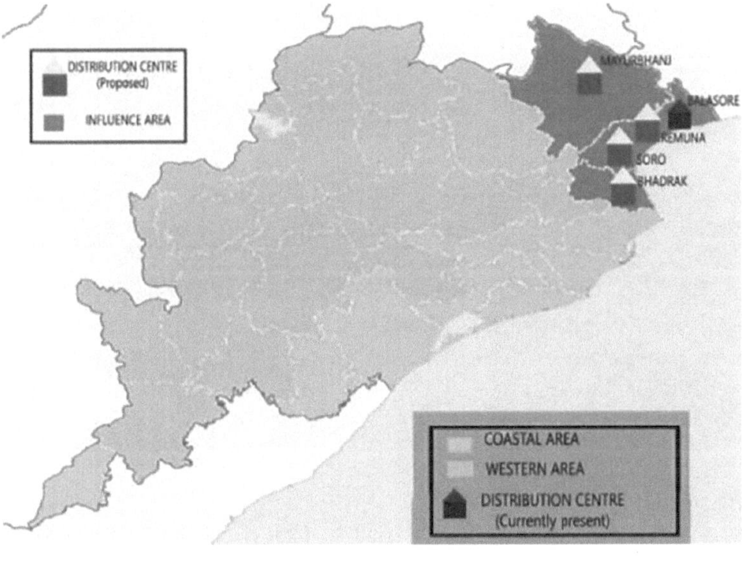

FIGURE 12.28
Influence area of proposed distribution centers: Remuna, Bhadrak, Soro, and Mayurbhanj.

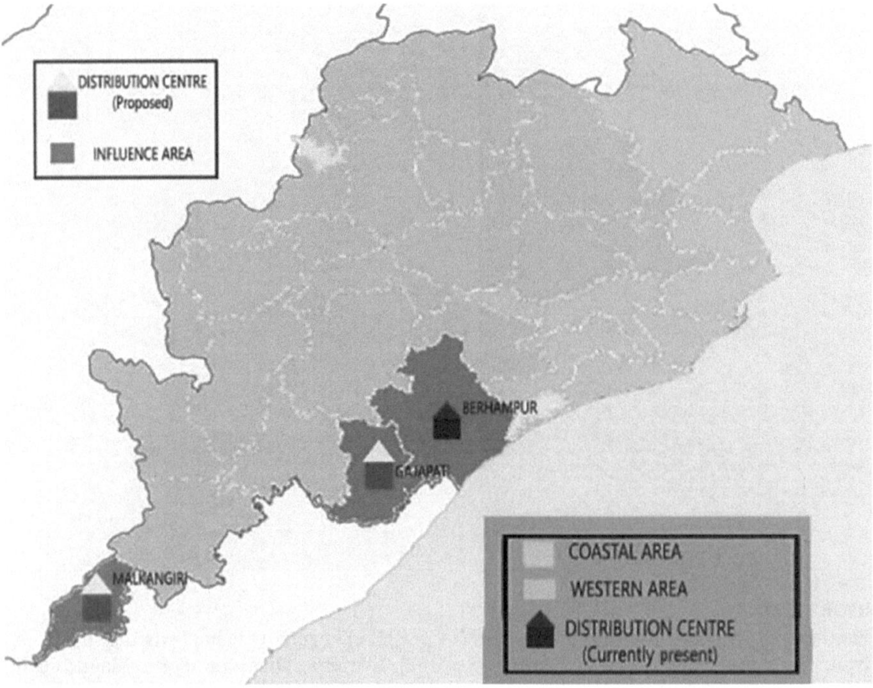

FIGURE 12.29
Influence area of proposed distribution centers: Malkanagiri and Gajapati.

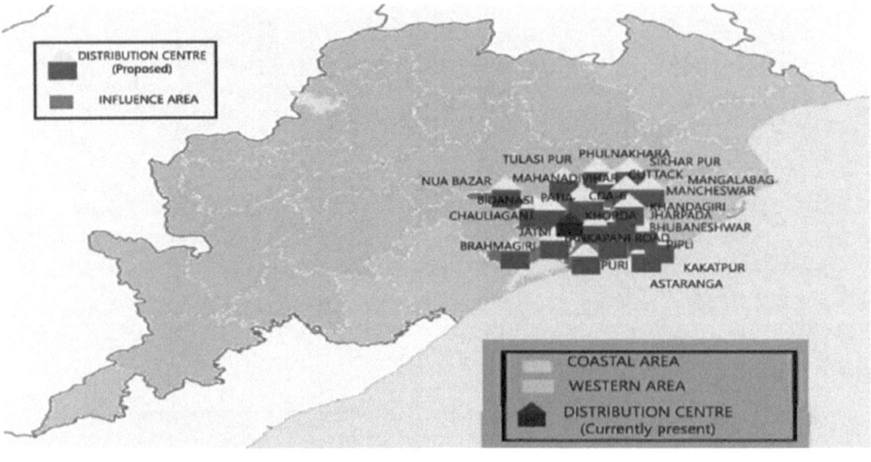

FIGURE 12.30
Influence area of proposed distribution centers (28) of coastal Odisha: Khandagiri, Bhubaneswar, Cuttack, Puri, Pipili, Kakatpur, Brahmagiri, Astaranga, Mancheswar, Patia, Tankapani Road, Mahanadi Vihar, Chauliaganj, Sikharpur, Phulnakhara, Tulasipur, CDA-6, Nua Bazar, Mangalabag, Jharpada, Bidanasi, and Jatni.

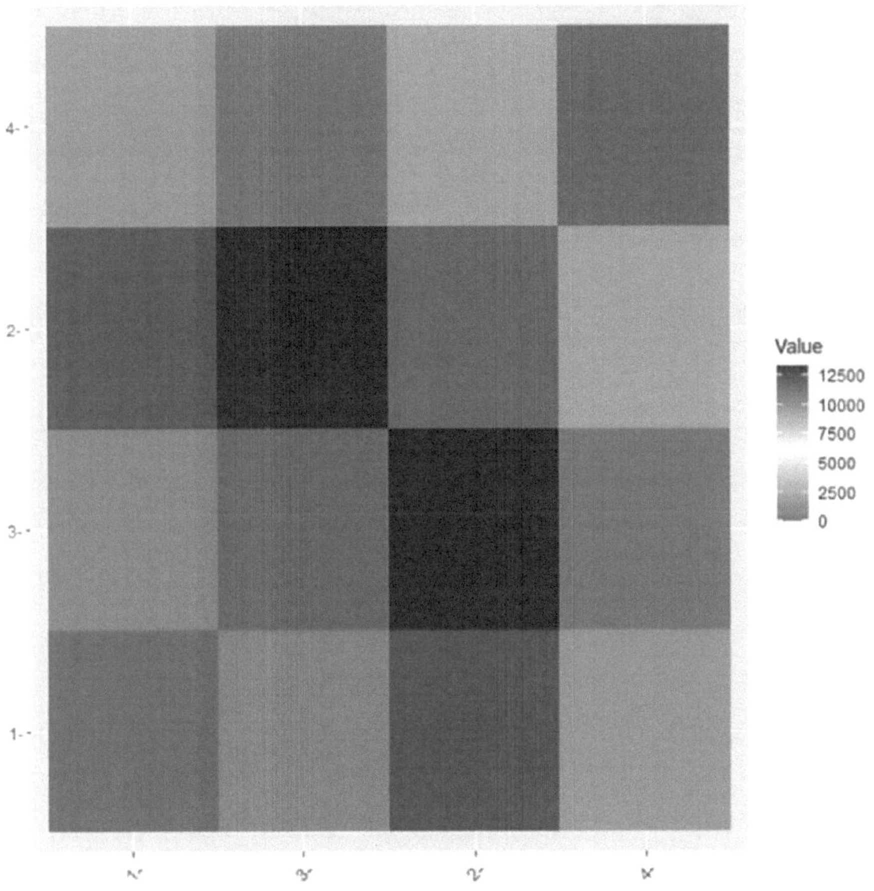

FIGURE 12.31
Visualization of the distance matrices of Case I and Case II.

of the model. Objects belonging to the same cluster are displayed in consecutive order. Figure 12.32 shows the estimate of the optimum number of clusters according to the given variables. The location of a bend in the plot is usually measured as a pointer of the suitable number of clusters applying K-means clustering for the clustering model. Figure 12.33 shows the scatter plot of all the variables and the cluster of the clusters for the SCND model. Figure 12.34 shows the bar plot of a number of clusters using K-means clustering for this model. Figure 12.35 visualizes the cut of the hierarchical tree at a given height to partition the clusters into clusters for the present SCND model.

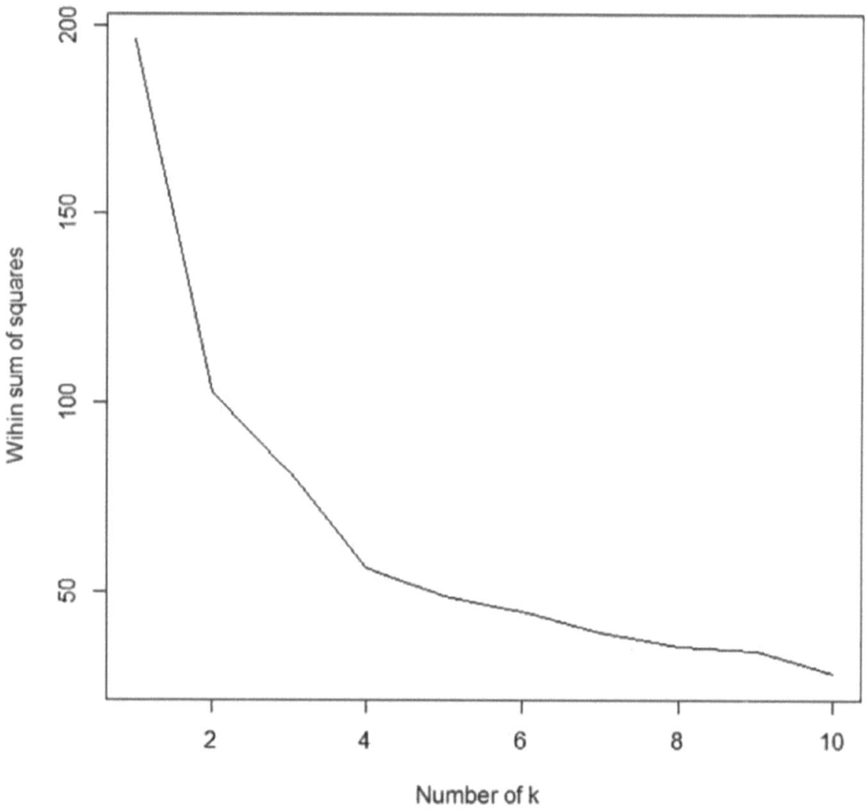

FIGURE 12.32
Estimation of the optimum number of clusters according to the given variables using K-means clustering for Case I and Case II.

12.5.2 Sensitive Analysis

It is motivating to investigate the influence of the key variables, $F, h, R, C_r, f_r, C_f, C_v, \alpha, \beta$, and b of the SCND model on the decision variables. The results of Table 12.8 under Case I are presented in Table 12.9, and the results of Table 12.10 under Case II are summarized in Table 12.11.

It is investigated that, for 4 numbers of DCs, A_i^* is sensitive with high degree to the constants of TFC, TOC, TOTC, C_r, and f_r. Q_i^* is sensitive with high degree to the constant α, and T^* is sensitive to C_v, α, and b. τ^* and TPEC* are sensitive with high degree only to the constants associated with deterioration and preservation function, respectively. Lastly, π_I^* is highly sensitive to α, β, and b. Hence, the manufacturer should be cautious to fix input

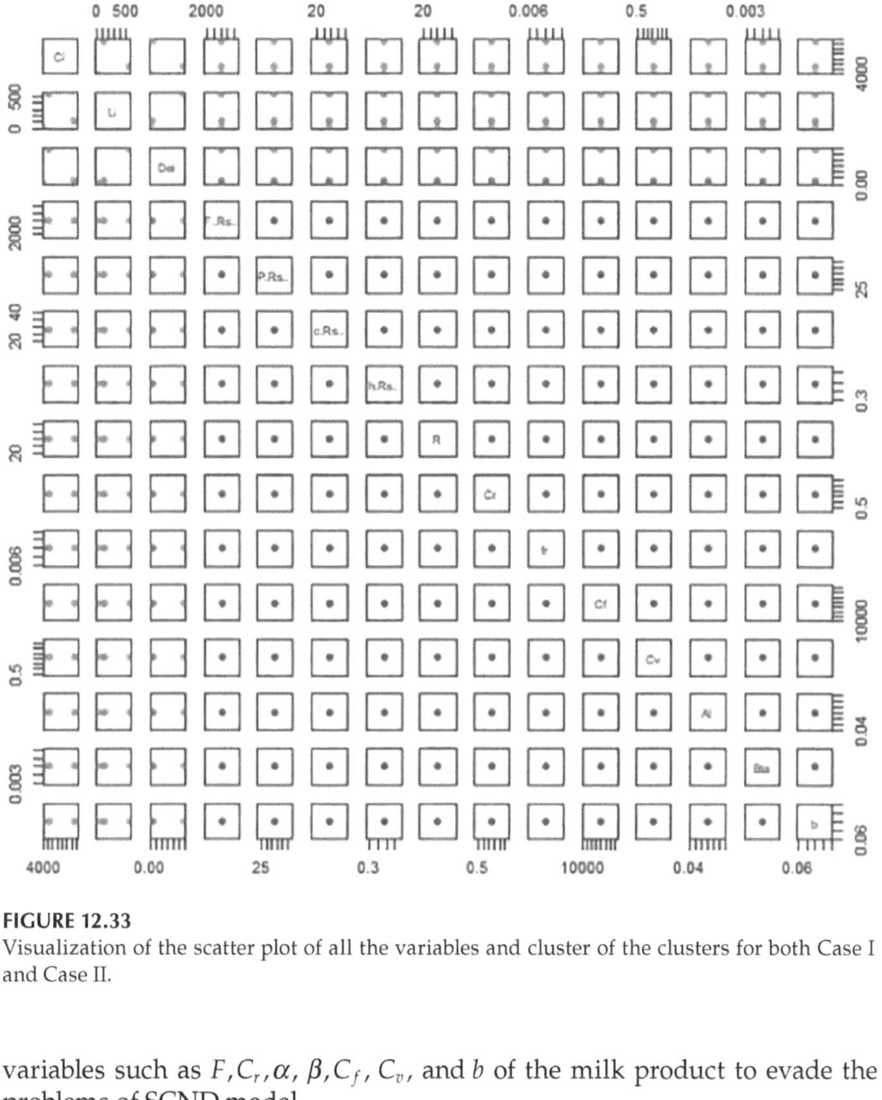

FIGURE 12.33
Visualization of the scatter plot of all the variables and cluster of the clusters for both Case I and Case II.

variables such as $F, C_r, \alpha, \beta, C_f, C_v,$ and b of the milk product to evade the problems of SCND model.

It is investigated that A_i^* are highly sensitive to the parameters of TFC, TOC, TOTC, and constant of distance metric for 32 numbers of DCs. Q_i^* and $T^*$are insensitive to almost all the input variables. τ^* and TPEC* are sensitive with a high degree only to the constants of preservation and deterioration functions, respectively. Finally, π_{II}^* is sensitive with a high degree to the input variables such as $C_f, C_v, \alpha,$ and β. Hence, the manufacturer should be cautious in fixing input variables such as $C_f, C_v, \alpha,$ and β of the milk product to evade the problems of sensitivity in the SCND model.

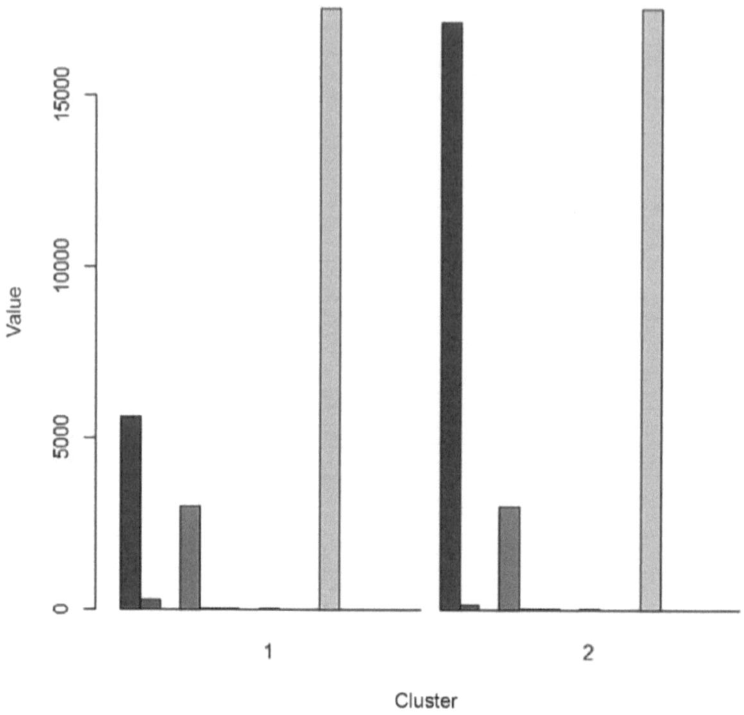

FIGURE 12.34
Bar plot of the number of clusters using K-means clustering for Case I and Case II.

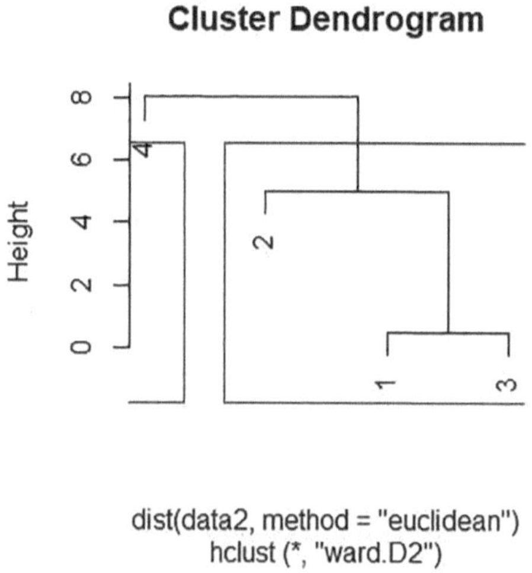

FIGURE 12.35
Visualization of the cut of the hierarchical tree for partitioning the clusters into clusters for Case I and Case II.

TABLE 12.8

Sensitivity Analysis under Case I

Variables	Value	Decision Variables	−50%	−30%	0%	+30%	50%
F	3,000	A_i	2144.109	2673.222	3381.226	4021.357	4420.883
			19425.76	24219.57	30634.13	36433.75	40053.49
			3847.605	4797.101	6067.615	7216.330	7933.281
			186.6542	232.7159	294.3508	350.0770	384.8575
		Q_i	70237.52	70237.52	70237.52	70237.52	70237.52
			684.9192	684.9192	684.9192	684.9192	684.9192
			31256.44	31256.44	31256.44	31256.44	31256.44
			1,129,189	1,129,189	1,129,189	1,129,189	1,129,189
		T	0.9976443	0.9976443	0.9976443	0.9976443	0.9976443
		τ	0.4114753	0.4114749	0.4114745	0.9976443	0.4114742
		TSR	0.4533088E+08	0.4533088E+08	4,533,0880	0.4533088E+08	0.4533088E+08
		TPC	0.3532724E+08	0.3532724E+08	35,377,240	0.3532724E+08	0.3532724E+08
		TFC	73772.06	82838.38	93560.90	102267.9	107337.3
		TITC	150502.0	168049.2	1,094,738	206112.9	216109.2
		TOTC	1094738	1094738	188997.4	1094738	1094738
		TPEC	20197.52	20197.48	20197.44	20197.42	20197.41
		TOC	1478.925	1186.200	937.8182	788.5337	717.2720
		THC	0.000000	0.000000	0.00	0.000000	0.000000
		π_I	8,662,948	8,636,627	8,605,204	8,579,531	8,564,537
H	0.47	A_i	3381.226	3381.226	3381.226	3381.226	3381.226
			30634.13	30634.13	30634.13	30634.13	30634.13
			6067.615	6067.615	6067.615	6067.615	6067.615
			294.3508	294.3508	294.3508	294.3508	294.3508
		Q_i	70237.52	70237.52	70237.52	70237.52	70237.52
			684.9192	684.9192	684.9192	684.9192	684.9192
			31256.44	31256.44	31256.44	31256.44	31256.44
			1,129,189	1,129,189	1,129,189	1,129,189	1,129,189

(Continued)

TABLE 12.8 (*Continued*)

Sensitivity Analysis under Case I

Variables	Value	Decision Variables	−50%	−30%	0%	+30%	50%
		T	0.9976443	0.9976443	0.9976443	0.9976443	0.9976443
		τ	0.4114745	0.4114745	0.4114745	0.4114745	0.4114745
		TSR	0.4533088E+08	0.4533088E+08	4,533,0880	0.4533088E+08	0.4533088E+08
		TPC	0.3532724E+08	0.3532724E+08	3,537,7240	0.3532724E+08	0.3532724E+08
		TFC	93560.90	93560.90	93560.90	93560.90	93560.90
		TITC	188997.4	188997.4	1,094,738	188997.4	188997.4
		TOTC	1,094,738	1,094,738	188997.4	1,094,738	1,094,738
		TPEC	20197.44	20197.44	20197.44	20197.44	20197.44
		TOC	937.8182	937.8182	937.8182	937.8182	937.8182
		THC		0.000000	0.00	0.000000	0.000000
		π_I	8,605,204	8,605,204	8,605,204	8,605,204	8,605,204
R	*30*	A_i	3370.032	3374.512	3381.226	3387.934	3392.403
			30532.70	30573.29	30634.13	30694.90	30735.38
			6047.526	6055.566	6067.615	6079.652	6087.670
			293.3762	293.7662	294.3508	294.9347	295.3237
		Q_i	70237.52	70237.52	70237.52	70237.52	70237.52
			684.9192	684.9192	684.9192	684.9192	684.9192
			31256.44	31256.44	31256.44	31256.44	31256.44
			1,129,189	1,129,189	1,129,189	1,129,189	1,129,189
		T	0.9976443	0.9976443	0.9976443	0.9976443	0.9976443
		τ	0.4114738	0.4114741	0.4114745	0.4114749	0.4114752

(Continued)

TABLE 12.8 (*Continued*)

Sensitivity Analysis under Case I

Variables	Value	Decision Variables	−50%	−30%	0%	+30%	50%
		TSR	0.4533088E+08	0.4533088E+08	45330880	0.4533088E+08	0.4533088E+08
		TPC	0.3532724E+08	0.3532724E+08	35377240	0.3532724E+08	0.3532724E+08
		TFC	93871.69	93747.07	93560.90	93375.66	93252.67
		TITC	188684.3	188809.7	1,094,738	189184.8	189309.5
		TOTC	1,094,738	1,094,738	188997.4	1,094,738	1094738
		TPEC	20197.37	20197.40	20197.44	20197.48	20197.51
		TOC	470.4667	657.7790	937.8182	1216.750	1402.093
		THC	0.000000	0.000000	0.00	0.000000	0.000000
		π_l	8,605,204	8,605,204	8,605,204	8,605,204	8,605,204
C_r	0.75	A_i	5367.362	4288.864	3381.226	2838.651	2580.359
			48628.65	38857.38	30634.13	25718.36	23378.21
			9631.738	7696.371	6067.615	5093.963	4630.457
			467.2527	373.3646	294.3508	247.1172	224.6317
		Q_i	70237.52	70237.52	70237.52	70237.52	70237.52
			684.9192	684.9192	684.9192	684.9192	684.9192
			31256.44	31256.44	31256.44	31256.44	31256.44
			1,129,189	1,129,189	1,129,189	1,129,189	1,129,189
		T	0.9976443	0.9976443	0.9976443	0.9976443	0.9976443
		τ	0.4114740	0.4114742	0.4114745	0.4114748	0.4114749
		TSR	0.4533088E+08	0.4533088E+08	4,533,0880	0.4533088E+08	0.4533088E+08
		TPC	0.3532724E+08	0.3532724E+08	3,537,7240	0.3532724E+08	0.3532724E+08
		TFC	58939.68	73760.94	93560.90	111444.0	122599.5
		TITC	119060.9	149000.6	1,094,738	225122.1	247656.7
		TOTC	1094738	1094738	188997.4	1,094,738	1094738
		TPEC	20197.39	20197.41	20197.44	20197.47	20197.48
		TOC	590.7884	739.3510	937.8182	1117.071	1228.889
		THC	0.000000	0.000000	0.00	0.000000	0.000000

(*Continued*)

TABLE 12.8 (*Continued*)

Sensitivity Analysis under Case I

Variables	Value	Decision Variables	−50%	−30%	0%	+30%	50%
		π_I	8,605,204	8,605,204	8,605,202	8,605,204	8,605,203
		T	0.9976443	0.9976443	0.9976443	0.9976443	0.9976443
		τ	0.4114745	0.4114745	0.4114745	0.4114745	0.4114745
f_r	0.01	A_i	5367.362	4288.864	3381.226	2838.651	2580.359
			48628.65	38857.38	30634.13	25718.36	23378.21
			9631.738	7696.371	6067.615	5093.963	4630.457
			467.2527	373.3646	294.3508	247.1172	224.6317
		Q_i	70237.52	70237.52	70237.52	70237.52	70237.52
			684.9192	684.9192	684.9192	684.9192	684.9192
			31256.44	31256.44	31256.44	31256.44	31256.44
			1,129,189	1,129,189	1,129,189	1,129,189	1,129,189
		T	0.9976443	0.9976443	0.9976443	0.9976443	0.9976443
		τ	0.4114740	0.4114742	0.4114745	0.4114748	0.4114749
		TSR	0.4533088E+08	0.4533088E+08	4,533,0880	0.4533088E+08	0.4533088E+08
		TPC	0.3532724E+08	0.3532724E+08	3,537,7240	0.3532724E+08	0.3532724E+08
		TFC	58939.68	73760.94	93560.90	111444.0	122599.5
		TITC	119060.9	149000.6	1,094,738	225122.1	247656.7
		TOTC	1,094,738	1,094,738	188997.4	1,094,738	1,094,738
		TPEC	20197.39	20197.41	20197.44	20197.47	20197.48
		TOC	590.7884	739.3510	937.8182	1117.071	1228.889
		THC	0.000000	0.000000	0.00	0.000000	0.000000
		π_1	8,605,204	8,605,204	8,605,204	8,605,204	8,605,204

(*Continued*)

TABLE 12.8 (*Continued*)

Sensitivity Analysis under Case I

Variables	Value	Decision Variables	−50%	−30%	0%	+30%	50%
C_f	17531	A_i	3381.226	3381.226	3381.226	3381.226	3381.226
			30634.13	30634.13	30634.13	30634.13	30634.13
			6067.615	6067.615	6067.615	6067.615	6067.615
			294.3508	294.3508	294.3508	294.3508	294.3508
		Q_i	70237.52	70237.52	70237.52	70237.51	70237.51
			684.9192	684.9192	684.9192	684.9191	684.9191
			31256.44	31256.44	31256.44	31256.44	31256.43
			1,129,189	1,129,189	1,129,189	1,129,189	1,129,189
		T	0.9976443	0.9976443	0.9976443	0.9976443	0.9976444
		τ	0.4114229	0.4114435	0.4114745	0.4115054	0.4115261
		TSR	0.4533088E+08	0.4533088E+08	4,533,0880	0.4533088E+08	0.4533088E+08
		TPC	0.3532725E+08	0.3532724E+08	3,537,7240	0.3532724E+08	0.3532724E+08
		TFC	93560.90	93560.90	93560.90	93560.90	93560.90
		TITC	188997.4	188997.4	1,094,738	188997.4	188997.4
		TOTC	1,059,594	1,073,652	188997.4	1,115,825	1,129,883
		TPEC	20192.38	20194.40	20197.44	20200.48	20202.51
		TOC	937.8182	937.8182	937.8182	937.8182	937.8182
		THC	0.000000	0.000000	0.00	0.000000	0.000000
		π_l	8,605,204	8,605,204	8,605,204	8,605,204	8,605,204
C_v	0.83	A_i	3381.227	3381.227	3381.226	3381.226	3381.226
			30634.13	30634.13	30634.13	30634.13	30634.13
			6067.615	6067.615	6067.615	6067.615	6067.615
			294.3508	294.3508	294.3508	294.3508	294.3508
		Q_i	70238.33	70238.01	70237.52	70237.03	70236.70
			684.9271	684.9239	684.9192	684.9144	684.9112
			31256.80	31256.66	31256.44	31256.22	31256.08
			1,129,202	1,129,197	1,129,189	1,129,181	1,129,176

(*Continued*)

TABLE 12.8 (*Continued*)

Sensitivity Analysis under Case I

Variables	Value	Decision Variables	−50%	−30%	0%	+30%	50%
		T	0.9976425	0.9976432	0.9976443	0.9976455	0.9976462
		τ	0.4060918	0.4082449	0.4114745	0.4147041	0.4168571
		TSR	0.4533088E+08	0.4533088E+08	45330880	0.4533088E+08	0.4533088E+08
		TPC	0.3532776E+08	0.3532755E+08	3,537,7240	0.3532693E+08	0.3532673E+08
		TFC	93560.90	93560.90	93560.90	93560.90	93560.90
		TITC	188997.4	188997.4	1,094,738	188997.4	188997.4
		TOTC	582521.0	787409.6	188997.4	1402062	1606942
		TPEC	19672.47	19881.63	20197.44	20515.74	20729.31
		TOC	937.8200	937.8193	937.8182	937.8171	937.8164
		THC	0.000000	0.000000	0.00	0.000000	0.000000
		π_I	8,605,201	8,605,211	8,605,204	8,605,123	8,605,205
α	0.07	A_i	3383.465	3381.199	3381.226	3381.265	3381.296
			29690.45	30633.88	30634.13	30634.47	30634.76
			6068.985	6067.565	6067.615	6067.683	6067.739
			294.4056	294.3483	294.3508	294.3541	294.3568
		Q_i	65892.34	69589.11	70237.52	70858.12	71255.19
			642.5473	678.5962	684.9192	690.9710	694.8430
			29322.79	30967.89	31256.44	31532.61	31709.31
			1,059,333	1,118,765	1,129,189	1,139,167	1145550.
		T	0.9981774	0.9988787	0.9976443	0.9959479	0.9945579
		τ	0.1920891	0.3975291	0.4114745	0.4259085	0.4358112

(*Continued*)

TABLE 12.8 (*Continued*)

Sensitivity Analysis under Case I

Variables	Value	Decision Variables	−50%	−30%	0%	+30%	50%
		TSR	0.4533088E+08	0.4533088E+08	45330880	0.4533088E+08	0.4533088E+08
		TPC	1.822501	0.3491188E+08	3,537,7240	0.3576553E+08	0.3607095E+08
		TFC	93549.92	93561.67	93560.90	93559.85	93558.98
		TITC	189008.9	188996.7	1,094,738	188998.5	188999.4
		TOTC	1030811	1,083,940	188997.4	1,105,670	1113025
		TPEC	0.4401657E+08	18851.61	20197.44	21639.29	22657.26
		TOC	937.2074	936.6670	937.8182	939.4050	940.7093
		THC	1.489643	0.000000	0.00	0.000000	0.000000
		π_1	8,605,192	9,030,655	8,605,204	8,152,488	7,838,694
β	0.01	A_i	3381.229	3381.228	3381.226	3381.224	3381.223
			30634.15	30634.14	30634.13	30634.11	30634.09
			6067.619	6067.618	6067.615	6067.611	6067.608
			294.3510	294.3509	294.3508	294.3506	294.3504
		Q_i	70284.24	70269.29	70237.52	70194.48	70159.51
			685.3747	685.2290	684.9192	684.4995	684.1585
			31277.23	31270.58	31256.44	31237.29	31221.73
			1,129,940	1,129,700	1,129,189	1,128,497	1,127,935
		T	0.9975351	0.9975704	0.9976443	0.9977424	0.9978203
		τ	0.2059296	0.2882154	0.4114745	0.5344547	0.6162440
		TSR	0.4533088E+08	0.4533088E+08	4,533,0880	0.4533088E+08	0.4533088E+08
		TPC	0.3535674E+08	0.3534729E+08	3,537,7240	0.3530018E+08	0.3527826E+08
		TFC	93560.83	93560.86	93560.90	93560.96	93561.01
		TITC	188997.5	188997.5	1,094,738	188997.4	188997.3
		TOTC	1,095,540	1,095,283	188997.4	1,094,003	1,093,408
		TPEC	5058.806	9909.343	20197.44	34074.72	45301.85
		TOC	937.9202	937.8873	937.8182	937.7266	937.6539
		THC	0.000000	0.000000	0.00	0.000000	0.000000

(Continued)

TABLE 12.8 (*Continued*)

Sensitivity Analysis under Case I

Variables	Value	Decision Variables	−50%	−30%	0%	+30%	50%
		π_I	8,587,988	8,592,843	8,603,151	8,617,075	8,628,358
b	0.1	A_i	3381.223	3381.225	3381.226	3381.227	3381.228
			30634.10	30634.11	30634.13	30634.13	30634.14
			6067.609	6067.612	6067.615	6067.616	6067.617
			294.3505	294.3506	294.3508	294.3507	294.3508
		Q_i	70175.13	70210.79	70237.52	70251.90	70258.29
			684.3108	684.6585	684.9192	685.0594	685.1217
			31228.67	31244.55	31256.44	31262.84	31265.68
			1128186	1128760	1129189	1129420	1129523.
		T	0.9977857	0.0.9977055	0.9976443	0.9976110	0.9975961
		τ	0.8219175	0.5875055	0.4114745	0.3166100	0.2744305
		TSR	0.4533088E+08	0.4533088E+08	4,533,0880	0.4533088E+08	0.4533088E+08
		TPC	0.3528804E+08	0.3531042E+08	3,537,7240	0.3533631E+08	0.3534034E+08
		TFC	93560.99	93560.94	93560.90	93560.88	93560.87
		TITC	188997.4	188997.4	1,094,738	188997.5	188997.5
		TOTC	1,093,673	1,094,281	188997.4	1,094,985	1,095,094
		TPEC	40293.69	28822.55	20197.44	15545.45	13476.17
		TOC	937.6862	937.7611	937.8182	937.8493	937.8632
		THC	0.000000	0.000000	0.00	0.000000	0.000000
		π_I	8,605,209	8,605,206	8,605,202	8,605,205	8,605,203

TABLE 12.9

Summary of Sensitivity Analysis for Table 12.8

Parameters	A_i^*	Q_i^*	T^*	τ^*	TPEC*	π_I^*
F	H sensitive	I sensitive	I sensitive	I sensitive	I sensitive	Sensitive
h	I sensitive	I sensitive	I sensitive	I sensitive	I sensitive	I sensitive
R	Sensitive	I sensitive	I sensitive	I sensitive	I sensitive	I sensitive
C_r	H sensitive	I sensitive	I sensitive	I sensitive	I sensitive	I sensitive
f_r	H sensitive	I sensitive	I sensitive	I sensitive	I sensitive	I sensitive
C_f	I sensitive	I sensitive	I sensitive	I sensitive	I sensitive	I sensitive
C_v	I sensitive	I sensitive	Sensitive	Sensitive	Sensitive	Sensitive
α	I sensitive	H sensitive	Sensitive	H sensitive	H sensitive	H sensitive
β	I sensitive	I sensitive	I sensitive	Sensitive	H sensitive	H sensitive
b	I sensitive	Sensitive	Sensitive	H sensitive	H sensitive	H sensitive

Note: H, highly; I, In.

12.5.3 Sensitive Analysis Using ANOVA Test under Case II

The sensitive analysis of π_{II}^* with the significant decision variables such as T and τ is presented using the analysis of variance (ANOVA) test.

H_{01}: π_{II}^* is not sensitive to different values of T.

H_{02}: π_{II}^* is not sensitive to different values of τ.

The ANOVA test in Table 12.13 is shown using data from Table 12.12; it is observed that p-values of T and τ are less than 0.05. So, the two null hypotheses may be rejected. Hence, the annual SCND profit is sensitive to the decision variables like T and τ.

12.6 Conclusion

This chapter investigated SCND problems of premium milk in coastal Odisha with deterioration effect and preservation effort. The key results impose for obtaining the optimum A_i, T, and τ to maximize π_j. With this, optimum m_i and Q_i are also evaluated for coastal Odisha. Numeric analysis on a real-life case data is conducted to justify and validate the present SCND model. In the present model there are four numbers of clusters like Balasore, Tirtol, Berhampur, and Khurda. Existing DCs in each of the four clusters is only 1, which is insufficient to meet the customer demand. It is observed that π_j under preservation is of Rs. 8,603,151 with 219 iterations. In π_{II}, the profit is more than that of without PT so, it may be finalized that the effects of PT for selling milk product is outperformed. This model is the best as the number of DCs is more

TABLE 12.10

Sensitivity Analysis under Case II

Variables		Decision Variables	−50%	−30%	0%	+30%	50%
F	3,000	Iteration	188	172	149	184	185
		A_i	2144.109	2673.222	3818.227	4021.357	4420.883
			4403.000	4403.000	4361.277	4403.00	4403.000
			3847.605	4797.101	5090.299	7216.330	7933.281
			186.6542.	232.7159	2889	350.0770	384.8575.
		Q_i	70237.52	70237.52	69592.41	70237.52	70237.52
			684.9192	684.9192	28373.57	684.91	684.9192
			31256.44	31256.44	18923.84	31256.44	31256.44
			1,129,189	1,129,189	2222.504	1,129,189	1,129,189
		T	0.99	0.99	0.99	0.99	0.99
		τ	0.411	0.411	0.413	0.411	0.411
		TSR	2144.109	0.45	45,330,880	0.45	0.45
		TPC	4403.000	0.35	35,327,240	0.35	0.35
		TFC	3847.605	84556.61	96129.72	105696.6	111342.7
		TITC	186.6542.	1,094,738	10947.38	1,094,738	1094738
		TOTC		167604.9	188456.6	205493.2	215443.5
		TPEC		20197.48	20197.45	20197.42	20197.41
		TOC		1210.804	963.567	814.9705	744..0372
		THC		0.0000	0.00	0.0000	0.0000.
		π_{II}	8,662,128	8,635,328	8,603,151	8,576,696	8,561,170
H	0.47	Iteration	220	186	149	218	205
		A_i	3381.226	3381.236	3818.227	3381.226	3381.226
			4403.000	4403.000	4361.277	4403.000	4403.000
			6067.615	6067.615	5090.299	6067.615	6067.615
			294.3508	294.3508	2,889	294.3508	294.3508
		Q_i	70237.52	70237.52	69592.41	70237.52	70237.52
			684.9192	684.9192	28373.57	684.9192	684.9192
			31256.44	31256.44	18923.84	31256.44	31256.44
			1,129,189	1,129,189	2222.504	1129189	1129189.
		T	0.99	0.99	0.99	0.99	0.99
		τ	0.411	0.411	0.413	0.411	0.411
		TSR	0.45	0.45	45,330,880	0.45	0.45
		TPC	0.35	0.35	35327240	0.35	0.35
		TFC	96129.72	96129.72	96129.72	96129.72	96129.72
		TITC	1,094,738	1,094,738	10947.38	1094738	1094738
		TOTC	188456.6	188456.6	188456.6	188456.6	188456.6
		TPEC	20197.45	20197.45	20197.45	20197.45	20197.45
		TOC	963.5670	963.5670	963.567	963.560	963.5670
		THC	0.0000	0.0000	0.00	0.0000	0.0000
		π_{II}	8,603,151	8,603,151	8,603,151	8,603,151	8,603,151
		Iteration	163	180	149	192	170

(Continued)

TABLE 12.10 (*Continued*)

Sensitivity Analysis under Case II

Variables		Decision Variables	−50%	−30%	0%	+30%	50%
R	30	A_i	3370.032	3374.512	3818.227	3384.934	3392.403
			4403.000	4403.000	4361.277	4403.000	4403.000
			6047.526	6055.566	5090.299	6079.652	6087.670
			293.37.	293.7662.	2889	294.9347.	295.3237
		Q_i	70237.52	70237.52	69592.41	70237.52	70237.52
			684.9192	684.9192	28373.57	684.9192	684.9192
			31256.44	31256.44	18923.84	31256.44	31256.44
			1,129,189.	1,129,189	2222.504	1,129,189.	1,129,189
		T	0.99	0.99	0.99	0.99	0.99
		τ	0.411	0.411	0.413	0.411	0.411
		TSR	0.45	0.45	45,330,880	0.45	0.45
		TPC	0.35	0.35	35,327,240	0.35	0.35
		TFC	96439.07	96135.02	96129.72	95945.33	95822.91
		TITC	1094738	1094738	10947.38	1094738	1094738
		TOTC	188145.0	188269.8	188456.6	188643.2	188767.3
		TPEC	20197.38	20197.40	20197.45	20197.49	20197.51
		TOC	483.3339	675.7971	963.567	1250.234	1440.738
		THC	0.0000	0.0000	0.00	0.0000	0.0000.
		π_{II}	8,603,633	8,603,440	8,603,151	8,602,862	8,602,670
		Iteration	181	162	149	181	211
C_r	0.75	A_i	5367.362	4288.864	3818.227	2838.651	2580.359
			4403.000	4403.000	4361.277	4403.000	4403.000
			9631.738	7696.371	5090.299	5093.963	4630.457
			467.2527.	373.3646.	2889	247.1172.	224.6317.
		Q_i	70237.52	70237.52	69592.41	70237.52	70237.52
			684.9192	684.9192	28373.57	684.9192	684.9192
			31256.44	31256.44	18923.84	31256.44	31256.44
			1129189.	119189	2222.504	1129181.	1129189.
		T	0.99	0.99	0.99	0.99	0.99
		τ	0.411	0.411	0.413	0.411	0.411
		TSR	0.45	0.45	4,5330,880	0.45	0.45
		TPC	0.35	0.35.76421.00	35327240	0.35	0.35
		TFC	61668.05	10947.38	96129.72	113930	125034.5
		TITC	1094738	148545.0	10947.38	1,094,738	1,094,738
		TOTC	118677.3	20197.42	188456.6	224513.90	247010.7
		TPEC	20197.40	766.0145	20197.45	20197.47	20197.49
		TOC	618.1366	0.0000	963.567	1141.994	1253.297
		THC	0.0000		0.00	0.0000	0.0000
		π_{II}	8,707,737	8,662,968	8,603,151	8,549,114	8,515,402

(Continued)

TABLE 12.10 (*Continued*)

Sensitivity Analysis under Case II

Variables		Decision Variables	−50%	−30%	0%	+30%	50%
		Iteration	181	162	149	181	211
f_r	0.01	A_i	5367.362	4288.864	3818.227	2838.651	2580.359
			4403.000	4403.000	4361.277	4403.000	4403.000
			9631.738	7696.371	5090.299	5093.963	4630.457
			467.2527	373.3664	2,889	247.1172	224.6317
		Q_i	70237.52	70237.52	69592.41	70237.52	70237.52
			684.9192	684.9192	28373.57	684.192	684.9192
			31256.44	31256.44	18923.84	31256.44	31256.44
			1,129,189.	1,129,189	2222.504	1,129,189	1129,189
		T	0.99	0.99	0.99	0.99	0.99
		τ	0.411	0.411	0.413	0.411	0.411
		TSR	0.45	0.45	45,330,880	0.45	0.45
		TPC	0.35	0.35	35,327,240	0.35	0.35
		TFC	61668.05	76421.00	96129.72	113930.41	125034.5
		TITC	1,094,738	1,094,738	10947.38	1,094,738	1,094,738
		TOTC	118677.3	148545.0	188456.6	224513.9	297010.7
		TPEC	20197.40	20197.42	20197.45	20197.47	20197.49
		TOC	618.1366	766.0145	963.567	1141.994	1253.297
		THC	0.0000	0.0000	0.00	0.0000	0.0000
		π_{II}	8,707,737	8,662,968	8,603,151	8,549,114	8,515,402
		Iteration	151	195	149	191	189
C_f	17531	A_i	3381.226	3381.226	3818.227	3381.226	3381.226
			4403.000	4403.000	4361.277	4403.000	4403.000
			6067.615	6067.615	5090.299	6067.615	6067.615
			294.3508.	294.3508.	2889	294.3508.	294.3508.
		Q_i	70237.52	70237.52	69592.41	70237.51	70237.51
			684.9192	684.9192	28373.57	684.9191	684.9191
			31256.44	31256.44	18923.84	31256.44	31256.43
			1,129,189	1,129,189	2222.504	1,129,189	1,129,189
		T	0.99	0.99	0.99	0.99	0.99
		τ	0.411	0.411	0.413	0.411	0.411
		TSR	0.45	0.45	45,330,880	0.45	0.45
		TPC	0.35	0.35	35,327,240	0.35	0.35
		TFC	96129.72	96129.72	96129.72	96129.72	96129.72
		TITC	1,059,594	1,073,652	10947.38	1,115,825	1,129,883
		TOTC	188,456	188456.6	188456.6	188456.6	188456.6
		TPEC	20192.38	963.5670	20197.45	20200.48	20202.51
		TOC	963.5670	0.0000	963.567	963.5670	963.5670
		THC	0.0000		0.00	0.0000	0.0000
		π_{II}	8,638,295	8,624,238	8,603,151	8,582,064	8,568,006
		Iteration	172	176	149	141	136

(Continued)

TABLE 12.10 (*Continued*)

Sensitivity Analysis under Case II

Variables		Decision Variables	−50%	−30%	0%	+30%	50%
C_v	0.83	A_i	3381.227	3381.227	3818.227	3381.226	3381.226
			4403.000	4403.000	4361.277	4403.000	4403.000
			6067.615	6067.615	5090.299	6067.615	6067.615
			24.3508	294.3508.	2,889	2,943,508	294.3508
		Q_i	70238.33	70238.01	69592.41	70237.03	70236.70
			684.9271	684.7239	28373.57	684.9144	684.9112
			31256.80	31256.66	18923.84	31256.22	31256.08
			1,129,202	1,129,197	2222.504	112,981.	1,129,176
		T	0.99	0.99	0.99	0.99	0.99
		τ	0.40	0.40	0.413	0.41	0.41
		TSR	0.45	0.45	45,330,880	0.45	0.45
		TPC	0.35	0.35	353,272,40	0.35	0.35
		TFC	96129.72	96129.72	96129.72	96129.72	96129.72
		TITC	582521.0	787409.6	10947.38	1,402,062	1,606,942
		TOTC	188456.6	188456.6	188456.6	188456.6	188456.6
		TPEC	19672.47	1988.64	20197.45	20515.74	20729.32
		TOC	963.5688	963.5681	963.567	963.5659	963.5652
		THC	0.0000.	0.0000	0.00	0.0000	0.0000
		π_{II}	9,115,378	8,910,487	8,603,151	8,295,817	8,090,930
		Iteration	1031	162	149	168	170
α	0.07	A_i	16556.94	3381.199	3818.227	3,381,265	3381.296
			4403.000	4403.000	4361.277	4403.000	4403.000
			17711.98	6067.565	5090.299	6067.683	606,773
			6835.917.	294.3483.	2889	294.3541.	294.3568.
		Q_i	68044.25	69589.11	69592.41	70858.12	71255.19
			663.5316	678.5962	28373.57	640.9710	694.8430
			30280.41	30967.89	18923.84	31532.61	31709.31
			1,093,929	1,118,765	2222.504	1,139,167	114,550
		T	1.0000	0.99	0.99	0.99	0.99
		τ	7.00014	0.39	0.413	0.42	0.43
		TSR	0.45	0.45	45,330,880	0.45	0.45
		TPC	0.34	0.34	35,327,240	0.35	0.36
		TFC	0.34	96130.48	96129.72	96128.67	96127.80
		TITC	12000.49	1,083,940	10947.38	1,105,670	1,113,025
		TOTC	1,060,245	188455.9	188456.6	188457.7	188456.6
		TPEC	774563.4	18851.61	20197.45	21639.30	226.5381
		TOC	5,845,321	962.3839	963.567	965.1977	0.0000
		THC	120.0049	0.0000	0.00	0.0000	
			0000				
		π_{II}	9,030,623	9,030,655	8,603,151	8,152,488	7,838,694
		Iteration	167	160	149	175	178

(*Continued*)

TABLE 12.10 (*Continued*)

Sensitivity Analysis under Case II

Variables		Decision Variables	−50%	−30%	0%	+30%	50%
β	0.01	A_i	3381.229	3381.228	3818.227	3381.224	3381.223
			4403.000	4403.000	4361.277	4403.000	4403.000
			6067.619	6067.618	5090.299	6067.611	6067.608
			294.3510	294.3509.	2889	294.3506.	294.3504.
		Q_i	70284.24	70269.29	69592.41	70194.48	70159.51
			685.3747	685.2290	28373.57	684.4995	684.1585
			31277.23	31270.58	18923.84	31237.29	31221.73
			1,129,940	1,129,700	2222.504	1,128,497	11,279,935
		T	0.99	0.99	0.99	0.99	0.99
		τ	0.20	0.28	0.413	0.53	0.61
		TSR	0.45	0.45	45,330,880	0.45	0.45
		TPC	0.35	0.35	35,327,240	0.35	0.35
		TFC	96129.65	96129.67	96129.72	96129.78	96129.83
		TITC	1,095,540	1,095,283	10947.38	109,4003	1,093,408
		TOTC	188456.7	188456.7	188456.6	188456.6	188456.5
		TPEC	5058.807	9909.345	20197.45	34074.73	45301.86
		TOC	0.0000	963.6380	963.567	963.4729	963.3982
		THC		0.0000	0.00	0.0000	0.0000
		π_{II}	8,587,988	8,592,843	8,603,151	8,617,075	8,628,358
b	0.1	Iteration	149	155	149	182	152
		A_i	3381.223	3381.225	3818.227	3381.227	3381.228
			4403.000	4403.000	4361.277	4403.000	4403.000
			6067.609	6067.612	5090.299	6067.616	6067.617
			294.3505.	294.3506.	2889	294.3508	294.3508
						294.3508	
		Q_i	70175.13	70210.79	69592.41	70251.90	70258.29
			684.3108	684.6585	28373.57	685.0594	685.1217
			31228.67	31244.55	18923.84	31262.84	31265.68
			1,128,186	1,128,760	2222.504	1,129,420	1,129,523
		T	0.99	0.99	0.99	0.99	0.99
		τ	0.82	0.58	0.413	0.31	0.27
		TSR	0.45	0.45	45,330,880	0.45	0.45
		TPC	0.35	0.35	35,327,240	0.35	0.35
		TFC	96129.80	96129.75	96129.72	96129.70	96129.69
		TITC	1,093,673	188456.6	10947.38	1,094,985	1,095,094
		TOTC	188456.6	1,094,281	188456.6	188456.7	188456.7
		TPEC	40293.6	28822.56	20197.45	15,545	13476.17
		TOC	963.4313	963.5083	963.567	963.5990	963.6133
		THC	0.0000	0.0000	0.00	0.0000	0.0000
		π_{II}	8,623,323	8,611,802	8,603,151	8,598,488	8,596,415

TABLE 12.11

Summary of Sensitivity Analysis for Table 12.10

Parameters	A_i^*	Q_i^*	T^*	τ^*	TPEC*	π^*
F	H sensitive	I sensitive	I sensitive	Sensitive	I sensitive	Sensitive
h	I sensitive	I sensitive	I sensitive	Sensitive	I sensitive	I sensitive
R	Sensitive	Sensitive	I sensitive	Sensitive	I sensitive	Sensitive
C_r	H sensitive	Sensitive	I sensitive	Sensitive	I sensitive	I sensitive
f_r	H sensitive	Sensitive	I sensitive	Sensitive	Sensitive	H sensitive
C_f	I sensitive	Sensitive	I sensitive	Sensitive	Sensitive	H sensitive
C_v	I sensitive	Sensitive	I sensitive	Sensitive	Sensitive	H sensitive
α	H sensitive	H sensitive	Sensitive	H sensitive	H sensitive	H sensitive
β	I sensitive	Sensitive	I sensitive	H sensitive	H sensitive	H sensitive
b	I sensitive	Sensitive	I sensitive	H sensitive	H sensitive	H sensitive

Note: H, highly and I, In.

TABLE 12.12

Impact of T and τ on π_{II}

τ /T	1.00	1.01	1.02	1.03	1.04	1.05
0.42	8,402,207	7,558,580	6,731,487	5,920,447	5,124,997	4,344,692
0.43	8,402,115	7,556,977	6,728,404	5,915,913	5,119,039	4,337,336
0.44	8,401,999	7,555,348	6,725,290	5,911,343	5,113,042	4,329,939
0.45	8,401,859	7,553,690	6,722,145	5,906,740	5,107,008	4,322,502
0.46	8,401,695	7,552,005	6,718,969	5,902,102	5,100,936	4,315,023
0.47	8,401,507	7,550,293	6,715,763	5,897,429	5,094,826	4,307,503

TABLE 12.13

ANOVA

Tests of between Subjects Effects						
Dependent Variable: Total Profit						
Source		**Sum of Squares**	**Degrees of Freedom**	**Mean Square**	***F***	***p*-Value**
T	Due to T effects	69766873557354.290	5	13953374711470.857	534715.924	0.000
τ	Due to τ effects	1545369156.222	5	309073831.244	11.844	0.000
Error	Due to errors	652373255.445	25	26094930.218	–	–
Total	Due to total	–	35	–	–	–

than one for each cluster. It is also investigated that the optimum A_i for three clusters such as Balasore, Berhampur, and Khurda is less than that of the service area, while that of Tirtol is the same. So, the optimum m_i for Balasore, Berhampur, and Khurda clusters are 5, 3, and 23, respectively, and there is only 1 DC for Tirtol. By adding the existing 4 with the proposed 28 numbers of DCs, the total DCs in the coastal region will be 4+28=32. Optimum Q_i is highest for Khurda and lowest for cluster Tirtol, where the optimal value of T and τ are 0.997645 and 0.4114745, respectively. From sensitive analysis, it is observed that almost all decision variables are not sensitive except the input variables of deterioration rate are sensitive to the decision variables and π_j. From the ANOVA test, it is investigated that π_j is sensitive with high degree to T and τ. Hence, the present model provides a significant upshot for analyzing prospective changes in maximum key variables for tactical decision-making principle in an SCND.

References

Abad, P. L., & Jaggi, C. K. (2003). A joint approach for setting unit price and the length of the credit period for a seller when end demand is price sensitive. *International Journal of Production Economics, 83*(2), 115–122. doi:10.1016/S0925-5273(02)00142-1

Ackerman, M. A. M., Pantel, C. A., & Lasserre, A. A. A. (2017). A green supply chain network design framework for the processed food industry: Application to the orange juice agro food cluster. *Computers & Industrial Engineering, 109*, 369–389. doi:10.1016/j.cie.2017.04.031

Agrawal, D. K. (2003). *Logistics and Supply Chain Management*. Macmillan, India.

Alejandro, G. H. J., Willmer, E. J., & Alvaro, F. C. (2013). A multi-product lot-sizing model for a manufacturing company. *Ingeniería, Investigación y Tecnología, 14*(3), 413–419. doi:10.1016/S1405-7743(13)72254-6

Anand, A., & Pandian, A. S. (2019). A customer-based supply chain network design. *International Journal of Enterprise Network Management, 10*(3/4), 224–240. doi:10.1504/IJENM.2019.103153

Asgharizadeh, E., Torabi, S. A., Mohaghar, A., & Zare-Shourijeh, M. A. (2019). Sustainable supply chain network design: A review on quantitative models using content analysis. *Environmental Energy and Economic Research, 3*(2), 143–176. doi:10.22097/eeer.2019.184458.1081

Ballou, R. H., & Srivastava, S. K. (2007). *Business Logistics/Supply Chain Management*. Pearson Education, India.

Bhatia, S., Sharma, M., & Bhatia, K. K. (2016). A novel approach for crawling the opinions from World Wide Web. *International Journal of Information Retrieval Research (IJIRR)*, 6(2), 1–23. 10.4018/IJIRR.2016040101

Bhatia, S., Sharma, M., & Bhatia, K. K. (2017). Opinion score mining: An algorithmic approach. *International Journal of Intelligent Systems and Applications, 10*(11), 34. https://doi.org/10.5815/ijisa.2017.11.05

Bhunia, A. K., Kundu, S., Sannigrahi, T., & Goyal, S. K. (2009). An application of tournament genetic algorithm in a marketing oriented economic production lot-size model for deteriorating items. *International Journal of Production Economics, 119*(1), 112–121. doi:10.1016/j.ijpe.2009.01.010

Biuki, M., Kazemi, A., & Alinezhad, A. (2020). An integrated location-routing-inventory model for sustainable design of a perishable products supply chain network. *Journal of Cleaner Production, 260,* 120842. Advance online publication. doi:10.1016/j.jclepro.2020.120842

Bourlakis, M. A., & Weightman, P. W. H. (2004). *Food Supply Chain Management.* Blackwell Publishing, USA.

Chang, C. T., Teng, J. T., & Chern, M. S. (2010). Optimal manufacturer's replenishment policies for deteriorating items in a supply chain with up-stream and down-stream trade credit. *International Journal of Production Economics, 127*(1), 197–202. doi:10.1016/j.ijpe.2010.05.014

Chen, J. M., & Chen, T. H. (2005a). The multi-item replenishment problem in a two-echelon supply chain: The effect of centralization versus decentralization. *Computers & Operations Research, 32*(12), 3191–3207. doi:10.1016/j.cor.2004.05.007

Chen, J. M., & Chen, T. H. (2005b). Effects of joint replenishment and channel coordination for managing multiple deteriorating products in a supply chain. *The Journal of the Operational Research Society, 56*(10), 1224–1234. doi:10.1057/palgrave.jors.2601921

Chen, J. M., & Chen, T. H. (2007). The profit maximization model for a multi-item distribution channel. *Transportation Research Part E, Logistics and Transportation Review, 43*(4), 338–354. doi:10.1016/j.tre.2005.12.001

Cortinhal, M. J., Lopes, M. J., & Melo, M. T. (2019). A multi-stage supply chain network design problem with in-house production and partial product outsourcing. *Applied Mathematical Modelling, 70,* 572–594. doi:10.1016/j.apm.2019.01.046

Coyle, J. J., Bardi, E. J., & Langley, C. J. (2007). *The Management of Business Logistics.* Thomson South Western, Ohio.

Dasci, A., & Verter, V. (2001). A continuous model for production-distribution system design. *European Journal of Operational Research, 129*(2), 287–298. doi:10.1016/S0377-2217(00)00226-5

Dave, U. (1979). On a discrete-in-time order-level inventory model for deteriorating items. *The Journal of the Operational Research Society, 30*(4), 349–354. doi:10.1057/jors.1979.75

Dave, U. & Patel, L.K. (1981). (T, Si) Policy inventory Model for Deteriorating items with Time-Proportional Demand. Journal of the Operational Research Society, 32, 137–142. https://doi.org/10.1057/jors.1981.27

Dye, C. Y. (2013). The effect of preservation technology investment on a non-instantaneous deteriorating inventory model. *Omega, 41*(5), 872–880. doi:10.1016/j.omega.2012.11.002

Dye, C. Y., & Hsieh, T. P. (2012). An optimal replenishment policy for deteriorating items with effective investment in preservation technology. *European Journal of Operational Research, 218*(1), 106–112. doi:10.1016/j.ejor.2011.10.016

Dye, C. Y., & Hsieh, T. P. (2013). A particle swarm optimization for solving lot-sizing problem with fluctuating demand and preservation technology cost under tradecredit. *Journal of Global Optimization, 55*(3), 655–679. doi:10.1007/s10898-012-9950-z

Eskandarpour, M., Dejax, P., Miemczyk, J., & Péton, O. (2015). Sustainable supply chain network design: An optimization-oriented review. *International Journal of Omega, 54*, 11–32.

Gahan, P., & Pattnaik, M. (2017). *Multi-Product Optimization Models: Retailers' Perspective*. Lambert Academic Publishing, Germany.

Govindan, K., Fattahi, M., & Keyvanshokooh, E. (2017). Supply chain network design under uncertainties: A comprehensive review and future research directions. *European Journal of Operational Research, 263*(1), 108–141. doi:10.1016/j.ejor.2017.04.009

Hsu, P. H., Wee, H. M., & Teng, H. M. (2010). Preservation technology investment for deteriorating inventory. *International Journal of Production Economics, 124*(2), 388–394. doi:10.1016/j.ijpe.2009.11.034

Imran, M., & Ramzan, M. B. (2017). Medicine supply chain model for integrated healthcare system with uncertain product complaints. *Journal of Manufacturing Systems, 46*(C), 13–28.

Javid, J., & Kannan, G. (2020). (in press). On the sustainable perishable food supply chain network design: A dairy products case to achieve sustainable development goals. *Journal of Cleaner Production*. doi:10.1016/j.jclepro.2020.123060

Jiang, Y., Zhao, Y., Dong, M., & Han, S. (2019). Sustainable supply chain network design with carbon footprint consideration: A case study in China. *Mathematical Problems in Engineering, 2019*, 1–19. doi:10.1155/2019/3162471

Kang, S., & Kim, I. (1983). A study on the price and production level of the deteriorating inventory system. *International Journal of Production Research, 21*(6), 899–908. doi:10.1080/00207548308942422

Khalaf, M. F., Naderi, B., Mohammandi, M., & Pishavee, M. S. (2020). (in press). Design of a sustainable and reliable hydrogen supply chain network under mixed uncertainties: A case study. *International Journal of Hydrogen Energy*. Advance online publication. doi:10.1016/j.ijhydene.2020.05.276

Lee, F. C., & Wen, U. P. (2007). A heuristic approach for solving serially distributed storage depots under general integer policy. *Asia-Pacific Journal of Operational Research, 24*(4), 479–497. doi:10.1142/S0217595907001346

Li, X., & Ouyang, Y. (2010). A continuum approximation approach to reliable facility location design under correlated probabilistic disruptions. *Transportation Research Part B: Methodological, 44*(4), 535–548. doi:10.1016/j.trb.2009.09.004

Lodree, E. J. Jr, & Uzochukwu, B. M. (2008). Production planning for a deteriorating item with stochastic demand and consumer choice. *International Journal of Production Economics, 116*(2), 219–232. doi:10.1016/j.ijpe.2008.09.010

Martel, A., & Klibi, W. (2016). *Designing Value-Creating Supply Chain Networks*. Springer.

Muratet, A., Verter, V., & Laporte, G. (2010). A continuous analysis framework for the solution of location-allocation problems with dense demand. *Computers & Operations Research, 37*(1), 123–136. doi:10.1016/j.cor.2009.04.001

Murr, D. P., & Morris, L. L. (1975). Effect of storage temperature on post change in mushrooms. *Journal of the American Society for Horticultural Science, 100*, 16–19. 10.21273/JASHS.100.1.16

Ouyang, L. Y., Teng, J. T., & Chen, L. H. (2006). Optimal ordering policy for deteriorating items with partial backlogging under permissible delay in payments. *Journal of Global Optimization, 34*(2), 245–271. doi:10.1007/s10898-005-2604-7

Paksoy, T., & Chang, C. T. (2010). Revised multi-choice goal programming for multi-period, multi-stage inventory controlled supply chain model with popup stores in Guerrilla marketing. *Applied Mathematical Modelling, 34*(11), 3586–3598. doi:10.1016/j.apm.2010.03.008

Pal, B., San, S. S., & Chaudhuri, K. (2015). Two-echelon manufacturer-retailer supply chain strategies with price, quality, and promotional effort sensitive demand. *International Transactions in Operational Research, 22*(6), 56–68. doi:10.1111/itor.12131

Pattnaik, M. (2014). *Inventory Models: A Management Perspective*. Lambert Academic Publishing, Germany.

Pattnaik, M. (2018). Impact of publicity effort and variable ordering cost in multi-product order quantity model of units lost sales due to deterioration. *Log Forum, 14*(3), 407–424.

Pattnaik, M. (2019). Preservation effort effects on retailers and manufacturers in integrated multi-deteriorating item discrete supply chain model. *OPSEARCH*. 10.1007/s12597-020-00477-2

Pattnaik, M. (2020). Milk supply chain network design (SCND): A case of the milk industry in western region of Odisha. *International Journal of Strategic Decision Sciences, 11*(3), 63–120. doi:10.4018/IJSDS.2020070104.

Pattnaik, M., & Gahan, P. (2018). Retail-chain multi-item optimization: A mixed integer non-linear (MINL) heuristic approach. *Singaporean Journal of Business Economics and Management Studies, 6*(8), 9–25. doi:10.12816/0048616

Pattnaik, M., & Gahan, P. (2019). *Multi-Product Discrete Supply Chain Models*. Lambert Academic Publishing, Germany.

Pham, T., & Yendradee, P. (2017). Optimal supply chain network design with process network and BOM under uncertainties: A case study in toothbrush industry. *Computers & Industrial Engineering, 108*, 177–191. doi:10.1016/j.cie.2017.04.012

Ramezaanian, R., & Behboodi, Z. (2017). Blood supply chain network design under uncertainties in supply and demand considering social aspects. *International Journal of Production Research, 55*, 2013–2033. doi:10.1080/00207543.2016.1262563

Reddy, C. K. K., Vijaya Babu, B., & Rupa, C. H. (2014). SLEAS: Supervised learning using entropy as attribute selection measure. *International Journal of Engineering and Technology, 6*, 2053–2060.

Reddy, C. K. K., Rupa, C. H., & Vijaya Babu, B. (2015). SLGAS: Supervised learning using gain ratio as attribute selection measure to nowcast snow/no-snow. *International Review on Computers and Software*. https://doi.org/10.15866/irecos.v10i2.4841

Sahay, B. S. (2000). *Supply Chain Management in the* Twenty-*First* Century. Macmillan, India.

Saravanan, S., & Anubama, B. (2017). Selection of cold chain logistics service providers in pharmaceutical industry with reference to India. *International Journal of Managing Value and Supply Chains, 8*(2), 1–12. doi:10.5121/ijmvsc.2017.8201

Sople, V. V. (2010). *Logistics Management*. Pearson.

Taxakis, K., & Papadopoulos, C. (2016). A design model and a production-distribution and inventory planning model in multi-product supply chain networks. *International Journal of Production Research, 54*(21), 6436–6457. doi:10.1080/00207543.2016.1158882

Teng, J. T., & Lou, K. R. (2012). Seller's optimal credit period and replenishment time in a supply chain with up-stream and down-stream trade credits. *Journal of Global Optimization, 53*(3), 417–430. doi:10.1007/s10898-011-9720-3

Teng, J. T., Min, J., & Pan, Q. (2012). Economic order quantity model with trade credit financing for nondecreasing demand. *Omega, 40*(3), 328–335. doi:10.1016/j.omega.2011.08.001

Tsao, Y. C. (2016). Designing a supply chain network for deteriorating inventory under preservation effort and trade credits. *International Journal of Production Research, 54*(13), 3837–3851. doi:10.1080/00207543.2016.1157272

Tsao, Y. C., & Lu, J. C. (2012). A supply chain network design considering transportation cost discounts. *Transportation Research Part E, Logistics and Transportation Review, 48*(2), 401–414. doi:10.1016/j.tre.2011.10.004

Tsao, Y. C., & Sheen, G. J. (2008). Dynamic pricing, promotion and replenishment policies for a deteriorating item under permissible delay in payments. *Computers & Operations Research, 35*(11), 3562–3580.

Tsao, Y. C., & Sheen, G. J. (2012). A multi-item supply chain with credit periods and weight freight cost discounts. *International Journal of Production Economics, 135*(1), 106–115. doi:10.1016/j.ijpe.2010.11.013

Tsao, Y. C., & Teng, W. G. (2013). Heuristics for the joint multi-item replenishment problem under trade credits. *IMA Journal of Management Mathematics, 24*(1), 63–77. https://doi.org/10.1093/imaman/dps004

Tsao, Y. C., Chen, T. H., & Zhang, Q. H. (2013). Effects of maintenance policy on an imperfect production system under trade credit. *International Journal of Production Research, 51*(5), 1549–1562. doi:10.1080/00207543.2012.714001

Tsao, Y. C., Mangotra, D., Lu, J. C., & Dong, M. (2012). A continuous approximation approach for the integrated facility-inventory allocation problem. *European Journal of Operational Research, 222*(2), 216–228. doi:10.1016/j.ejor.2012.04.033

Varsei, M., & Polyakovskiy, S. (2017). Sustainable supply chain network design: A case of the wine industry in Australia. *Omega, 66*(Part B), 236–247.

Wang, K. J., Wang, S. M., & Yang, S. J. (2007). Are source portfolio model for equipment investment and allocation of semiconductor testing industry. *European Journal of Operational Research, 179*(2), 390–403. doi:10.1016/j.ejor.2006.04.006

Wang, S. J., Huang, C. T., Wang, W. L., & Chem, Y. H. (2010). Incorporating ARIMA forecasting and service level based replenishment in RFID-enabled supply chain. *International Journal of Production Research, 48*(9), 2655–2677. doi:10.1080/00207540903564983

Wang, S. M., Chen, J. C., & Wang, K. J. (2007). Resource portfolio planning of make-to-stock products using a constraint programming-based genetic algorithm. *Omega, 35*(2), 237–246. doi:10.1016/j.omega.2005.06.001

Wee, H. (1998). Optimal Buyers-Sellers Discount Pricing and Ordering Policy for Deteriorating Items. Journal of Devoted to the problems of Capital Investment, 43, 151–168.

Yang, P.C. & Wee, H.M. (2006). A collaborative inventory system with permissible delay in payment for deteriorating items, *Mathematical and Computer Modelling*, 43(3-4), 209–221.

Zauberman, G., Ronen, R., Akerman, M. & Fuchs, Y. (1990). Low PH treatment protects litchi fruit colour. *Acta Hortic*, 269, 309–314.

Zhong, Y., Shu, J., Xie, W., & Zhou, Y.W., (2017). Optimal trade credit and replenishment policies or supply chain network design. International Journal Production, 46, 453–460.

13

Integrated Multi-Item Multi-Echelon Supply Chain Model: A Mixed-Integer Heuristic Approach

Monalisha Pattnaik, Aryan Pattnaik, Monee Kieran Roul, and P. Abhilash Kumar

13.1 Introduction

An organization may be happy to have created a genuinely optimal leanness company of operations by redesigning procedures, lowering overhead, and eliminating duplicate tasks. It should improve the standard of the products and services by eliminating errors and misinformation inside the organization, as well as breaking down divisional barriers and encouraging individuals to collaborate and exchange information. Most high-tech businesses have taken a more active approach to reforming their operations through cross-company operations. It entails altering the economics of their product's supply chain. Whenever, a multitude of individuals are scattered around multiple organizations with employing an assortment of disconnected information systems, they are in a burdensome set of processes in tandem which creates a greater value for the organization. In a multi-echelon supply chain (SC), the hub is a computer system developed by the company to exchange information between all participants, and this multi-echelon procedure has significantly improved supply chain efficiency. The newly established relationship additionally benefits the supplier, because dealing with one major customer instead of a slew of little ones is easier and more secure.

In modern times, all firms are constantly striving for quicker and less expensive supply chain management. The industrial cycle influenced companies' quests. When a company is crashing, the managerial focus is on growing speed, and, when the economy is tanking, businesses scramble to increase profits. As a result, the efficiency of the supply chain is required to optimize its performance when maximizing its interests. Hence flexible, versatile, and coordinated supply networks can give firms a durable competitive edge.

DOI: 10.1201/9781003503934-13

Price negotiation has traditionally been a major activity between suppliers and merchants. Retailers select the best supplier based on negotiating strength. Organizations rarely give their suppliers either genuine demand or pertinent sales information; as a result, suppliers make stocking decisions based on incomplete or null information. The "Win–Win" approach is currently gaining popularity in the commercial sector, replacing friction between suppliers and merchants vying for negotiating leverage. The present-day commercial climate has grown increasingly competitive in the quickly expanding worldwide market. A "Win–Win" supply chain management (SCM) system must be important regardless of whether the role is on the supplier or purchasing side. As a result, a multi-echelon SC model is proposed to address stock issues in supplier–retailer collaborative operations. In the business world, sourcing and inventory management are massive tasks that frequently involve hundreds of items. With a multi-item SC, synchronizing the purchase of multiple items can result in significant cost savings during replenishment. Multi-echelon multi-item restocking approaches have been extensively implemented in real-life situations, such as delivering computer and automotive components (Hahm and Yano, 1995a,b) or chilled items to retailers (Hammer, 2001). In such industries, a manufacturer typically manufactures numerous products for an individual consumer and transports them all at the same time in one vehicle. In the perishable and fast-moving consumer product businesses, several types of chilled items (derived milk products, general mills yogurt, etc.) can be delivered in a single vehicle to a single supermarket or retailer. Kao (1979), Graves (1979), Hammer (2001), Goyal (1988), Ben-Khedher and Yano (1994), Van Eijs (1994), Rempala (2003), Chen and Chen (2005a,b), Nilsson et al. (2007), Tsao and Sheen (2012), and others have created models and algorithms to solve multi-item replenishment challenges under a variety of constraints. Because the coordination of multi-echelon is commonly used in modern business practices, it is an integral component of this model. The current study focuses on the multi-echelon supply chain.

Efficient collaboration between channel participants such as distributors, wholesalers, retailers, manufacturers, and suppliers is critical for increasing a channel's efficiency. Narayanan and Raman (2004) found that connecting with channel partners can help an organization improve its market share. However, enterprises within the same channel frequently fail to behave in ways that maximize channel profit, resulting in poor overall supply chain performance. Recently, Khouja (2003) proposed a three-stage integrated SC with numerous buyers and vendors. A corporation frequently believes that any policy that maximizes its own profit will likewise maximize the channel profit. Hsu and Wee (2005) explored the challenge of horizontal supplier coordination in the face of unpredictable deliveries. Chen and Chen (2005c) examined the impact of channel coordination and joint replenishment on governing multiple perishable items.

Li and Wang (2007) extended a comprehensive literature assessment on the coordination of the SC system. The current work will focus on channel coordination of several items, using joint replenishment as a collaboration mechanism.

Nowadays, store publicity activities are becoming increasingly prevalent in the business sector. Wal-Mart, Costco, Pantaloons, Big Bazaar, Reliance Fresh, The World., for example, frequently offer discounted prices to boost consumer interest in different kind of products; clothing retailers Baleno, NET, and Reliance Trends create shelf space for particular clothing products for extended periods of time; and McDonald's, Pizza Hut, Burger King, and KFC frequently use coupons to entice customers. Freebies, advertisements, and exhibits are all examples of additional publicity methods. The retail industry places a high value on public relations. The amount of promotional activity the retailer does has a significant collision on the yearly earnings. Too much publicity might entail additional expenditures, but too little can result in reduced sales revenue. Multi-item prices, such as deteriorating items, first-moving consumer goods (FMCG) products, automobile parts, and electrical parts, will be supplied to a supermarket or retailer at one time. Price promotion, rebates, and marketing are standard marketing tactics used to accelerate product flow. In the multi-item sector, it is critical to make joint optimum retail price and publicity effort judgments while taking demand into account. The model with the PR effort component offers the decision maker meaningful and practical information. Tsao and Sheen (2008) introduced promotion, dynamic pricing, and replenishment strategies for a degrading item with an acceptable payment delay. This study approaches the issue by providing a multi-echelon, multi-item SC model with a promotional campaign. As a result, the proposed model was optimized by incorporating a publicity effort cost and a publicity constraint. The goal is to estimate the retailer's ideal publicity effort and replenishment strategies for multi-items using the mixed-integer issue, in which the multi-item demand rates are presumed to be integers.

The remainder of the present chapter is as follows: In Section 13.2, assumptions and notations are explained. Section 13.3 explains supply chain models for both decentralized and centralized policies; two different models are derived for each policy such as noncooperative and cooperative replenishment with individual and joint items, respectively. In this section some mathematical properties and solution procedures are also presented. The algorithm of the models is developed in Section 13.4. A numerical analysis is conducted in Section 13.5, and comparative and sensitive analyses with significance testing have been performed in Sections 13.6 and 13.7, respectively, to justify the results. Finally, in Section 13.8, conclusions are drawn. Major contributions of related research are summarized in Table 13.1. Figure 13.1 depicts the setting of the integrated multi-item and multi-echelon SC with publicity effort effects.

TABLE 13.1

Major Contributions of Related Researches

	Research Parameters							
Authors	**Demand**	**Forward Financing**	**Models**	**Items**	**Echelons**	**Benefits**	**Publicity Effort Effects**	**Replenishment**
Cohen (1977)	Price dependent	NA	NLP	S	1	NA	NA	FI
Khender and Yano	Constant	NA	NLP	MI	2	Container effect	NA	FI
Chang et al. (2001)	Linear	Y	NLP	MI	1	NA	NA	IF
Abad and Jaggi (2003)	Price dependent and exponential	Y	NLP	S	2	NA	NA	FI
Chen and Chen (2005d)	constant	NA	NLP	MI	M	Quantity	NA	FI
Chen and Chen (2005e)	Constant	NA	NLP	MI	2	Centralization	NA	FI
Tsao and Sheen (2008)	Time and price dependent	Y	NLP	MI	1	NA	Y	FI
Chung (1998)	Constant	Y	NLP	S	1	NA	NA	FI
Tsao and Sheen (2012)	Price dependentand linear	Y	NLP	MI	M	Freight	NA	FI
Chen and Chen (2007)	price dependent	NA	NLP	MI	M	Demand and revenue increments	NA	FI
Present Study	Retail price dependentand linear decreasing function	NA	MINLP	MI	M	Integration and publicity effort	Y	FI

Note: Y, Yes; NA, Not Applicable; NLP, Nonlinear Programming; MINLP, Mixed Integer NLP; S, Single; MI, Multi-item; M, Multi; FI, Finite; IF, Infinite.

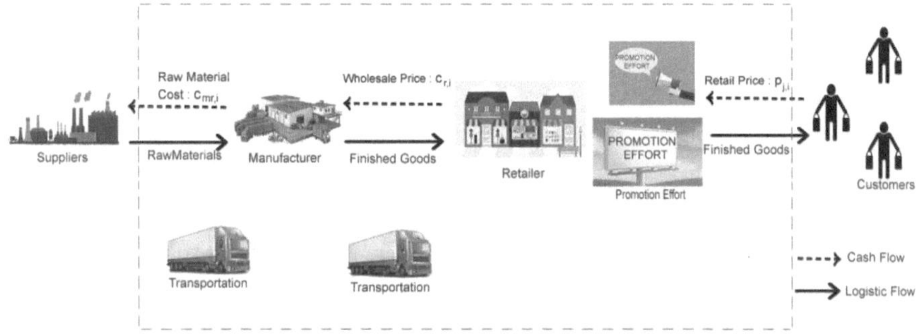

FIGURE 13.1
Environment of the multi-item multi-echelon integrated supply chain with publicity effort.

13.2 Assumptions and Notations

In Table 13.2 notations are shown and assumptions are also defined.

The assumptions are adopted to develop an SC model:

- The present SC comprising a supplier (vendor), manufacturer, and retailer, where the stock of multiple (k) deteriorated items with a short life-time and a stable decaying rate per unit time are sold to the end customers.
- A lot-for-lot strategy is adopted by the supplier where the supplier's ordered quantity and retailer's demand quantity are equal.

13.3 Mathematical Model

The aim of this chapter is to develop a procedure of the SC model under four policies to find the optimal selling price of each item, individual replenishment time for decentralized and common replenishment time for centralized policies, order quantity of each item for manufacturer and retailer, discrete demand rate for each item, discrete demand with publicity, publicity effort effects for each item, total SC profit per unit of time under the decentralized and centralized policies for individual and joint items, respectively. Maximizing the total SC profit comprising the cost of purchasing, holding cost of inventory, and setup cost under major and minor and publicity effort cost for individual perspective, namely retailer, manufacturer, and the chain perspective.

TABLE 13.2
Summary of Notations

Supply Chain	
k	No. of finished items
i	Indicator of finished items, $i = 1, 2, \ldots k$
j	Indicator of the policies, $j \in [I, II, III, IV]$
π_j	Total supply chain profit per unit item and per unit time under j policy
Manufacturer	
	Raw Material
$a_{mr,i}$	Ordering cost for ith item per lot
$h_{mr,i}$	Holding cost of inventory for ith item
$c_{mr,i}$	Purchasing cost per unit item i
$I_{mr,i}(t)$	Stock level for ith item at time t
	Finished Item
A_m	Setup major cost per lot
$a_{m,i}$	Setup minor cost of ith item
$h_{m,i}$	Holding cost of inventory for ith item
μ_i	ith item production rate
u_i	ith item usage rate of raw material
$\theta_{m,i}$	Raw material deterioration rate for ith item
$t_{m,i}$	Production time for ith item
$I_{m,i}(t)$	Finished item i inventory level at time t
$\pi_{j,m}$	Under policy j manufacturer's total profit per unit time
Retailer	
A_r	Setup major cost per order
$a_{r,i}$	Setup minor cost for ith item into the order
$h_{r,i}$	Holding cost of inventory for ith item
$c_{r,i}$	Purchasing cost of per unit item i
$D_i(p_{j,i})$	Under policy j for ith item demand rate, $D_i(p_{j,i}) = D_i = \alpha_i - \beta_i p_{j,i}, \textit{ where } \alpha_i > 0, \beta_i > 0, p_{j,i} \le \alpha_i / \beta_i$
θ_i	Finished item i deterioration rate in supply chain
$T_{j,i}$	Under policy j individual cycle time of replenishment of ith item, $j \in [I, III]$ (decision variable)
T_j	Under policy j common cycle time of replenishment of all items, $j \in [II, IV]$ (decision variable)
$I_{r,i}(t)$	Finished item i inventory level at time t
$p_{j,i}$	Under policy j retail selling price of item i(decision variable)
$PE_j(\rho_i, D_i)$	Under policy j publicity effort cost, $PE_j(\rho_i, D_i) = \sum_{i=1}^{k} \tau_i (\rho_i - 1)^2 D_i(p_{j,i})^{\gamma_i}$, where $\tau_i > 0$, ρ_i is a publicity effort factor and γ_i is a constant
$\pi_{j,r}$	Under policy j the total profit per unit time of the retailer

BOX 13.1 SUPPLY CHAIN MODELS UNDER FOUR POLICIES

Decentralized Policy		Centralized Policy	
Individual Item	Joint Items	Individual Item	Joint Items
Policy I	Policy II	Policy III	Policy IV

13.3.1 Mathematical Model for the Decentralized Policy

Four supply chain models are formulated from the perspectives of the players, namely supplier, manufacturer, retailer, and supply chain.

Under decentralized production and replenishment managerial policy, the objective of each player within the SC is to optimize its own profit, without consideration being given to its counterpart's reaction. All the replenishment decisions on an economic order quantity policy are taken by the retailer, which includes all the above-discussed costs. Box 13.1 shows the description of four different policies.

13.3.1.1 Policy I: Uncooperative Replenishment for Individual Items

13.3.1.1.1 Retailer's Model

Under this situation, the retailer has to maximize their profit with the three decision variables, namely retailer selling price, the cycle time of replenishment, and factor of publicity effort for each individual item. In this policy, the stock level for each item changes due to the joint effects, like effects of publicity effort on demand requisite and deterioration over the cycle time. The model can be developed using the differential equation.

$$\frac{dI_{r,i}(t)}{dt} = -\theta_i I_{r,i}(t) - D_i \rho_i, \quad 0 \le t \le T_{I,i} \tag{13.1}$$

$$\frac{dI_{r,i}(t)}{dt} + \theta_i I_{r,i}(t) = -D_i \rho_i \tag{13.2}$$

This is a linear differential equation of first order, i.e. $\frac{dy}{dx} + Py = Q$, where $P = Q = f(x)$.

The integrating factor $IF = e^{\int Pdx}$ is introduced, and then both sides of the above equation are multiplied with $\frac{dy.e^{\int Pdx}}{dx} = Q.e^{\int Pdx}$. Hence, the value is $dy.e^{\int Pdx} = Q.e^{\int Pdx}dx$ and then both sides are integrated with $\int dy.e^{\int Pdx} = \int Q.e^{\int Pdx}dx \Rightarrow y.e^{\int Pdx} = \int Q.e^{\int Pdx}dx$ so that y can be evaluated. In this way, the above differential equation can be solved as:

In this way the above differential equation can be solved by multiplying the integrating factor (IF) i.e.$e^{\int \theta_i dt} = e^{t\theta_i}$ with Equation (13.2). The equation becomes:

$$\frac{dI_{r,i}(t)e^{t\theta_i}}{dt} = -D_i\rho_i e^{t\theta_i} \Rightarrow \int dI_{r,i}(t)e^{t\theta_i} = -\int D_i\rho_i e^{t\theta_i}\,dt \Rightarrow e^{t\theta_i} I_{r,i}(t)$$

$$= \frac{e^{t\theta_i} D_i\rho_i}{\theta_i} + c_1 \Rightarrow I_{r,i}(t) = \frac{-D_i\rho_i}{\theta_i} + \frac{c_1}{e^{t\theta_i}} \text{ where } c_1 \text{ is an arbitrary constant} \tag{13.3}$$

Given that $I_{r,i}(T_{I,i}) = 0$, for $t = T_{I,i}$, $I_{r,i}(t) = 0$.

So, $0 = \frac{-D_i\rho_i}{\theta_i} + \frac{c_1}{e^{T_{I,i}\theta_i}} \Rightarrow c_1 = \frac{D_i\rho_i e^{T_{I,i}\theta_i}}{\theta_i}$; putting the value of c_1 in Equation (13.3):

$$I_{r,i}(t) = \frac{-D_i\rho_i}{\theta_i} + \frac{D_i\rho_i e^{T_{I,i}\theta_i}}{\theta_i e^{t\theta_i}} \Rightarrow I_{r,i}(t) = \frac{D_i\rho_i}{\theta_i}\left[e^{\theta_i(T_{I,i}-t)} - 1\right], \text{ for } 0 \le t \le T_{I,i}$$

Purchasing Cost (PuC):

$$\text{PuC} = \sum_{i=1}^{k} c_{r,i}.I_{r,i}(0)$$

$I_{r,i}(0)$ is the initial inventory level of item i, and it is also equivalent to the purchasing quantity.

$I_{r,i}(t) = \frac{D_i\rho_i}{\theta_i}\left[e^{\theta_i(T_{I,i}-t)} - 1\right]$, for $0 \le t \le T_{I,i}$; putting $t = 0$ in $I_{r,i}(t)$ it can be obtained: $I_{r,i}(0) = \frac{D_i\rho_i}{\theta_i}\left[e^{\theta_i(T_{I,i})} - 1\right]$I

Total purchasing cost of all k items is: $\text{PuC} = \sum_{i=1}^{k} c_{r,i}.I_{r,i}(0) \Rightarrow \text{PuC} = \sum_{i=1}^{k} \frac{c_{r,i}D_i\rho_i}{\theta_i}\left[e^{\theta_i(T_{I,i})} - 1\right]$

Total holding cost (HoC): $\text{HoC} = \sum_{i=1}^{k} h_{r,i}.\int_0^{T_{I,i}} I_{r,i}(t)dt$

$$\text{HoC} = \sum_{i=1}^{k} h_{r,i}.\int_0^{T_{I,i}} I_{r,i}(t)dt = \sum_{i=1}^{k} h_{r,i}.\int_0^{T_{I,i}} \frac{D_i\rho_i}{\theta_i}\left[e^{\theta_i(T_{I,i}-t)} - 1\right]dt$$

$$\text{HoC} = \sum_{i=1}^{k} \frac{h_{r,i}D_i\rho_i}{\theta_i T_{I,i}}\left[\frac{e^{\theta_i T_{I,i}} - 1}{\theta_i} - T_{I,i}\right]$$

Setup cost major (MaSC):

$$\text{MaSC} = \sum_{i=1}^{k} \frac{A_r}{T_{I,i}}$$

Setup cost minor (MSC):

$$\text{MSC} = \sum_{i=1}^{k} \frac{a_r}{T_{I,i}}$$

Publicity effort cost (PEC):

$$\text{PEC} = \sum_{i=1}^{k} \frac{\tau_i (\rho_i - 1)^2 D_i (p_i)^{\gamma_i}}{T_{I,i}}$$

Sales revenue (SR):

$$\text{SR} = \sum_{i=1}^{k} p_{I,i} D_i \rho_i$$

The approximation of an exponential term $e^{\theta_i T_{I,i}}$ using the expansion of the Taylor series is $e^{\theta_i T_{I,i}} = 1 + \theta_i T_{I,i} + \frac{(\theta_i T_{I,i})^2}{2!} + \frac{(\theta_i T_{I,i})^3}{3!} + \ldots$; it is known that for a deteriorated perishable item the third or the terms of a higher order of this expansion may be neglected. Hence, revised PuC and HoC are:

$$\text{PuC} = \sum_{i=1}^{k} c_{r,i} D_i \rho_i + \frac{c_{r,i} D_i \rho_i \theta_i T_{I,i}}{2} \quad \text{and} \quad \text{HoC} = \sum_{i=1}^{k} \frac{h_{r,i} D_i \rho_i T_{I,i}}{2}$$

The total profit per unit time of the retailer is $\pi_{I,r}$

$$\pi_{I,r} = \text{SR} - \text{PuC} - \text{HoC} - \text{MaSC} - \text{MSC} - \text{PEC}$$

$$\pi_{I,r} = \sum_{i=1}^{k} \left[p_{I,i} D_i \rho_i - \left\{ c_{r,i} D_i \rho_i + \frac{c_{r,i} \theta_i D_i \rho_i T_{I,i}}{2} \right\} - \frac{h_{r,i} D_i \rho_i T_{I,i}}{2} - \frac{A_r}{T_{I,i}} - \frac{a_{r,i}}{T_{I,i}} - \frac{\tau_i (\rho_i - 1)^2 D_i^{\gamma_i}}{T_{I,i}} \right] \Rightarrow$$

$$\pi_{I,r} = \sum_{i=1}^{k} \left[(p_{I,i} - c_{r,i}) D_i \rho_i - \frac{(A_r + a_{r,i})}{T_{I,i}} - \frac{(h_{r,i} + c_{r,i} \theta_i) D_i \rho_i T_{I,i}}{2} - \frac{\tau_i (\rho_i - 1)^2 D_i^{\gamma_i}}{T_{I,i}} \right] \tag{13.4}$$

13.3.1.1.2 Optimization-I

The retailer has to optimize their profit with three decision variables. Under the individual uncooperative policy, the following propositions are framed for optimality testing.

Proposition I: $\pi_{I,r}$ is concave in $p_{I,i}$.

Proof:

$$\frac{\partial \pi_{I,r}}{\partial p_{I,i}} = D_i\rho_i - (p_{I,i} - c_{r,i})\beta_i\rho_i + \frac{T_{I,i}\beta_i\rho_i}{2}(h_{r,i} + c_{r,i}\theta_i) + \frac{\tau_i\beta_i\gamma_i}{T_{I,i}}(\rho_i - 1)^2 D_i^{\gamma_i - 1}$$

$$\frac{\partial \pi_{I,r}}{\partial p_{I,i}} = 0 \Rightarrow \frac{T_{I,i}\beta_i\rho_i}{2}(h_{r,i} + c_{r,i}\theta_i) + \frac{\tau_i\beta_i\gamma_i}{T_{I,i}}(\rho_i - 1)^2 D_i^{\gamma_i - 1} = D_i\rho_i + (p_{I,i} - c_{r,i})\beta_i\rho_i$$

Since the above equation is an implicit function, $p_{I,i}^*$ is difficult to obtain, but $\pi_{I,r}''$ should satisfy the concavity property $\frac{\partial^2 \pi_{I,r}}{\partial p_{I,i}^2} = -\left[2\beta_i\rho_i + \frac{\tau_i\beta_i^2\gamma_i(\gamma_i - 1)}{T_{I,i}}(\rho_i - 1)^2 D_i^{\gamma_i - 2}\right] < 0.$

For $\gamma_i > 1$, $\pi_{I,r}$ is concave.

Proposition II: $\pi_{I,r}$ is concave in $T_{I,i}$.

Proof: $$\frac{\partial \pi_{I,r}}{\partial T_{I,i}} = \frac{A_r + a_{r,i} + (\rho_i - 1)^2 \tau_i D_i^{\gamma_i}}{T_{I,i}^2} - \frac{D_i\rho_i}{2}(h_{r,i} + c_{r,i}\theta_i)$$

$$\frac{\partial \pi_{I,r}}{\partial T_{I,i}} = 0 \Rightarrow T_{I,i}^* = \sqrt{\frac{2\left(A_r + a_{r,i} + (\rho_i - 1)^2 \tau_i D_i^{\gamma_i}\right)}{D_i\rho_i(h_{r,i} + c_{r,i}\theta_i)}}$$

$T_{I,i}^*$ is obtained and $\pi_{I,r}''$ should satisfy the concavity property $\frac{\partial^2 \pi_{I,r}}{\partial T_{I,i}^2} = -\left[\frac{2\left(A_r + a_{r,i} + (\rho_i - 1)^2 \tau_i D_i^{\gamma_i}\right)}{T_{I,i}^3}\right] < 0.$

Hence, $\pi_{I,r}$ is concave.

Proposition III: $\pi_{I,r}$ is concave in ρ_i.

Proof: $$\frac{\partial \pi_{I,r}}{\partial \rho_i} = (p_{I,i} - c_{r,i})D_i - \frac{D_i T_{I,i}}{2}(h_{r,i} + c_{r,i}\theta_i) - \frac{2\tau_i(\rho_i - 1)D_i^{\gamma_i}}{T_{I,i}}$$

$$\frac{\partial \pi_{I,r}}{\partial \rho_i} = 0 \Rightarrow \rho_i^* = 1 + \frac{T_{I,i}}{2\tau_i D_i^{\gamma_i}}\left[(p_{I,i} - c_{r,i})D_i - \frac{D_i T_{I,i}}{2}(h_{r,i} + c_{r,i}\theta_i)\right]$$

ρ_i^* is obtained and $\pi_{I,r}''$ should satisfy concavity property $\frac{\partial^2 \pi_{I,r}}{\partial \rho_i^2} = -\left[\frac{2\tau_i D_i^{\gamma_i}}{T_{I,i}}\right] < 0.$

Hence, $\pi_{I,r}$ is concave.

13.3.1.1.3 Manufacturer's Model

A make-to-order policy is adopted by the upstream manufacturer, where the production quantity and demand by the retailer, who is the downstream player of supply chain, is the same. Hence, for each production run, the major setup cost and minor setup cost are incurred due to some factors. To produce the finished goods for the manufacturer, the publicity effort cost is not required to meet the demand of the market, but for raw materials supplementary holding and ordering costs are required.

$$\frac{dI_{mr,i}(t)}{dt} = -I_{mr,i}(t)\theta_{m,i} - \mu_i u_i, \quad t_{m,i} \le t \le T_{I,i} \tag{13.5}$$

$$\frac{dI_{mr,i}(t)}{dt} + I_{mr,i}(t)\theta_{m,i} = -\mu_i u_i \tag{13.6}$$

This is a linear differential equation of first order, i.e. $\frac{dy}{dx} + Py = Q$, where $P = Q = f(x)$.

Integrating factor $IF = e^{\int Pdx}$ is introduced, and both sides of the above equation is multiplied with $\frac{dye^{\int Pdx}}{dx} = Qe^{\int Pdx}$. Hence, the value is $dye^{\int Pdx} = Qe^{\int Pdx}dx$, and then both sides are integrated with $\int dye^{\int Pdx} = \int Qe^{\int Pdx}dx \Rightarrow ye^{\int Pdx} = \int Qe^{\int Pdx}dx$ so that y can be evaluated. $IF = e^{\int \theta_{m,i}dt} = e^{t\theta_{m,i}}$, and upon multiplying this integrating factor with Equation (13.6), the equation becomes:

$$\frac{dI_{mr,i}(t)e^{t\theta_{m,i}}}{dt} = -\mu_i u_i e^{t\theta_{m,i}} \Rightarrow \int dI_{mr,i}(t)e^{t\theta_{m,i}}$$

$$= -\int \mu_i u_i e^{t\theta_{m,i}}dt \Rightarrow e^{t\theta_{m,i}} I_{mr,i}(t) = \frac{-e^{t\theta_{m,i}}\mu_i u_i}{\theta_{m,i}} + c_2,$$

$$\Rightarrow I_{mr,i}(t) = \frac{-\mu_i u_i}{\theta_{m,i}} + \frac{c_2}{e^{t\theta_{m,i}}} \text{ where } c_2 \text{ is an arbitrary constant.} \tag{13.7}$$

As $I_{mr,i}(T_{I,i}) = 0$ for $t = T_{I,i}$, $I_{mr,i}(t) = 0$.

So, $\frac{-\mu_i u_i}{\theta_{m,i}}+\frac{c_2}{e^{T_{I,i}\theta_{m,i}}}=0 \Rightarrow c_2=\frac{\mu_i u_i e^{T_{I,i}\theta_{m,i}}}{\theta_{m,i}}$, putting the value of c_2 in Equation (13.7):

$$I_{mr,i}(t)=\frac{-\mu_i u_i}{\theta_{m,i}}+\frac{\mu_i u_i e^{T_{I,i}\theta_{m,i}}}{\theta_{m,i}e^{t\theta_{m,i}}} \Rightarrow I_{mr,i}(t)=\frac{\mu_i u_i}{\theta_{m,i}}\left[e^{\theta_{m,i}(T_{I,i}-t)}-1\right], \text{ for } t_{m,i}\le t\le T_{I,i}$$

Finished product:

$$\frac{dI_{m,i}(t)}{dt}=\mu_i-\theta_i I_{m,i}(t), \quad t_{m,i}\le t\le T_{Ii} \tag{13.8}$$

$$\frac{dI_{m,i}(t)}{dt}+I_{m,i}(t)\theta_i=\mu_i \tag{13.9}$$

This is a linear differential equation of first order, i.e. $\frac{dy}{dx}+Py=Q$, where $P=Q=f(x)$.

Integrating factor $IF=e^{\int Pdx}$ is introduced, and then both sides of the above equation are multiplied with $\frac{dye^{\int Pdx}}{dx}=Qe^{\int Pdx}$. The value is $dye^{\int Pdx}=Qe^{\int Pdx}dx$, and then both sides are integrated with $\int dye^{\int Pdx}=\int Qe^{\int Pdx}dx \Rightarrow ye^{\int Pdx}$ so that y can be evaluated.

$IF=e^{\int \theta_i dt}=e^{t\theta_i}$, and upon multiplying this integrating factor with Equation (13.9), the equation becomes: $\frac{dI_{m,i}(t)e^{t\theta_i}}{dt}=\mu_i e^{t\theta_i} \Rightarrow \int dI_{m,i}(t)e^{t\theta_i}=\int \mu_i e^{t\theta_i}dt \Rightarrow e^{t\theta_i}I_{m,i}(t)=\frac{\mu_i e^{t\theta_i}}{\theta_i}+c_3$,

$$\Rightarrow I_{m,i}(t)=\frac{\mu_i}{\theta_i}+\frac{c_3}{e^{t\theta_i}} \text{ where } c_3 \text{ is an arbitrary constant.} \tag{13.10}$$

As, $I_{m,i}(t_{m,i})=0$ for $t=t_{m,i}$, $I_{m,i}(t)=0$.

So, $\frac{\mu_i}{\theta_i}+\frac{c_3}{e^{t_{m,i}\theta_i}}=0 \Rightarrow c_3=\frac{-\mu_i e^{t_{m,i}\theta_i}}{\theta_i}$, putting the value of c_3 in Equation (13.10)

$$I_{m,i}(t)=\frac{\mu_i}{\theta_i}-\frac{\mu_i e^{t_{m,i}\theta_i}}{\theta_i e^{t\theta_i}} \Rightarrow I_{m,i}(t)=\frac{\mu_i}{\theta_i}\left[1-e^{\theta_i(t_{m,i}-t)}\right], \text{ for } t_{m,i}\le t\le T_{I,i}$$

A similar procedure of retailer is used to develop the profit function per unit time; the manufacturer profit function is as follows:

$$\pi_{I,m}=\text{SR}-\text{PuC}-\text{HoC}-\text{MaSC}-\text{MSC}$$

$$\pi_{I,m}=\sum_{i=1}^{k}\left[\begin{array}{l} c_{r,i}D_i\rho_i-\left\{c_{mr,i}\mu_iu_i+\dfrac{c_{mr,i}\theta_{m,i}\mu_iu_iT_{I,i}}{2}\right\} \\ -\dfrac{(h_{m,i}+h_{mr,i}u_i)D_i\rho_iT_{I,i}}{2}-\dfrac{(A_m+a_{m,i}+a_{mr,i})}{T_{I,i}}\end{array}\right]\Rightarrow$$

$$\pi_{I,m}=\sum_{i=1}^{k}\left[\begin{array}{l} c_{r,i}D_i\rho_i-c_{mr,i}\mu_iu_i-\left\{\dfrac{c_{mr,i}\theta_{m,i}\mu_iu_i+(h_{m,i}+h_{mr,i}u_i)D_i\rho_i}{2}\right\} \\ T_{I,i}-\dfrac{(A_m+a_{m,i}+a_{mr,i})}{T_{I,i}}\end{array}\right] \tag{13.11}$$

13.3.1.1.4 Supply Chain Model

Hence, after summing up Equations (13.4) and (13.11), the profit model under Policy I is:

$$\pi_I=\pi_{I,r}+\pi_{I,m}$$

13.3.1.2 Policy II: Uncooperative Replenishment of Retailer and Manufacturer for Joint Items

13.3.1.2.1 Retailer's Model

Under the joint uncooperative replenishment policy, the retailer has to optimize their profit with the three decision variables such as T_{II}^*, $p_{II,i}^*$, and ρ_i^* for each item. The profits and the related costs measured by the retailer are the same as that of individual replenishment, only for the number of main replenishment setups that are reduced to single over the cycle time. Hence, the profit per unit time of the retailer $\pi_{II,r}$is:

$$\pi_{II,r}=[\text{SR}-\text{PuC}-\text{HoC}-\text{MSC}-\text{PEC}]-\text{MaSC}$$

$$\pi_{II,r}=\sum_{i=1}^{k}\left[\begin{array}{l} p_{II,i}D_i\rho_i-\left\{c_{r,i}D_i\rho_i+\dfrac{c_{r,i}\theta_iD_i\rho_iT_{II}}{2}\right\} \\ -\dfrac{h_{r,i}D_i\rho_iT_{II}}{2}-\dfrac{a_{r,i}}{T_{II}}-\dfrac{\tau_i(\rho_i-1)^2D_i^{\gamma_i}}{T_{II}}\end{array}\right]-\frac{A_r}{T_{II}}\Rightarrow$$

$$\pi_{II,r}=\sum_{i=1}^{k}\left[\begin{array}{l} (p_{II,i}-c_{r,i})D_i\rho_i-\dfrac{a_{r,i}}{T_{II}} \\ -\dfrac{(h_{r,i}+c_{r,i}\theta_i)D_i\rho_iT_{II}}{2}-\dfrac{\tau_i(\rho_i-1)^2D_i^{\gamma_i}}{T_{II}}\end{array}\right]-\frac{A_r}{T_{II}}$$

13.3.1.2.2 Optimization II

The retailer has to optimize his profit for the decision variables such as T_{II}^*, $p_{II,i}^*$, and ρ_i^* of *i*th item. For optimization under joint replenishment uncooperative policy, the following propositions are discussed.

Proposition I: $\pi_{II,r}$ is concave in $p_{II,i}$.

Proof:

$$\frac{\partial \pi_{II,r}}{\partial p_{II,i}} = \sum_{i=1}^{k}\left[D_i\rho_i - \left(p_{II,i} - c_{r,i}\right)\beta_i\rho_i + \frac{T_{II}\beta_i\rho_i}{2}\left(h_{r,i} + c_{r,i}\theta_i\right) + \frac{\tau_i\beta_i\gamma_i}{T_{II}}\left(\rho_i - 1\right)^2 D_i^{\gamma_i - 1}\right]$$

$$\frac{\partial \pi_{I,r}}{\partial p_{II,i}} = 0 \Rightarrow \sum_{i=1}^{k}\left[\frac{T_{II}\beta_i\rho_i}{2}\left(h_{r,i} + c_{r,i}\theta_i\right) + \frac{\tau_i\beta_i\gamma_i}{T_{II}}\left(\rho_i - 1\right)^2 D_i^{\gamma_i - 1}\right]$$

$$= \sum_{i=1}^{k}\left[D_i\rho_i + \left(p_{II,i} - c_{r,i}\right)\beta_i\rho_i\right]$$

Since the above equation is an implicit function, $p_{II,i}^*$ is difficult to obtain, but $\pi_{II,r}''$ should satisfy the concavity property $\frac{\partial^2 \pi_{II,r}}{\partial p_{II,i}^{\ 2}} = -\sum_{i=1}^{k}\left[\left[2\beta_i\rho_i + \frac{\tau_i\beta_i^{\ 2}\gamma_i(\gamma_i - 1)}{T_{II}}\left(\rho_i - 1\right)^2 D_i^{\gamma_i - 2}\right]\right] < 0.$

For $\gamma_i > 1$, the $\pi_{II,r}$ is concave.

Proposition II: $\pi_{II,r}$ is concave in T_{II}.

Proof: $$\frac{\partial \pi_{II,r}}{\partial T_{II}} = \sum_{i=1}^{k}\left[\frac{a_{r,i} + \left(\rho_i - 1\right)^2 \tau_i D_i^{\gamma_i}}{T_{II}^2} - \frac{D_i\rho_i}{2}\left(h_{r,i} + c_{r,i}\theta_i\right)\right] + \frac{A_r}{T_{II}^2}$$

$$\frac{\partial \pi_{II,r}}{\partial T_{II}} = 0 \Rightarrow T_{II}^* = \sqrt{\frac{2\left[\sum_{i=1}^{k}\left(a_{r,i} + \left(\rho_i - 1\right)^2 \tau_i D_i^{\gamma_i}\right) + A_r\right]}{\sum_{i=1}^{k} D_i\rho_i(h_{r,i} + c_{r,i}\theta_i)}}$$

T_{II}^* is obtained and $\pi_{II,r}''$ should satisfy the concavity property

$$\frac{\partial^2 \pi_{II,r}}{\partial T_{II}^{\ 2}} = -\left[\sum_{i=1}^{k}\left[\frac{2\left(a_{r,i} + \left(\rho_i - 1\right)^2 \tau_i D_i^{\gamma_i}\right)}{T_{II}^3}\right] + \frac{2A_r}{T_{II}^3}\right] < 0.$$

Hence, $\pi_{II,r}$ is concave.

Proposition III: $\pi_{II,r}$ **is concave in** ρ_i.

Proof: $$\frac{\partial \pi_{II,r}}{\partial \rho_i} = \sum_{i=1}^{k} \left[(p_{II,i} - c_{r,i}) D_i - \frac{D_i T_{II}}{2} (h_{r,i} + c_{r,i}\theta_i) - \frac{2\tau_i(\rho_i - 1) D_i^{\gamma_i}}{T_{II}} \right]$$

$$\frac{\partial \pi_{II,r}}{\partial \rho_i} = 0 \Rightarrow \rho_i^* = 1 + \sum_{i=1}^{k} \left[\frac{T_{II}}{2\tau_i D_i^{\gamma_i}} \left[(p_{II,i} - c_{r,i}) D_i - \frac{D_i T_{II}}{2} (h_{r,i} + c_{r,i}\theta_i) \right] \right]$$

ρ_i^* is obtained and $\pi_{II,r}''$ should satisfy the concavity property $\frac{\partial^2 \pi_{II,r}}{\partial \rho_i^2} = -\sum_{i=1}^{k} \frac{2\tau_i D_i^{\gamma_i}}{T_{II}} < 0.$

Hence, $\pi_{II,r}$ is concave.

13.3.1.2.3 Manufacturer's Model

Hence, the total SC profit per unit time of the manufacturer is:

$$\pi_{II,m} = [\text{SR} - \text{PuC} - \text{HoC} - \text{MSC}] - \text{MaSC}$$

$$\pi_{II,m} = \sum_{i=1}^{k} \left[\begin{array}{l} c_{r,i} D_i \rho_i - \dfrac{(a_{m,i} + a_{mr,i})}{T_{II}} - u_i \mu_i c_{mr,i} \\ -\left\{ \theta_{m,i} u_i \mu_i c_{mr,i} + (h_{m,i} + h_{mr,i} u_i) D_i \rho_i \right\} \dfrac{T_{II}}{2} \end{array} \right] - \frac{A_m}{T_{II}}$$

13.3.1.2.4 Supply Chain Model

The total SC profit under Policy II is:

$$\pi_{II} = \pi_{II,r} + \pi_{II,m}$$

13.3.2 Mathematical Model for the Centralized Policy

Under this policy, the three decision variables, namely retail selling price, cycle time for replenishment, and publicity effort factor of *i*th item, are evaluated simultaneously where the total supply chain optimized profit earned by both the retailer and manufacturer.

13.3.2.1 Policy III: Cooperative Replenishment of Retailer and Manufacturer for Individual Items

Under this policy, the decision variables for *i*th item are evaluated jointly by each player in SC. The system profit of Policy III is:

$$\pi_{III} = \pi_{I,r} + \pi_{I,m}$$

$$= \left[\sum_{i=1}^{k} \left[\left(p_{III,i} - c_{r,i}\right) \rho_i D_i - \frac{(A_r + a_{r,i})}{T_{III,i}} - \frac{\left(h_{r,i} + c_{r,i}\theta_i\right)}{2} D_i \rho_i T_{III,i} - \frac{(\rho_i - 1)^2 \tau_i D_i^{\gamma_i}}{T_{III,i}} \right] \right]$$

$$+ \left[\sum_{i=1}^{k} \left[\begin{array}{l} \left(c_{r,i} \rho_i D_i\right) - \frac{(A_m + a_{m,i} + a_{mr,i})}{T_{III,i}} - c_{mr,i} u_i \mu_i \\ -\left[c_{mr,i} u_i \mu_i \theta_{m,i} + \left(h_{m,i} + h_{mr,i} u_i\right) \rho_i D_i \right] \frac{T_{III,i}}{2} \end{array} \right] \right]$$

$$\pi_{III} = \pi_{I,r} + \pi_{I,m}$$

$$= \sum_{i=1}^{k} \left[\begin{array}{l} \left(p_{III,i} \rho_i D_i\right) - \frac{(A_r + A_m + a_{r,i} + a_{m,i} + a_{mr,i})}{T_{III,i}} - c_{mr,i} u_i \mu_i \\ -\left[\left[c_{mr,i} u_i \mu_i \theta_{m,i} + \left(h_{r,i} + c_{r,i}\theta_i + h_{m,i} + h_{mr,i} u_i\right) \rho_i D_i \right] \frac{T_{III,i}}{2} - \frac{(\rho_i - 1)^2 \tau_i D_i^{\gamma_i}}{T_{III,i}} \right] \end{array} \right]$$

13.3.2.1.1 Optimization I

The retailer is to individually determine an individual replenishment cycle time, the retail selling price, and the factor of publicity effort for *i*th item to maximize their profit. The following propositions are developed for optimality testing of the model under Policy III.

Proposition I: π_{III} is concave in $p_{III,i}$.

Proof: $$\frac{\partial \pi_{III}}{\partial p_{III,i}} = D_i \rho_i - p_{III,i} \beta_i \rho_i - \frac{T_{III,i} \beta_i \rho_i}{2} \left(h_{r,i} + c_{r,i}\theta_i + h_{m,i} + h_{mr,i} u_i\right) + \frac{\tau_i \beta_i \gamma_i}{T_{III,i}} (\rho_i - 1)^2 D_i^{\gamma_i - 1}$$

$$\frac{\partial \pi_{III}}{\partial p_{III,i}} = 0 \Rightarrow p_{III,i} \beta_i \rho_i + \frac{T_{III,i}}{2} \left(h_{r,i} + c_{r,i}\theta_i + h_{m,i} + h_{mr,i} u_i\right) \beta_i \rho_i$$

$$= D_i \rho_i + \frac{\tau_i \beta_i \gamma_i}{T_{III,i}} (\rho_i - 1)^2 D_i^{\gamma_i - 1}$$

Since the above equation is an implicit function, $p^*_{III,i}$ is difficult to obtain, but π_{III}'' should satisfy the concavity property

$$\frac{\partial^2 \pi_{III}}{\partial p_{III,i}{}^2} = -\left[2\beta_i\rho_i + \frac{\tau_i\beta_i^2\gamma_i(\gamma_i-1)}{T_{III,i}}(\rho_i-1)^2 D_i^{(\gamma_i-2)}\right]$$

$$= -\left[2\beta_i\rho_i + \frac{\tau_i\beta_i^2\gamma_i(\gamma_i-1)}{T_{III}}(\rho_i-1)^2 D_i^{(\gamma_i-2)}\right] < 0.$$

For, $\gamma_i > 1$, π_{III} is concave in $p_{III,i}$.

Proposition II: π_{III} is concave in $T_{III,i}$.

$$\text{Proof: } \frac{\partial \pi_{III}}{\partial T_{III,i}} = \frac{\left(A_r + A_m + a_{m,i} + a_{mr,ir} + (\rho_i-1)^2\tau_i D_i^{\gamma_i}\right)}{T^2_{III,i}}$$

$$-\frac{1}{2}\left[D_i\rho_i\left(h_{r,i} + c_{r,i}\theta_i + h_{m,i} + h_{mr,i}u_i\right) + c_{mr,i}u_i\mu_i\theta_{m,i}\right]$$

$$\frac{\partial \pi_{III}}{\partial T_{III,i}} = 0 \Rightarrow \frac{\left(A_r + A_m + a_{m,i} + a_{mr,ir} + (\rho_i-1)^2\tau_i D_i^{\gamma_i}\right)}{T^2_{III,i}}$$

$$= \frac{1}{2}\left[D_i\rho_i\left(h_{r,i} + c_{r,i}\theta_i + h_{m,i} + h_{mr,i}u_i\right) + c_{mr,i}u_i\mu_i\theta_{m,i}\right] \quad \Rightarrow T^*_{III,i}$$

$$= \sqrt{\frac{2\left[A_r + A_m + a_{m,i} + a_{mr,ir} + (\rho_i-1)^2\tau_i D_i^{\gamma_i}\right]}{\left[D_i\rho_i\left(h_{r,i} + c_{r,i}\theta_i + h_{m,i} + h_{mr,i}u_i\right) + c_{mr,i}u_i\mu_i\theta_{m,i}\right]}}$$

$T^*_{III,i}$ is obtained, and π_{III}'' should satisfy the concavity property

$$\frac{\partial^2 \pi_{III}}{\partial T_{III,i}{}^2} = -\left[\frac{2\left(A_r + A_m + a_{r,i} + a_{m,i} + a_{mr,i} + (\rho_i-1)^2\tau_i D_i^{\gamma_i}\right)}{T^3_{III,i}}\right] < 0.$$

Hence, π_{III} is concave in $T_{III,i}$.

Proposition III: π_{III} is concave in ρ_i.

$$\text{Proof: } \frac{\partial \pi_{III}}{\partial \rho_i} = p_{III,i}D_i - \frac{D_i T_{III,i}}{2}\left(h_{r,i} + c_{r,i}\theta_i + h_{m,i} + h_{mr,i}u_i\right) - \frac{2\tau_i(\rho_i-1)D_i^{\gamma_i}}{T_{III,i}}$$

$$\frac{\partial \pi_{III}}{\partial \rho_i} = 0 \Rightarrow \rho_i^* = 1 + \frac{T_{III,i}}{2D_i^{\gamma_i}\tau_i}\left[p_{III,i}D_i - \frac{D_i T_{III,i}}{2}\left(h_{r,i} + c_{r,i}\theta_i + h_{m,i} + h_{mr,i}u_i\right)\right]$$

ρ_i^* is obtained, and π_{III}'' should satisfy the concavity property $\frac{\partial^2 \pi_{III}}{\partial \rho_i^2} = -\left[\frac{2\tau_i D_i^{\gamma_i}}{T_{III,i}}\right] < 0.$
Hence, π_{III} is concave in ρ_i.

13.3.2.2 Policy IV: Cooperative Replenishment of Retailer and Manufacturer for Joint Items

The retailer and manufacturer are to jointly decide $T_{IV}^*, p_{IV,i}^*, \rho_i^*$ for the ith item. The total SC profit per unit time is:

$$\pi_{IV} = \pi_{II,r} + \pi_{II,m}$$

$$= \left[\sum_{i=1}^{k}\left[\begin{array}{l}(p_{IV,i} - c_{r,i})\rho_i D_i - \frac{a_{r,i}}{T_{IV}} \\ -\frac{(h_{r,i} + c_{r,i}\theta_i)}{2} D_i \rho_i T_{IV} - \frac{(\rho_i - 1)^2 \tau_i D_i^{\gamma_i}}{T_{IV}}\end{array}\right] - \frac{A_r}{T_{IV}}\right]$$

$$+ \left[\sum_{i=1}^{k}\left[\begin{array}{l}(c_{r,i}\rho_i D_i) - \frac{(a_{m,i} + a_{mr,i})}{T_{IV}} - c_{mr,i}u_i\mu_i \\ -\left[c_{mr,i}u_i\mu_i\theta_{m,i} + (h_{m,i} + h_{mr,i}u_i)\rho_i D_i\right]\frac{T_{IV}}{2}\end{array}\right] - \frac{A_m}{T_{IV}}\right]$$

After simplifying:

$$\pi_{IV} = \pi_{II,r} + \pi_{II,m}$$

$$= \sum_{i=1}^{k}\left[\begin{array}{l}(p_{IV,i}\rho_i D_i) - \frac{(a_{r,i} + a_{m,i} + a_{mr,i})}{T_{IV}} - c_{mr,i}u_i\mu_i \\ -\left[\left[c_{mr,i}u_i\mu_i\theta_{m,i} + (h_{r,i} + c_{r,i}\theta_i + h_{m,i} + h_{mr,i}u_i)\rho_i D_i\right]\frac{T_{IV}}{2} - \frac{(\rho_i - 1)^2 \tau_i D_i^{\gamma_i}}{T_{IV}}\right] \\ -\frac{(A_r + A_m)}{T_{IV}}\end{array}\right]$$

13.3.2.2.1 Optimization II

The retailer determines $T_{IV}^*, p_{IV,i}^*, \rho_i^*$ for ith item to optimize their profit. The following propositions are developed for optimality testing of the model under Policy IV.

Proposition I: π_{IV} is concave in $p_{IV,i}$.

$$\text{Proof: } \frac{\partial \pi_{IV}}{\partial p_{IV,i}} = \sum_{i=1}^{k} \left[\begin{array}{l} D_i\rho_i - p_{IV,i}\beta_i\rho_i + \frac{T_{IV}\beta_i\rho_i}{2}\left(h_{r,i} + c_{r,i}\theta_i + h_{m,i} + h_{mr,i}u_i\right) \\ + \frac{\tau_i\beta_i\gamma_i}{T_{IV}}\left(\rho_i - 1\right)^2 D_i^{\gamma_i - 1} \end{array} \right]$$

$$\frac{\partial \pi_{IV}}{\partial p_{IV,i}} = 0 \Rightarrow \sum_{i=1}^{k} \left[\begin{array}{l} D_i\rho_i - p_{IV,i}\beta_i\rho_i + \frac{T_{IV}\beta_i\rho_i}{2}\left(h_{r,i} + c_{r,i}\theta_i + h_{m,i} + h_{mr,i}u_i\right) \\ + \frac{\tau_i\beta_i\gamma_i}{T_{IV}}\left(\rho_i - 1\right)^2 D_i^{\gamma_i - 1} \end{array} \right] = 0$$

Since the above equation is an implicit function, $p^*_{IV,i}$ is difficult to obtain, but π_{IV}''should satisfy the concavity property $\frac{\partial^2 \pi_{IV}}{\partial p_{IV,i}{}^2} = -\left[2\beta_i\rho_i + \frac{\tau_i\beta_i^2\gamma_i(\gamma_i - 1)}{T_{IV,i}}\left(\rho_i - 1\right)^2 D_i^{(\gamma_i - 2)} \right]$

$$= -\left[2\beta_i\rho_i + \frac{\tau_i\beta_i^2\gamma_i(\gamma_i - 1)}{T_{IV}}\left(\rho_i - 1\right)^2 D_i^{(\gamma_i - 2)} \right] < 0.$$

For, $\gamma_i > 1$, π_{IV} is concave in $p_{IV,i}$.

Proposition-II: π_{IV} is concave in T_{IV}.

$$\text{Proof: } \frac{\partial \pi_{IV}}{\partial T_{IV}} = \sum_{i=1}^{k} \left[\begin{array}{l} \frac{A_r + A_m + a_{r,i} + a_{m,i} + a_{mr,i} + \left(\rho_i - 1\right)^2 \tau_i D_i^{\gamma_i})}{T_{IV}^2} \\ -\frac{1}{2}\left[D_i\rho_i\left(h_{r,i} + c_{r,i}\theta_i + h_{m,i} + h_{mr,i}u_i\right) + c_{mr,i}u_i\mu_i\theta_{m,i} \right] \end{array} \right]$$

$$\frac{\partial \pi_{IV}}{\partial T_{IV}} = 0 \Rightarrow \sum_{i=1}^{k} \left[\frac{A_r + A_m + a_{r,i} + a_{m,i} + a_{mr,i} + \left(\rho_i - 1\right)^2 \tau_i D_i^{\gamma_i}}{T_{IV}^2} \right]$$

$$= \sum_{i=1}^{k} \left[\frac{1}{2}\left[D_i\rho_i\left(h_{r,i} + c_{r,i}\theta_i + h_{m,i} + h_{mr,i}u_i\right) + c_{mr,i}u_i\mu_i\theta_{m,i} \right] \right] \quad \Rightarrow T_{IV}^*$$

$$= \sqrt{\frac{2\sum_{i=1}^{k}\left[A_r + A_m + a_{r,i} + a_{m,i} + a_{mr,i} + \left(\rho_i - 1\right)^2 \tau_i D_i^{\gamma_i} \right]}{\sum_{i=1}^{k}\left[D_i\rho_i\left(h_{r,i} + c_{r,i}\theta_i + h_{m,i} + h_{mr,i}u_i\right) + c_{mr,i}u_i\mu_i\theta_{m,i} \right]}}$$

T_{IV}^* is obtained, and π_{IV}'' should satisfy the property of concavity $\frac{\partial^2 \pi_{IV}}{\partial T_{IV}^2} = -\left[\frac{2\left(A_r + A_m + a_{r,i} + a_{m,i} + a_{mr,i} + (\rho_i - 1)^2 \tau_i D_i^{\gamma_i}\right)}{T_{IV}^3}\right] < 0.$

Hence, π_{IV} is concave in T_{IV}.

Proposition III: π_{IV} is concave in ρ_i.

Proof: $\frac{\partial \pi_{IV}}{\partial \rho_i} = p_{IV,i} D_i - \frac{D_i T_{IV}}{2}\left(h_{r,i} + c_{r,i}\theta_i + h_{m,i} + h_{mr,i} u_i\right) - \frac{2\tau_i(\rho_i - 1) D_i^{\gamma_i}}{T_{IV}}$

$$\frac{\partial \pi_{IV}}{\partial \rho_i} = 0 \Rightarrow \rho_i^* = 1 + \frac{T_{IV}}{2 D_i^{\gamma_i} \tau_i}\left[p_{IV,i} D_i - \frac{D_i T_{IV}}{2}\left(h_{r,i} + c_{r,i}\theta_i + h_{m,i} + h_{mr,i} u_i\right)\right]$$

ρ_i^* is obtained, and π_{IV}'' should satisfy the concavity property $\frac{\partial^2 \pi_{IV}}{\partial \rho_i^2} = -\left[\frac{2\tau_i D_i^{\gamma_i}}{T_{IV}}\right] < 0.$

Hence, π_{IV} is concave in ρ_i.

13.4 Algorithm

Table 13.3 presents the algorithm of the SC problem for determining the optimum solutions.

13.5 Numerical Analysis

These four models are applied to solve three-item problems in the supply chain with and without publicity effort under a mixed-integer heuristic approach, and it is solved by a personal computer using Lingo 14.0 software package.

Three items, like 1, 2, and 3, are replenished jointly. In Table 13.4, the dataset of the model is illustrated. Figure 13.2 shows the retail price and linear decreasing demand function, and Figure 13.3 shows the convexity of the publicity effort cost with respect to the retail selling price and publicity

TABLE 13.3

Algorithm

Policies I and II	
Step 1	Obtain critical points after solving $\frac{d\pi_{Ir}}{dp_{Ii}} = \frac{d\pi_{Ir}}{d\rho_i} = \frac{d\pi_{Ir}}{dT_{Ii}} = 0$ and $\frac{d\pi_{IIr}}{dp_{IIi}} = \frac{d\pi_{IIr}}{d\rho_i} = \frac{d\pi_{IIr}}{dT_{II}} = 0$. Verify that $\frac{d^2\pi_{Ir}}{dp_{Ii}^2} < 0, \frac{d^2\pi_{Ir}}{d\rho_i^2} < 0$ and $\frac{d^2\pi_{Ir}}{dT_{Ii}^2} < 0$ and $\frac{d^2\pi_{IIr}}{dp_{IIi}^2} < 0, \frac{d^2\pi_{IIr}}{d\rho_i^2} < 0$ and $\frac{d^2\pi_{IIr}}{dT_{II}^2} < 0$ satisfying the property of concavity.
Step 2	$\pi_{Ir}^*\left(p_{Ii}^*, \rho_i^*, T_{I,i}^*\right)$ and $\pi_{IIr}^*\left(p_{IIi}^*, \rho_i^*, T_{II}^*\right)$ are optimum solutions; it gives the maximum values of π_{Ir} and π_{IIr}, respectively, after putting the optimum values in Equation (13.8) to obtain the value of π_{Im}.
Step 3	Determining $Q_r^*, Q_f^*, \& D_i^*$ for $i = 1,2,\ldots k$ items, where $Q_r^* = \mu_i u_i T_{Ii}\left[1 + \frac{\theta_{mi} T_{Ii}}{2}\right]$ and $Q_f^* = D_i^* T_{Ii}\left[1 + \frac{\theta_i T_{Ii}}{2}\right]$ under policy I and $Q_r^* = \mu_i u_i T_{II}\left[1 + \frac{\theta_{mi} T_{II}}{2}\right]$ and $Q_f^* = D_i^* T_{II}\left[1 + \frac{\theta_i T_{II}}{2}\right]$ under Policy II.
Step 4	Under Policies I and II after adjusting $Q_r^*, Q_f^*, \& D_i^*$ to the close integer values for $i = 1,2,\ldots k$ items .
Step 5	Now $\pi_I^* = \pi_{Ir}^* + \pi_{Im}^*$ and $\pi_{II}^* = \pi_{IIr}^* + \pi_{IIm}^*$ be calculated.
Policies III and IV	
Step 1	Obtain critical points after solving the Equations (13.10) and (13.11) for $\frac{d\pi_{III}}{dp_{IIIi}} = \frac{d\pi_{III}}{d\rho_i} = \frac{d\pi_{III}}{dT_{IIIi}} = 0$ and $\frac{d\pi_{IV}}{dp_{IVi}} = \frac{d\pi_{IV}}{d\rho_i} = \frac{d\pi_{IV}}{dT_{IV}} = 0$. Verify that $\frac{d^2\pi_{III}}{dp_{IIIi}^2} < 0, \frac{d^2\pi_{III}}{d\rho_i^2} < 0$ and $\frac{d^2\pi_{III}}{dT_{IIIi}^2} < 0$ and $\frac{d^2\pi_{IV}}{dp_{IVi}^2} < 0, \frac{d^2\pi_{IV}}{d\rho_i^2} < 0$ and $\frac{d^2\pi_{IV}}{dT_{IV}^2} < 0$ satisfying the property of concavity.
Step 2	$\pi_{III}^*\left(p_{IIIi}^*, \rho_i^*, T_{III,i}^*\right)$ and $\pi_{IV}^*\left(p_{IVi}^*, \rho_i^*, T_{IV}^*\right)$ are optimum solutions; it gives the maximum values of π_{III} and π_{IV}, respectively.
Step 3	Determining Q_r^*, Q_f^*, D_i^* for $i = 1,2,\ldots k$ items, where $Q_r^* = \mu_i u_i T_{IIIi}\left[1 + \frac{\theta_{mi} T_{IIIi}}{2}\right]$ and $Q_f^* = D_i^* T_{IIIi}\left[1 + \frac{\theta_i T_{IIIi}}{2}\right]$ and under Policy III and $Q_r^* = \mu_i u_i T_{IV}\left[1 + \frac{\theta_{mi} T_{IV}}{2}\right]$ and $Q_f^* = D_i^* T_{IV}\left[1 + \frac{\theta_i T_{IV}}{2}\right]$ under Policy IV, respectively.
Step 4	Under Policies III and IV after adjusting Q_r^*, Q_f^*, D_i^* to the close integer values for $i = 1,2,\ldots k$ items.
Step 5	Now $\pi_{III}^* = \pi_{IIIr}^* + \pi_{IIIm}^*$ and $\pi_{IV}^* = \pi_{IVr}^* + \pi_{IVm}^*$ be calculated.

TABLE 13.4
Dataset for Numerical Analysis

Item	1	2	3
Retailer			
A_r	250	250	250
$a_{r,i}$	45	60	40
$h_{r,i}$	5.2	5.1	4.8
$c_{r,i}$	47	45	40
α_i, β_i	(350,2.0)	(500,2.5)	(450,3.0)
τ_i, γ_i	(2,2)	(2,2)	(2,2)
θ_i	0.08	0.07	0.075
Manufacturer			
$h_{m,i}$	3.8	3.5	2.6
$h_{mr,i}$	1.8	1.7	1.2
$c_{mr,i}$	20	18	16
μ_i	250	350	300
$\theta_{m,i}$	0.04	0.02	0.03
u_i	0.3	0.2	0.25
A_m	450	450	450
$a_{m,i}$	65	80	55
$a_{mr,i}$	35	50	30

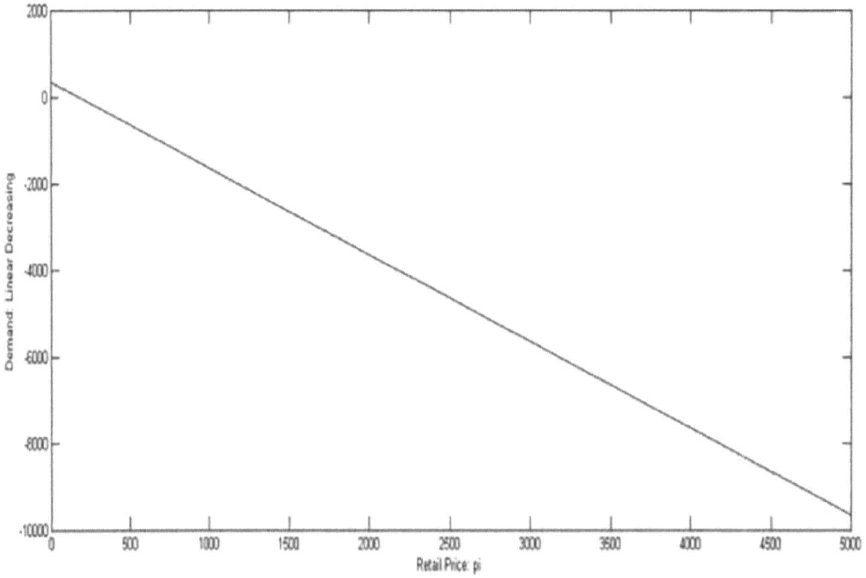

FIGURE 13.2
Two-dimensional plot of the retailing price and demand rate function.

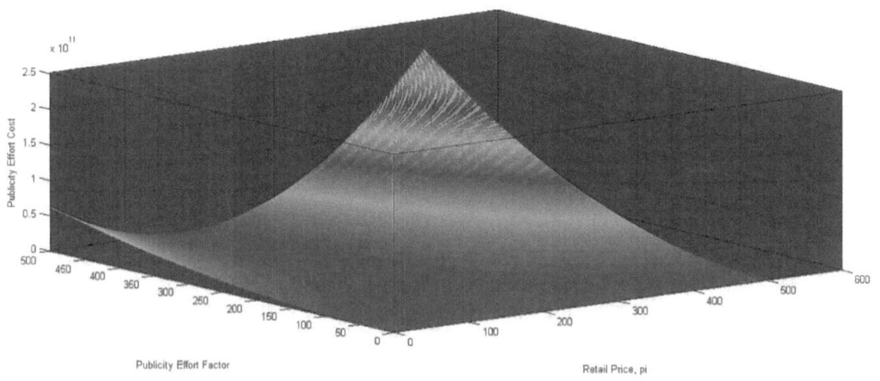

FIGURE 13.3
Three-dimensional mesh plot with retail price, publicity effort factor, and publicity effort cost.

effort factor. Figures 13.4–13.6 show the 3-D plot of retail price, publicity effort factor, and the total SC profit of three items. Table 13.5 presents the optimal values of the decision variables and total SC profits of four different models.

13.6 Comparative Analysis

From Table 13.6 it is seen that under Policy I there are three types of models: with publicity effort, without publicity effort, and mixed integer without publicity effort. Under this policy, it is investigated that the total profit with publicity effort is the highest at Rs. 44,267.10, whereas it is the lowest at Rs. 43,437.38 for the model of without publicity effort. It may be interpreted that the publicity effort effect of selling deteriorating items enables to increase the total profit. In the model with publicity effort, the replenishment cycle time of individual and retailer selling prices of each item is comparatively more than that of the other models. After imposing promotional activities with the deteriorated item, the demand rate is increased. For the decentralized model with individual replenishment, the publicity attempt can be an instrument for revenue growth. Under Policy II there are four different models: with publicity effort, mixed integer with publicity effort, without publicity effort, and mixed integer without publicity effort. From this policy, it is investigated that the total profit with publicity effort is the highest at Rs. 46,009.30, whereas it is the lowest at Rs. 45,409.00 for the model of mixed integer without publicity effort. The replenishment cycle time of common and retail selling prices of the supply chain model with publicity effort is more

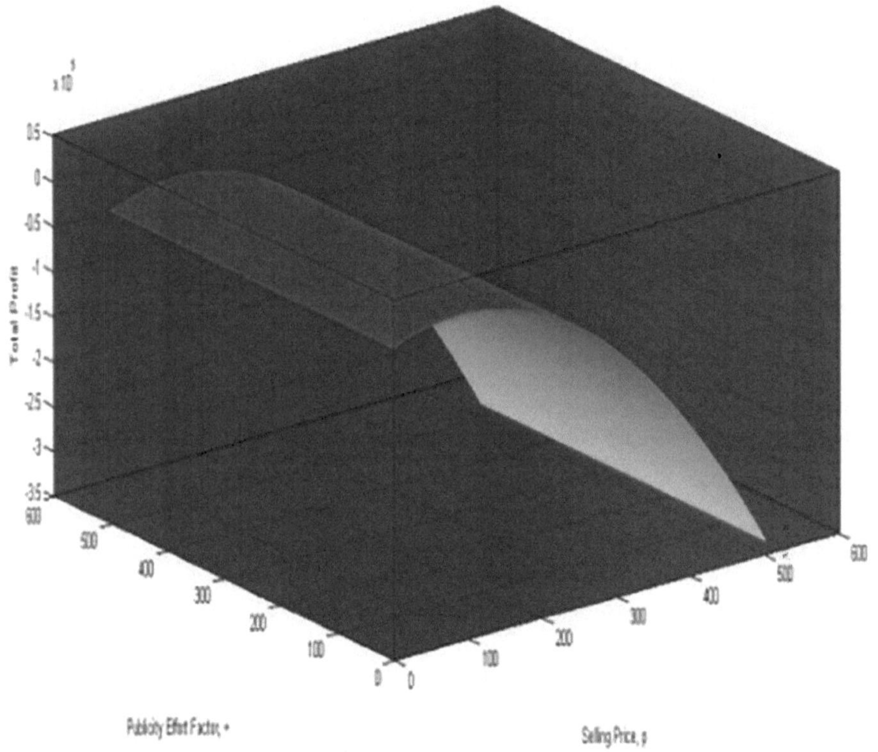

FIGURE 13.4
Three-dimensional mesh plot with retail price, publicity effort factor, and total supply chain channel profit of item no. 1.

than that of the other models. Through publicity efforts, the demand rate for the deteriorated items is increased. It is concluded that publicity effort can be the key parameter for competitive advantage in the decentralized supply chain under joint replenishment.

Under Policy III there are three different models: with publicity effort, without publicity effort, and mixed integer without publicity effort. From this policy, it is investigated that the total profit with publicity effort is the highest at Rs. 47,736.29, whereas it is the lowest at Rs. 46,925.75 for the model of mixed integer without publicity effort. The cycle time of individual replenishment and retail selling price of each deteriorating item of the model with publicity effort is more than that of the other models. Publicity effort affects the demand rate more for each deteriorated item. It is concluded that publicity effort is also the key parameter for competitive advantage in the centralized supply chain under individual replenishment.

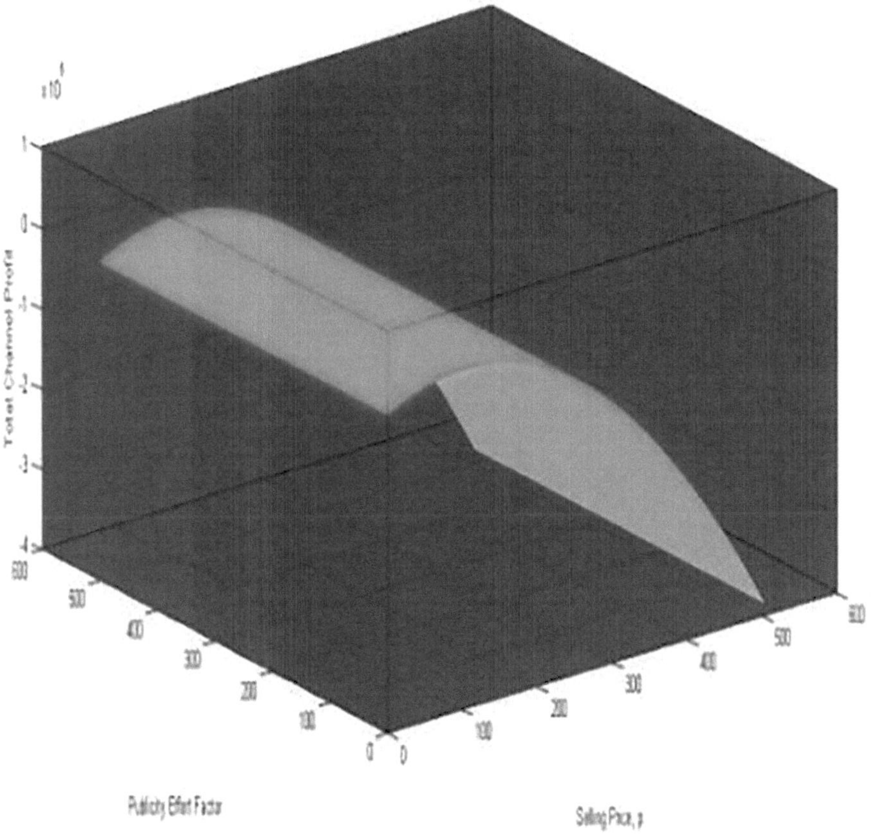

FIGURE 13.5
Three-dimensional mesh plot with retail price, publicity effort factor, and total supply chain channel profit of item no. 2.

Under Policy IV there are four different models: with publicity effort, mixed integer with publicity effort, without publicity effort, and mixed integer without publicity effort. From this policy, it is investigated that the total profit with publicity effort is the highest at Rs. 48,796.94 where it is the lowest at Rs. 48,141.38 for the model of mixed integer without publicity effort. The retail selling price and common replenishment cycle time of the model with publicity effort is more than that of the other models. Through publicity efforts, the demand for deteriorated items is increased. It is concluded that publicity efforts can play a significant role in the centralized supply chain with joint replenishment.

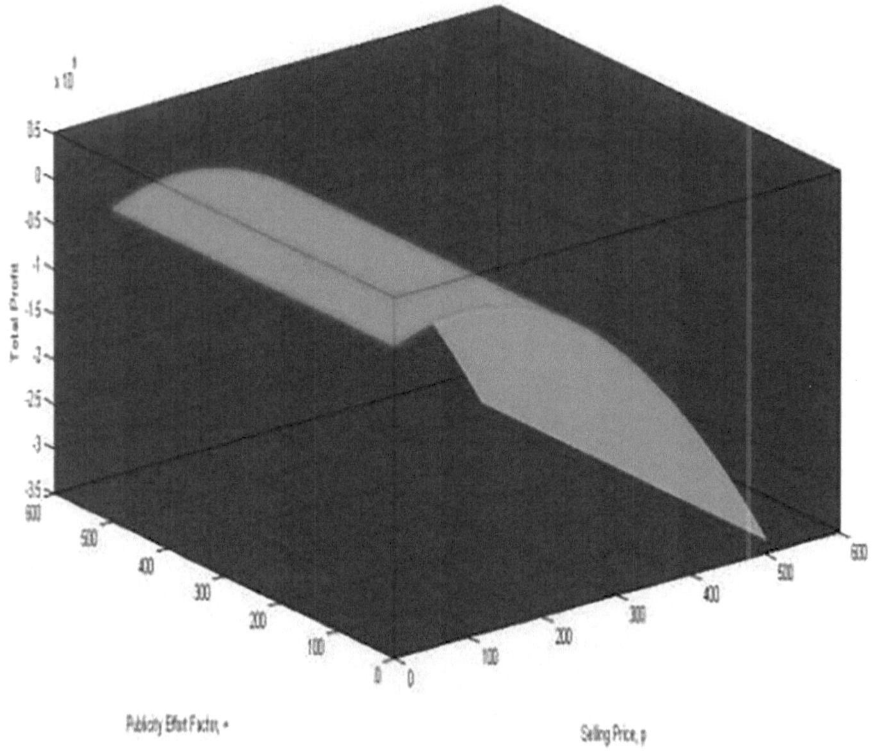

FIGURE 13.6
Three-dimensional mesh plot with retail price, publicity effort factor, and total supply chain channel profit of item no. 3.

Overall, from the four different policies, it is seen that the total profit of Policy IV (centralized) model with publicity effort, Rs. 48,796.94, is the highest, whereas that of Rs. 43,437.38 is the lowest for Policy I (decentralized) model without publicity effort. If it is compared under the individual replenishment or joint replenishment, the total profit of the joint replenishment model is more than that of individual replenishment. So, it is investigated that channel coordination total profit for ordering of joint items with publicity effort is higher than any other total profit for the supply chain model. Under Policy IV the total profit of the model under mixed-integer programming and publicity effort is Rs. 48,794.00, which is very close to the supply chain profit of Rs. 48,796.94 of the model with publicity effort.

TABLE 13.5

Results of Supply Chain Models under Four Different Policies

	Joint Replenishment	
Channel Coordination	**No**	**Yes**
No	Policy I	Policy II
	Iteration – 8,409	Iteration – 571
	$p^*_{I,1} = 113.93, T^*_{I,1} = 0.85, \rho^*_1 = 1.03$	$p^*_{II,1} = 112.85, T^*_{II} = 0.52, \rho^*_1 = 1.02$
	$p^*_{I,2} = 125.01, T^*_{I,2} = 0.73, \rho^*_2 = 1.03$	$p^*_{II,2} = 124.24, \rho^*_2 = 1.02$
	$p^*_{I,3} = 97.14, T^*_{I,3} = 0.80, \rho^*_3 = 1.02$	$p^*_{II,3} = 96.41, \rho^*_3 = 1.01$
	$Q^*_{r,1} = 110.52, Q^*_{f,1} = 64.57$	$Q^*_{r,1} = 39.4, Q^*_{f,1} = 67.42$
	$Q^*_{r,2} = 143.64, Q^*_{f,2} = 51.30$	$Q^*_{r,2} = 36.59, Q^*_{f,2} = 102.03$
	$Q^*_{r,3} = 132.86, Q^*_{f,3} = 60.40$	$Q^*_{r,3} = 29.3, Q^*_{f,3} = 86.47$
	$D^*_1 = 122.14, \rho^*_1 D^*_1 = 126.27$	$D^*_1 = 124.29, \rho^*_1 D^*_1 = 127.05$
	$D^*_2 = 187.47, \rho^*_2 D^*_2 = 192.52$	$D^*_2 = 189.4, \rho^*_2 D^*_2 = 192.74$
	$D^*_3 = 158.57, \rho^*_3 D^*_3 = 162.12$	$D^*_3 = 160.78, \rho^*_3 D^*_3 = 163.14$
	PEC = 110.4	PEC = 48.56
	$\pi^*_{I,r} = 29999.85$	$\pi^*_{II,r} = 30761.04$
	$\pi^*_{I,m} = 14267.25$	$\pi^*_{II,m} = 15248.26$
	$\pi^*_I = 44267.10$	$\pi^*_{II} = 46009.30$
Yes	Policy III	Policy IV
	Iteration – 175	Iteration – 149
	$p^*_{III,1} = 92.42, T^*_{III,1} = 0.98, \rho^*_1 = 1.04$	$p^*_{IV,1} = 91.35, T^*_{IV} = 0.75, \rho^*_1 = 1.03$
	$p^*_{III,2} = 104.20, T^*_{III,2} = 0.88, \rho^*_2 = 1.03$	$p^*_{IV,2} = 103.51, \rho^*_2 = 1.03$
	$p^*_{III,3} = 78.52, T^*_{III,3} = 0.96, \rho^*_3 = 1.03$	$p^*_{IV,3} = 77.78, \rho^*_3 = 1.02$
	$Q^*_{r,1} = 74.71, Q^*_{f,1} = 174.47$	$Q^*_{r,1} = 57.17, Q^*_{f,1} = 133.60$
	$Q^*_{r,2} = 62.08, Q^*_{f,2} = 224.01$	$Q^*_{r,2} = 52.97, Q^*_{f,2} = 190.70$
	$Q^*_{r,3} = 72.95, Q^*_{f,3} = 218.71$	$Q^*_{r,3} = 56.96, Q^*_{f,3} = 170.87$
	$D^*_1 = 165.15, \rho^*_1 D^*_1 = 171.86$	$D^*_1 = 167.30, \rho^*_1 D^*_1 = 172.71$
	$D^*_2 = 239.50, \rho^*_2 D^*_2 = 247.22$	$D^*_2 = 241.23, \rho^*_2 D^*_2 = 247.42$
	$D^*_3 = 214.44, \rho^*_3 D^*_3 = 220.17$	$D^*_3 = 216.66, \rho^*_3 D^*_3 = 221.28$
	PEC = 274.56	PEC = 177.77
	$\pi^*_{III,r} = 26613.71$	$\pi^*_{IV,r} = 26597.76$
	$\pi^*_{III,m} = 21122.58$	$\pi^*_{IV,m} = 22199.18$
	$\pi^*_{III} = 47736.29$	$\pi^*_{IV} = 48796.94$

TABLE 13.6
Comparative Analysis

Model	Iteration	$T^*_{I,i}$	$p^*_{I,i}$	ρ^*_i	Q_r	Q_f	$D^*_i(p_{I,i})$	$\rho^*_i D^*_i(p_{I,i})$	PEC	$\pi^*_{I,r}$	$\pi^*_{I,m}$	π^*_I
Policy I												
With Publicity	8,409	0.85	113.93	1.03	110.52	64.57	122.14	126.27	110.41	29999.85	14267.25	44267.10
Effort (PE)		0.73	125.01	1.03	143.64	51.30	187.47	192.52				
Effect		0.80	97.14	1.02	132.86	60.40	158.57	162.12				
Without PE	147	0.73	112.63	–	55.28	93.27	124.75	–	–	29615.14	13822.24	43437.38
Effect		0.63	123.79		44.24	122.28	190.51					
		0.68	96.33		51.49	112.21	161.02					
Mixed Integer,	461	0.73	112.5	–	55.23	93.36	(Integer)	–	–	29615.01	13852.87	43467.89
Without PE		0.63	123.6		44.18	122.44	125					
		0.68	96.33		51.49	112.20	191					
							161					
Policy II												
Model	Iteration	T^*_{II}	$p^*_{II,i}$	ρ^*_i	Q_r	Q_f	$D^*_i(p_{II,i})$	$\rho^*_i D^*_i(p_{II,i})$	PEC	$\pi^*_{II,r}$	$\pi^*_{II,m}$	π^*_{II}
With Publicity	571	0.52	112.85	1.02	39.40	67.42	124.29	127.05	48.56	30761.04	15248.26	46009.30
Effort Effect			124.24	1.02	36.59	102.03	189.4	192.74				
			96.41	1.01	29.30	86.47	160.78	163.14				
Mixed Integer,	6,356	0.51	113	1.02	38.89	66.52	(Integer)	(Integer)	44	30757.64	15239.40	45997.04
With PE			124	1.02	36.10	100.83	124	127				
			96	1.01	38.79	85.27	190	193				
							161	163				
Without	99	0.4457985	111.99	–	33.73	57.17	126.00	–	–	30504.15	14939.36	45443.50
Publicity			123.42		31.34	86.68	191.45					
Effort Effect			95.87		33.65	73.60	162.39					
Mixed Integer,	392	0.45	112	–	33.76	57.22	(Integer)126	–	–	30504.02	14904.98	45409.00
Without PE			123.6		31.37	86.55	191					
			96		33.69	73.49	162					

(Continued)

TABLE 13.6 (*Continued*)

Comparative Analysis

Model	Iteration	$T^*_{I,i}$	$p^*_{I,i}$	ρ^*_i	Q_r	Q_f	$D^*_i(p_{I,i})$	$\rho^*_i D^*_i(p_{I,i})$	PEC	$\pi^*_{I,r}$	$\pi^*_{I,m}$	π^*_I
Policy III												
Model	Iteration	$T^*_{III,i}$	$p^*_{III,i}$	ρ^*_i	Q_r	Q_f	$D^*_i(p_{III,i})$	$\rho^*_i D^*_i(p_{III,i})$	PEC	$\pi^*_{III,ri}$	$\pi^*_{III,m}$	π^*_{III}
With Publicity	175	0.98	92.42	1.04	74.71	174.47	165.15	171.86	274.56	26613.71	21122.58	47736.29
Effort Effect		0.88	104.20	1.03	62.08	224.01	239.50	247.22				
		0.96	78.52	1.03	72.95	218.71	214.44	220.17				
Without	100	0.85	90.34	–	65.22	149.71	169.31	–	–	26234.58	20691.26	46925.84
Publicity		0.77	102.34		54.54	193.90	244.15					
Effort Effect		0.83	77.23		63.33	187.77	218.30					
Mixed Integer,		0.86	90.5	–	65.28	149.57	(Integer)	–	–	26266.77	20658.98	46925.75
Without PE	391	0.77	102.4		54.56	193.84	169					
		0.83	77.33		63.38	187.65	244					
							218					
Policy IV												
Model	Iteration	T^*_{IV}	$p^*_{IV,i}$	ρ^*_i	Q_r	Q_f	$D^*_i(p_{IV,i})$	$\rho^*_i D^*_i(p_{IV,i})$	PEC	$\pi^*_{IV,r}$	$\pi^*_{IV,m}$	π^*_{IV}
With Publicity	149	0.75	91.35	1.03	57.17	133.60	167.30	172.71	177.77	26597.76	22199.18	48796.94
Effort Effect			103.51	1.03	52.97	190.70	241.23	247.42				
			77.78	1.02	56.96	170.87	216.66	221.28				
Mixed Integer,	5737	0.75	91	1.03	56.96	133.32	Integer	(Integer)	172	26611.25	22182.74	48794.00
With PE			103.6	1.02	52.77	189.66	168	173				
			78	1.02	56.75	170.00	241	247				
							216	221				
Without	78	0.661	89.68	–	49.88	114.95	170.63	–	–	26231.69	21909.69	48141.38
Publicity			101.98		46.25	164.55	245.04					
Effort Effect			76.75		49.71	147.79	219.73					
Mixed Int	372	0.66	89.5	–	49.86	115.15	Integer	–	–	26206.39	21934.90	48141.29
Without PE			102		46.23	164.45	171					
			76.67		49.70	147.90	245					
							220					

13.7 Sensitive Analysis

It is remarkable to study the control of the important variables, $\gamma_i, \tau_i, \alpha_i, \beta_i, \theta_i, A_r, a_{r,i}, h_{r,i}, c_{r,i}, \mu_i, \theta_{m,i}, u_i, A_m, a_{m,i}, a_{mr,i}, h_{m,i}, h_{mr,i},$ and $c_{mr,i}$ on cooperative replenishment for joint items supply chain model. In Table 13.7 the computational results of sensitive analysis are presented, and its summary is shown in Table 13.8.

13.7.1 Sensitive Analysis Using ANOVA Test

The sensitive analysis using the two-way ANOVA test of the total SC profit under publicity effort with respect to significant variables A_m and A_r has been presented.

Null Hypothesis: H_{01} : under joint replenishment and cooperative policy, the optimal total SC profit with publicity effort is insignificant for dissimilar values of A_m and A_r.

Two-way ANOVA results are presented in Table 13.9 using the data values of Table 13.10; it is concluded that the p-values of F values are greater than 0.05. Hence, the null hypothesis may not be rejected, and the total SC profit is highly insignificant for dissimilar values of A_m and A_r.

The sensitive analysis using the two-way ANOVA test of the total SC profit without publicity effort with respect to significant variables A_m and A_r has been presented.

Null Hypothesis: H_{01} : under joint replenishment and cooperative policy, the optimal total SC profit with publicity effort is insignificant for dissimilar values of A_m and A_r.

Two-way ANOVA results are presented in Table 13.11 using the data values of Table 13.12; it is concluded that the p-values of F values are less than 0.05. Hence, the null hypothesis may be rejected and the total supply chain profit is highly significant for dissimilar values of A_m and A_r.

13.8 Conclusions

This chapter considers four different profit optimization supply chain models, under mixed integer and publicity effort, for the individual and joint items. It is investigated that for the mixed integer multi-item supply chain model, publicity effort plays the economies of scale which benefits the retailers. An example is provided to justify these models. It indicates the managerial insights into the impact of a publicity effort, multi-item replenishment,

TABLE 13.7

Policy IV Sensitive Analysis with Respect to the Variables

Variables	Decision Variables	−50%	−30%	0%	+30%	+50%
γ_i	Iteration	176	184	149	331	202
	$T^*_{IV,i}$	0.9118654	0.8079256	0.7510187	0.7252874	0.8660130
	$p^*_{IV,i}$	93.86802	92.27598	91.34825	90.91221	91.40970
		105.8045	104.3554	103.5087	103.1102	103.5637
		79.32494	78.34785	77.77836	77.51069	150.0000
	ρ^*_i	1.082238	1.050520	1.032303	1.023800	1.024649
		1.064726	1.040012	1.025677	1.018950	1.019622
		1.053362	1.033122	1.021306	1.015742	0.000000
	Q_r	69.63715	61.57354	133.6034	55.18562	66.07594
		64.41262	57.01172	190.7030	51.13835	61.14589
		69.32534	61.32876	170.8667	54.98835	65.79470
	Q_f	165.9718	144.9608	57.17245	128.5015	153.4881
		235.9296	206.5953	52.96613	183.5557	219.3373
		210.6194	184.8571	56.96093	164.5670	0.000000
	$D^*_i(p_{IV,i})$	162.2640	165.4480	167.3035	168.1756	167.1806
		235.4886	239.1114	241.2281	242.2245	241.0908
		212.0252	214.9565	216.6649	217.4679	0.000000
	$\rho^*_i D^*_i(p_{IV,i})$	175.6082	173.8065	172.7080	172.1781	171.3014
		250.7308	248.6788	247.4221	246.8146	245.8216
		223.3394	222.0763	221.2812	220.8912	0.000000
	PE	538.4042	296.9276	177.7666	126.8994	118.0839
	$\pi^*_{IV,r}$	27051.20	26777.91	26597.76	26507.36	19018.37
	$\pi^*_{IV,m}$	22534.09	22336.04	22199.18	22128.98	13486.64
	π^*_{IV}	49585.29	49113.95	48796.94	48636.34	32505.02

(Continued)

TABLE 13.7 (*Continued*)

Policy IV Sensitive Analysis with Respect to the Variables

Variables	Decision Variables	−50%	−30%	0%	+30%	+50%
τ_i	Iteration	115	401	149	194	N/A
	$T^*_{IV,i}$	6.438246	7.433919	0.7510187	1.128731	N/A
	$p^*_{IV,i}$	130.3298	141.7649	91.34825	174.8677	N/A
		200.0000	159.1419	103.5087	78.82993	
		150.0000	119.8688	77.77836	150.0000	
	ρ^*_i	47.95383	11.67396	1.032303	0.3885545E-03	N/A
		9571.007	12.11808	1.025677	1.000748	
		0.000000	9.188401	1.021306	63374.96	
	Q_r	545.0450	640.4387	133.6034	86.56589	N/A
		479.6930	559.0586	190.7030	79.90300	
		529.5009	619.7150	170.8667	86.08812	
	Q_f	34686.24	7483.804	57.17245	0.1212406E-03	N/A
		0.000000	11595.92	52.96613	355.6946	
		0.000000	7895.673	56.96093	18.61614	
	$D^*_i(p_{IV,i})$	89.34044	66.47020	167.3035	1.531117	N/A
		0.000000	102.1453	241.2281	302.9252	
		0.000000	90.39372	216.6649	0.1285471E-03	
	$\rho^*_i D^*_i(p_{IV,i})$	4284.216	775.9704	172.7080	0.1027730E-03	N/A
		0.000000	1237.805	247.4221	303.1516	
		0.000000	830.5738	221.2812	15.82323	
	PE	393931.0	315310.7	177.7666	3.838374	N/A
	$\pi^*_{IV,r}$	49734.46	65910.95	26597.76	9711.905	N/A
	$\pi^*_{IV,m}$	137034.4	81743.49	22199.18	8885.728	N/A
	π^*_{IV}	186768.9	147654.4	48796.94	18597.63	N/A

(*Continued*)

TABLE 13.7 (*Continued*)

Policy IV Sensitive Analysis with Respect to the Variables

Variables	Decision Variables	−50%	−30%	0%	+30%	+50%
α_i	Iteration	291	162	149	147	109
	$T^*_{IV,i}$	0.4392586	0.8588249	0.7510187	0.6935511	0.6707949
	$p^*_{IV,i}$	87.50000	65.16943	91.34825	117.6899	135.3090
		99.99997	73.57462	103.5087	133.5920	153.7005
		75.00000	55.40341	77.77836	100.2855	115.3355
	ρ^*_i	534.2915	1.037112	1.032303	1.029758	1.028753
		1,937,836	1.029471	1.025677	1.023665	1.022871
		8.579761	1.024439	1.021306	1.019644	1.018987
	Q_r	33.23382	65.51824	133.6034	52.73785	50.98457
		30.88316	60.63405	190.7030	48.88529	47.27062
		33.16146	65.24165	170.8667	52.55747	50.81583
	Q_f	57.87342	105.6369	57.17245	161.2018	180.2548
		0.000000	151.2359	52.96613	229.8092	256.8455
		0.000000	135.1233	56.96093	206.1652	230.5342
	$D^*_i(p_{IV,i})$	0.000000	114.6611	167.3035	219.6202	254.3820
		0.7112373E-06	166.0635	241.2281	316.0200	365.7487
		0.000000	148.7898	216.6649	284.1434	328.9936
	$\rho^*_i D^*_i(p_{IV,i})$	0.000000	118.9165	172.7080	226.1556	261.6964
		129.7576	170.9575	247.4221	323.4987	374.1139
		0.000000	152.4261	221.2812	289.7251	335.2402
	PE	3.799205	110.5658	177.7666	259.5971	324.9902
	$\pi^*_{IV,r}$	4838.080	6391.511	26597.76	57287.34	83545.65
	$\pi^*_{IV,m}$	1.468075	13783.08	22199.18	30608.21	36214.54
	π^*_{IV}	4839.548	20174.59	48796.94	87895.55	119760.2

(Continued)

TABLE 13.7 (*Continued*)

Policy IV Sensitive Analysis with Respect to the Variables

Variables	Decision Variables	−50%	−30%	0%	+30%	+50%
β_i	Iteration	994	166	149	191	127
	$T^*_{IV,i}$	21.39483	1.160271	0.7510187	0.7088855	0.6963947
	$p^*_{IV,i}$	350.0000	133.1908	91.34825	70.41108	61.19595
		400.0000	150.3230	103.5087	79.75154	69.27498
		300.0000	214.2857	77.77836	60.01438	52.17526
	ρ^*_i	0.1869445E	1.074187	1.032303	1.023249	1.019726
		5,704,238	1.058482	1.025677	1.018513	1.015719
		12269.25	0.000000	1.021306	1.015380	1.013065
	Q_r	2291.221	89.03963	133.6034	53.92019	52.95705
		1818.056	82.16130	190.7030	49.97375	49.08711
		2119.569	88.53480	170.8667	53.73174	52.77519
	Q_f	366579.2	213.2783	57.17245	124.5197	121.4664
		0.000000	302.8023	52.96613	178.1790	174.0582
		0.000000	0.000000	56.96093	159.5658	155.7951
	$D^*_i(p_{IV,i})$	0.1320722E	163.5329	167.3035	166.9312	166.4121
		0.8148728E	236.9347	241.2281	240.8075	240.2188
		0.000000	0.000000	216.6649	215.9439	215.2113
	$\rho^*_i D^*_i(p_{IV,i})$	9232.711	175.6650	172.7080	170.8122	169.6948
		0.000000	250.7911	247.4221	245.2656	243.9948
		0.000000	0.000000	221.2812	219.2652	218.0230
	PE	126241.9	678.3773	177.7666	91.93408	65.87745
	$\pi^*_{IV,r}$	1,894,823	36915.98	26597.76	13393.25	7594.983
	$\pi^*_{IV,m}$	0.6294259	13851.23	22199.18	21931.65	21772.79
	π^*_{IV}	1,894,823	50767.21	48796.94	35324.90	29367.77

(*Continued*)

TABLE 13.7 (*Continued*)

Policy IV Sensitive Analysis with Respect to the Variables

Variables	Decision Variables	−50%	−30%	0%	+30%	+50%
θ_i	Iteration	7,113	396	149	925	222
	$T^*_{IV,i}$	0.9753830	0.7942131	0.7510187	0.4031860	0.9345633
	$p^*_{IV,i}$	175.0000	91.34827	91.34825	0.1612211	91.55129
		104.1742	103.5246	103.5087	197.9068	104.7244
		78.24255	77.76065	77.77836	150.0000	150.0000
	ρ^*_i	0.000000	1.034225	1.032303	0.6278095	1.040518
		1.033605	1.027194	1.025677	0.000000	1.032154
		1.027848	1.022559	1.021306	0.5477670E+10	0.000000
	Q_r	74.58078	60.51215	133.6034	3.026334	71.40236
		68.94277	56.03646	190.7030	2.823440	66.03082
		74.22402	60.27561	170.8667	3.025724	71.07483
	Q_f	0.000000	140.4782	57.17245	8.869736	163.2061
		245.6422	200.5928	52.96613	0.000000	241.0332
		219.7674	179.6725	56.96093	940.6219	0.000000
	$D^*_i(p_{IV,i})$	0.000000	167.3035	167.3035	349.6776	166.8974
		239.5646	241.1884	241.2281	5.232990	238.1889
		215.2724	216.7180	216.6649	0.000000	0.000000
	$\rho^*_i D^*_i(p_{IV,i})$	0.000000	173.0294	172.7080	219.5309	173.6598
		247.6152	247.7472	247.4221	0.000000	245.8476
		221.2674	221.6070	221.2812	23283.96	0.000000
	PE	201.5065	199.4124	177.7666	33931.13	208.7688
	$\pi^*_{IV,r}$	20090.98	26796.71	26597.76	0.000000	19195.61
	$\pi^*_{IV,m}$	14413.31	22242.74	22199.18	917359.5	13596.63
	π^*_{IV}	34504.29	49039.45	48796.94	917359.5	32792.24

(*Continued*)

TABLE 13.7 (*Continued*)

Policy IV Sensitive Analysis with Respect to the Variables

Variables	Decision Variables	−50%	−30%	0%	+30%	+50%
A_r	Iteration	142	151	149	292	183
	$T^*_{IV,i}$	0.7814031	0.6980460	0.7510187	0.8005388	1.731234
	$p^*_{IV,i}$	175.0000	91.07801	91.34825	91.60071	158.4818
		103.6501	103.2621	103.5087	103.7392	174.4945
		77.89021	77.58324	77.77836	77.96063	149.9999
	ρ^*_i	0.000000	1.029957	1.032303	1.034507	0.9121005
		1.026743	1.023823	1.025677	1.027416	0.9912551
		1.022187	1.019774	1.021306	1.022743	77746.40
	Q_r	59.52112	53.08435	133.6034	61.00170	134.3383
		55.12563	49.20431	190.7030	56.48632	123.2844
		59.29215	52.90163	170.8667	60.76138	133.2144
	Q_f	0.000000	124.0420	57.17245	142.5597	55.77908
		198.5391	177.0632	52.96613	203.4789	116.0549
		177.8542	158.6976	56.96093	182.2576	28.03873
	$D^*_i(p_{IV,i})$	0.000000	167.8440	167.3035	166.7986	33.03649
		240.8746	241.8447	241.2281	240.6520	63.76367
		216.3294	217.2503	216.6649	216.1181	0.1397249E-04
	$\rho^*_i D^*_i(p_{IV,i})$	0.000000	172.8720	172.7080	172.5543	30.13260
		247.3164	247.6061	247.4221	247.2498	63.20606
		221.1291	221.5461	221.2812	221.0333	15.20845
	PE	129.0686	153.8564	177.7666	201.6344	19.84709
	$\pi^*_{IV,r}$	20251.72	26894.56	26597.76	26329.77	11661.26
	$\pi^*_{IV,m}$	14366.45	22212.93	22199.18	22177.14	0.5745472
	π^*_{IV}	34618.17	49107.49	48796.94	48506.91	11661.84

(*Continued*)

TABLE 13.7 (*Continued*)

Policy IV Sensitive Analysis with Respect to the Variables

Variables	Decision Variables	−50%	−30%	0%	+30%	+50%
$a_{r,i}$	Iteration	11,612	389	149	173	191
	$T^*_{IV,i}$	0.2017088	1.809603	0.7510187	0.7608403	0.7673192
	$p^*_{IV,i}$	145.8287	174.9999	91.34825	91.39834	91.43137
		200.0000	200.0000	103.5087	103.5545	103.5846
		150.0000	150.0000	77.77836	77.81452	77.83837
	ρ^*_i	1.002789	360625.3	1.032303	1.032740	1.033028
		3,986,728	1779.300	1.025677	1.026021	1.026249
		0.000000	6775.877	1.021306	1.021591	1.021779
	Q_r	1.513426	140.6322	133.6034	57.93134	58.43211
		1.412246	128.9645	190.7030	53.66404	54.12449
		1.513274	139.4042	170.8667	57.71426	58.21132
	Q_f	1.181058	202.3979	57.17245	135.3784	136.5496
		17.57177	0.000000	52.96613	193.2350	194.9058
		0.000000	0.000000	56.96093	173.1248	174.6146
	$D^*_i(p_{IV,i})$	58.34267	0.8959681E-04	167.3035	167.2033	167.1373
		0.000000	0.000000	241.2281	241.1139	241.0385
		0.000000	0.000000	216.6649	216.5564	216.4849
	$\rho^*_i D^*_i(p_{IV,i})$	58.50542	104.2971	172.7080	172.6775	172.6574
		870.5308	0.000000	247.4221	247.3879	247.3654
		0.000000	0.000000	221.2812	221.2321	221.1996
	PE	0.5295819	2087.974	177.7666	182.3843	185.4618
	$\pi^*_{IV,r}$	99852.08	8572.442	26597.76	26543.94	26508.62
	$\pi^*_{IV,m}$	0.1815350	0.000000	22199.18	22195.46	22192.82
	π^*_{IV}	99852.27	8572.442	48796.94	48739.40	48701.44

(*Continued*)

TABLE 13.7 (*Continued*)

Policy IV Sensitive Analysis with Respect to the Variables

Variables	Decision Variables	−50%	−30%	0%	+30%	+50%
$h_{r,i}$	Iteration	6,385	349	149	265	288
	$T^*_{IV,i}$	18.66491	0.5195049	0.7510187	0.6965608	0.6662867
	$p^*_{IV,i}$	174.9998	175.0000	91.34825	91.33809	91.34289
		200.0000	199.9999	103.5087	103.5185	103.5342
		135.1211	150.0000	77.77836	77.82608	77.86224
	ρ^*_i	17.45174	66.78124	1.032303	1.029891	1.028555
		2,680,756	873625.7	1.025677	1.023771	1.022714
		2.080824	22.07804	1.021306	1.019731	1.018857
	Q_r	1922.437	39.36770	133.6034	52.96986	50.63741
		1550.409	36.55426	190.7030	49.09889	46.95082
		1791.794	39.26649	170.8667	52.78791	50.47093
	Q_f	0.2831383	0.000000	57.17245	123.3795	117.7184
		13236.32	65.68184	52.96613	176.2003	168.1668
		2947.038	0.9952558E	56.96093	157.8137	150.5830
	$D^*_i\left(p_{IV,i}\right)$	0.4976701E-03	0.000000	167.3035	167.3238	167.3142
		0.5343604E-04	0.5269037E-05	241.2281	241.2037	241.1646
		44.63672	0.8509021E-04	216.6649	216.5218	216.4133
	$\rho^*_i D^*_i\left(p_{IV,i}\right)$	0.8685207E-02	0.000000	172.7080	172.3253	172.0918
		428.9404	124.1737	247.4221	246.9373	246.6424
		92.88117	0.1878625E	221.2812	220.7939	220.4942
	PE	45695.58	42.37821	177.7666	152.2787	138.9714
	$\pi^*_{IV,r}$	40430.48	17062.86	26597.76	26318.98	26148.06
	$\pi^*_{IV,m}$	0.1731957	0.7392934	22199.18	22129.02	22081.10
	π^*_{IV}	40430.50	17062.86	48796.94	48448.00	48229.15

(*Continued*)

TABLE 13.7 (*Continued*)

Policy IV Sensitive Analysis with Respect to the Variables

Variables	Decision Variables	−50%	−30%	0%	+30%	+50%
$c_{r,i}$	Iteration	6,836	439	149	336	500
	$T^*_{IV,i}$	7.048093	2.178639	0.7510187	0.9012220	22.03043
	$p^*_{IV,i}$	174.9491	175.0000	91.34825	92.36279	160.7122
		200.0000	200.0000	103.5087	200.0000	200.0000
		149.9971	149.9999	77.77836	78.53109	149.9671
	ρ^*_i	3.153377	2.296087	1.032303	1.039016	0.8587452
		0.1552705E+08	12.97090	1.025677	0.000000	1520184.
		30.47721	1105927.	1.021306	1.025677	3.940832
	Q_r	603.1204	170.5176	133.6034	68.80995	2380.293
		528.1394	155.8273	190.7030	63.65408	1881.868
		584.4920	168.7377	170.8667	68.50538	2198.290
	Q_f	2.900155	0.000000	57.17245	160.3393	1017.002
		4399.819	0.000000	52.96613	0.000000	7372.553
		2.376051	421.5171	56.96093	204.8875	15.62473
	$D^*_i\left(p_{IV,i}\right)$	0.1017915	0.000000	167.3035	165.2744	28.57561
		0.1003474E-07	0.000000	241.2281	0.000000	0.6145298E-08
		0.8748984E-02	0.2744780E-05	216.6649	214.4067	0.9855243E-01
	$\rho^*_i D^*_i\left(p_{IV,i}\right)$	0.3209869	0.000000	172.7080	171.7228	24.53917
		500.7339	0.000000	247.4221	0.000000	188.9558
		0.2666446	178.8643	221.2812	219.9120	0.3883785
	PE	0.2615574	18.42879	177.7666	143.7782	32.75389
	$\pi^*_{IV,r}$	77043.76	20040.85	26597.76	8088.595	3820.700
	$\pi^*_{IV,m}$	0.000000	0.000000	22199.18	16441.02	0.1674728
	π^*_{IV}	77043.76	20040.85	48796.94	24529.61	3820.702

(*Continued*)

TABLE 13.7 (*Continued*)

Policy IV Sensitive Analysis with Respect to the Variables

Variables	Decision Variables	−50%	−30%	0%	+30%	+50%
μ_i	Iteration	136	122	149	147	350
	$T^*_{IV,i}$	0.7549328	0.7533598	0.7510187	0.7486994	0.9080405
	$p^*_{IV,i}$	91.36821	91.36019	91.34825	91.33642	92.14822
		103.5270	103.5196	103.5087	103.4979	104.2390
		77.79277	77.78698	77.77836	77.76982	150.0000
	ρ^*_i	1.032477	1.032407	1.032303	1.032200	1.039323
		1.025814	1.025759	1.025677	1.025595	1.031213
		1.021420	1.021374	1.021306	1.021239	0.000000
	Q_r	28.73742	40.14732	133.6034	74.09126	104.0098
		26.62212	37.19273	190.7030	68.64175	96.21002
		28.63056	39.99833	170.8667	73.81800	103.5460
	Q_f	134.3107	134.0264	57.17245	133.1843	162.0623
		191.7119	191.3064	52.96613	190.1052	231.2969
		171.7665	171.4049	56.96093	170.3335	0.000000
	$D^*_i\left(p_{IV,i}\right)$	167.2636	167.2796	167.3035	167.3272	165.7036
		241.1826	241.2009	241.2281	241.2551	239.4024
		216.6217	216.6391	216.6649	216.6906	0.000000
	$\rho^*_i D^*_i\left(p_{IV,i}\right)$	172.6958	172.7007	172.7080	172.7151	172.2195
		247.4085	247.4140	247.4221	247.4302	246.8748
		221.2616	221.2695	221.2812	221.2928	0.000000
	PE	179.5999	178.8620	177.7666	176.6846	196.5885
	$\pi^*_{IV,r}$	26599.14	26598.60	26597.76	26596.91	19139.59
	$\pi^*_{IV,m}$	24200.61	23400.02	22199.18	20998.40	11568.91
	π^*_{IV}	50799.76	49998.62	48796.94	47595.31	30708.50

(*Continued*)

TABLE 13.7 (*Continued*)

Policy IV Sensitive Analysis with Respect to the Variables

Variables	Decision Variables	−50%	−30%	0%	+30%	+50%
$\theta_{m,i}$	Iteration	1,418	251	149	253	422
	$T^*_{IV,i}$	3.987185	0.9518642	0.7510187	0.9107982	0.7471651
	$p^*_{IV,i}$	175.0000	92.37119	91.34825	92.16225	91.32860
		200.0000	200.0000	103.5087	104.2518	103.4908
		150.0000	78.51693	77.77836	150.0000	77.76417
	ρ^*_i	0.000000	1.041299	1.032303	1.039447	1.032132
		1,466,775	0.000000	1.025677	1.031311	1.025542
		8.705450	1.027159	1.021306	0.000000	1.021194
	Q_r	310.9621	72.34116	133.6034	69.92749	57.29346
		284.6671	67.07445	190.7030	64.51077	52.88773
		307.9813	72.10332	170.8667	69.52309	56.97944
	Q_f	0.000000	170.0358	57.17245	162.5637	132.9072
		519.8432	0.000000	52.96613	232.0119	189.7098
		0.000000	217.1545	56.96093	0.000000	169.9809
	$D^*_i\left(p_{IV,i}\right)$	0.000000	165.2576	167.3035	165.6755	167.3428
		0.000000	0.000000	241.2281	239.3704	241.2730
		0.000000	214.4492	216.6649	0.000000	216.7075
	$\rho^*_i D^*_i\left(p_{IV,i}\right)$	0.000000	172.0826	172.7080	172.2109	172.7199
		114.4121	0.000000	247.4221	246.8652	247.4355
		0.000000	220.2734	221.2812	0.000000	221.3005
	PE	0.000000	161.0024	177.7666	197.7671	175.9706
	$\pi^*_{IV,r}$	15627.66	13292.62	26597.76	19140.42	26596.33
	$\pi^*_{IV,m}$	0.000000	11435.29	22199.18	13559.11	22177.91
	π^*_{IV}	15627.66	24727.90	48796.94	32699.53	48774.25

(*Continued*)

TABLE 13.7 (*Continued*)

Policy IV Sensitive Analysis with Respect to the Variables

Variables	Decision Variables	−50%	−30%	0%	+30%	+50%
u_i	Iteration	967	102	149	178	202
	$T^*_{IV,i}$	6.935777	0.7580712	0.7510187	0.7441605	1.525338
	$p^*_{IV,i}$	174.9998	91.35401	91.34825	91.34294	174.7912
		173.2354	103.5225	103.5087	103.4956	200.0000
		149.9999	77.78745	77.77836	77.76967	116.8896
	ρ^*_i	9.321804	1.032617	1.032303	1.031999	25.20313
		1.010956	1.025924	1.025677	1.025436	1161105.
		9489.885	1.021510	1.021306	1.021107	1.153655
	Q_r	296.1704	40.40214	133.6034	73.63551	176.8356
		259.5889	37.42708	190.7030	68.22254	162.6035
		287.1507	40.25129	170.8667	73.36554	234.0358
	Q_f	0.2743299E	134.9266	57.17245	132.3175	17.03549
		583.0588	192.5591	52.96613	188.8988	93.37200
		33.64373	172.5284	56.96093	169.2514	184.7928
	$D^*_i\left(p_{IV,i}\right)$	0.3321549	167.2920	167.3035	167.3141	0.4176506
		66.91148	241.1938	241.2281	241.2611	0.3742565E
		0.2228501	216.6377	216.6649	216.6910	99.33120
	$\rho^*_i D^*_i\left(p_{IV,i}\right)$	0.3096283	172.7485	172.7080	172.6680	10.52610
		67.64453	247.4465	247.4221	247.3979	58.11156
		3.849523	221.2976	221.2812	221.2648	114.5939
	PE	10.01780	181.1707	177.7666	174.4872	670.2609
	$\pi^*_{IV,r}$	6925.354	26593.23	26597.76	26602.02	16139.24
	$\pi^*_{IV,m}$	0.1062922	23433.02	22199.18	20966.00	5.190582
	π^*_{IV}	6925.355	50026.25	48796.94	47568.02	16144.43

(*Continued*)

TABLE 13.7 (*Continued*)

Policy IV Sensitive Analysis with Respect to the Variables

Variables	Decision Variables	−50%	−30%	0%	+30%	+50%
A_m	Iteration	1,540	199	149	229	514
	$T^*_{IV,i}$	1.100706	0.8768263	0.7510187	0.1591771	0.8005388
	$p^*_{IV,i}$	93.12630	91.98932	91.34825	129.1676	91.60071
		105.1341	104.0940	103.5087	200.0000	103.7392
		150.0000	150.0000	77.77836	111.1195	77.96063
	ρ^*_i	1.048064	1.037920	1.032303	1.005810	1.034507
		1.038084	1.030108	1.025677	63135.51	1.027416
		198455.7	0.000000	1.021306	1.004753	1.022743
	Q_r	84.37032	66.91521	133.6034	11.97629	61.00170
		77.89754	61.91602	190.7030	11.16014	56.48632
		83.91598	66.62690	170.8667	11.96679	60.76138
	Q_f	197.2177	156.3912	57.17245	14.76916	142.5597
		281.4305	223.2080	52.96613	2.369080	203.4789
		15.15049	0.000000	56.96093	18.76624	182.2576
	$D^*_i\left(p_{IV,i}\right)$	163.7474	166.0214	167.3035	91.66478	166.7986
		237.1647	239.7651	241.2281	0.000000	240.6520
		0.6660787	0.000000	216.6649	116.6414	216.1181
	$\rho^*_i D^*_i\left(p_{IV,i}\right)$	171.6177	172.3169	172.7080	92.19738	172.5543
		246.1970	246.9838	247.4221	14.80083	247.2498
		13.21871	0.000000	221.2812	117.1957	221.0333
	PE	636.5129	183.4868	177.7666	1.181971	201.6344
	$\pi^*_{IV,r}$	19606.18	19128.95	26597.76	12411.52	26610.83
	$\pi^*_{IV,m}$	14205.04	13737.86	22199.18	0.000000	21896.08
	π^*_{IV}	33811.22	32866.81	48796.94	12411.52	48506.91

(*Continued*)

TABLE 13.7 (*Continued*)

Policy IV Sensitive Analysis with Respect to the Variables

Variables	Decision Variables	−50%	−30%	0%	+30%	+50%
$a_{m,i}$	Iteration	6,761	275	149	4,899	210
	$T^*_{IV,i}$	0.4273148	5.015897	0.7510187	0.7936546	0.9754519
	$p^*_{IV,i}$	174.9999	174.9999	91.34825	137.7994	92.49115
		0.000000	199.8721	103.5087	152.3104	200.0000
		149.9777	150.0000	77.77836	150.0000	78.60355
	ρ^*_i	0.8445191	1.826302	1.032303	1.021105	1.042366
		1.128208	80.94885	1.025677	1.019172	0.000000
		124.0625	389982.8	1.021306	325758.1	1.027851
	Q_r	0.3204889	413.9311	133.6034	5.961858	74.58615
		0.2991217	368.7243	190.7030	5.559991	68.94769
		0.3204882	404.4964	170.8667	5.959496	74.22934
	Q_f	390.7220	0.2209858	57.17245	6.048643	174.3330
		0.2410536	152.6381	52.96613	9.670463	0.000000
		0.3545593	636.2427	56.96093	10.68669	222.6057
	$D^*_i(p_{IV,i})$	0.7495643	0.2009240	167.3035	74.40113	165.0177
		500.0000	0.3197866	241.2281	119.2240	0.000000
		0.6687959	0.000000	216.6649	0.1053898E-06	214.1894
	$\rho^*_i D^*_i(p_{IV,i})$	914349.6	0.3669478	172.7080	75.97136	172.0088
		564.1040	25.88636	247.4221	121.5097	0.000000
		8.297246	106.7634	221.2812	134.2521	220.1546
	PE	16368.45	1307.298	177.7666	15.38298	168.9201
	$\pi^*_{IV,r}$	0.2937416	12168.66	26597.76	22738.28	13297.38
	$\pi^*_{IV,m}$	0.4143911	5.125470	22199.18	1.864357	11309.23
	π^*_{IV}	0.4143911	12173.78	48796.94	22740.22	24606.62

(*Continued*)

TABLE 13.7 (*Continued*)

Policy IV Sensitive Analysis with Respect to the Variables

Variables	Decision Variables	−50%	−30%	0%	+30%	+50%
$a_{mr,i}$	Iteration	380	438	149	219	1,241
	$T^*_{IV,i}$	3.649351	0.2619691	0.7510187	0.7588185	0.1471092
	$p^*_{IV,i}$	174.9999	175.0000	91.34825	91.38803	175.0000
		200.0000	199.9999	103.5087	103.5450	199.9999
		149.9999	149.9999	77.77836	77.80707	150.0000
	ρ^*_i	1.856381	2.447085	1.032303	1.032650	2.152448
		180.3586	13.83545	1.025677	1.025950	1648331.
		452604.4	1058863.	1.021306	1.021532	0.000000
	Q_r	293.6779	19.75062	133.6034	57.77510	11.06565
		264.7770	18.38588	190.7030	53.52036	10.31279
		288.6838	19.72489	170.8667	57.55917	11.05754
	Q_f	0.1987945E-02	0.2923184E-04	57.17245	135.0129	0.000000
		0.5793304E-01	0.5808887E-03	52.96613	192.7137	31.61187
		523.3081	55.72999	56.96093	172.6599	0.000000
	$D^*_i(p_{IV,i})$	0.2560632E-03	0.4512631E-04	167.3035	167.2239	0.000000
		0.7804945E-04	0.1588129E-03	241.2281	241.1374	0.2741667E-05
		0.7038858E-04	0.2646821E-04	216.6649	216.5788	0.000000
	$\rho^*_i D^*_i(p_{IV,i})$	0.4753507E-03	0.1104279E-03	172.7080	172.6838	0.000000
		0.1407689E-01	0.2197248E-02	247.4221	247.3950	213.7863
		126.1358	210.6654	221.2812	221.2422	0.000000
	PE	2029.879	1570.935	177.7666	181.4290	40.84581
	$\pi^*_{IV,r}$	10167.79	1551.948	26597.76	26600.46	26090.25
	$\pi^*_{IV,m}$	3.510304	1582.326	22199.18	22150.79	0.5780599E
	π^*_{IV}	10171.30	3134.274	48796.94	48751.24	26090.25

(*Continued*)

TABLE 13.7 (*Continued*)

Policy IV Sensitive Analysis with Respect to the Variables

Variables	Decision Variables	−50%	−30%	0%	+30%	+50%
$h_{m,i}$	Iteration	157	172	149	193	198
	$T^*_{IV,i}$	0.9753073	0.7943388	0.7510187	0.7141699	0.8781748
	$p^*_{IV,i}$	175.0000	91.34654	91.34825	91.36075	92.40555
		104.1319	103.5046	103.5087	103.5224	200.0000
		78.29040	77.78465	77.77836	77.78051	78.52806
	ρ^*_i	0.000000	1.034230	1.032303	1.030670	1.037980
		1.033603	1.027198	1.025677	1.024386	0.000000
		1.027846	1.022563	1.021306	1.020240	1.025004
	Q_r	74.57489	60.52187	133.6034	54.32780	67.01990
		68.93737	56.04540	190.7030	50.34892	62.01207
		74.21818	60.28526	170.8667	54.13654	66.73070
	Q_f	0.000000	141.8147	57.17245	126.6466	155.8636
		249.8543	202.3093	52.96613	180.8647	0.000000
		223.5469	181.2150	56.96093	162.0904	199.3585
	$D^*_i(p_{IV,i})$	0.000000	167.3069	167.3035	167.2785	165.1889
		239.6703	241.2384	241.2281	241.1940	0.000000
		215.1288	216.6460	216.6649	216.6585	214.4158
	$\rho^*_i D^*_i(p_{IV,i})$	0.000000	173.0339	172.7080	172.4089	171.4628
		247.7239	247.7997	247.4221	247.0758	0.000000
		221.1194	221.5342	221.2812	221.0435	219.7770
	PE	201.4928	199.4840	177.7666	160.2906	136.2090
	$\pi^*_{IV,r}$	19739.98	26547.34	26597.76	26637.11	13341.19
	$\pi^*_{IV,m}$	14763.87	22492.71	22199.18	21929.65	11090.39
	π^*_{IV}	34503.85	49040.05	48796.94	48566.75	24431.58

(*Continued*)

TABLE 13.7 (*Continued*)

Policy IV Sensitive Analysis with Respect to the Variables

Variables	Decision Variables	−50%	−30%	0%	+30%	+50%
$h_{mr,i}$	Iteration	646	140	149	207	194
	$T^*_{IV,i}$	0.7588475	0.7556862	0.7510187	0.7464379	0.7434309
	$p^*_{IV,i}$	91.33777	91.34194	91.34825	91.35464	91.35895
		103.5133	103.5114	103.5087	103.5062	103.5046
		77.77903	77.77872	77.77836	77.77811	77.77800
	ρ^*_i	1.032651	1.032511	1.032303	1.032100	1.031966
		1.025951	1.025840	1.025677	1.025516	1.025411
		1.021533	1.021441	1.021306	1.021173	1.021086
	Q_r	57.77733	57.53306	133.6034	56.81860	56.58636
		53.52242	53.29778	190.7030	52.64067	52.42705
		57.56139	57.31891	170.8667	56.60966	56.37910
	Q_f	135.0995	134.4952	57.17245	132.7286	132.1546
		192.7848	191.9440	52.96613	189.4855	188.6865
		172.7339	171.9798	56.96093	169.7746	169.0580
	$D^*_i(p_{IV,i})$	167.3245	167.3161	167.3035	167.2907	167.2821
		241.2166	241.2214	241.2281	241.2345	241.2384
		216.6629	216.6638	216.6649	216.6657	216.6660
	$\rho^*_i D^*_i(p_{IV,i})$	172.7878	172.7557	172.7080	172.6607	172.6295
		247.4765	247.4547	247.4221	247.3898	247.3685
		221.3283	221.3094	221.2812	221.2532	221.2347
	PE	181.6002	180.0472	177.7666	175.5427	174.0906
	$\pi^*_{IV,r}$	26588.82	26592.45	26597.76	26602.92	26606.28
	$\pi^*_{IV,m}$	22254.13	22232.04	22199.18	22166.65	22145.13
	π^*_{IV}	48842.96	48824.49	48796.94	48769.57	48751.41

(*Continued*)

TABLE 13.7 (*Continued*)

Policy IV Sensitive Analysis with Respect to the Variables

Variables	Decision Variables	−50%	−30%	0%	+30%	+50%
$c_{mr,i}$	Iteration	290	243	149	346	384
	$T^*_{IV,i}$	0.7549328	0.7533598	0.7510187	0.8848852	0.7471651
	$p^*_{IV,i}$	91.36821	91.36019	91.34825	175.0000	91.32860
		103.5270	103.5196	103.5087	104.1314	103.4908
		77.79277	77.78698	77.77836	78.27084	77.76417
	ρ^*_i	1.032477	1.032407	1.032303	0.000000	1.032132
		1.025814	1.025759	1.025677	1.030393	1.025542
		1.021420	1.021374	1.021306	1.025200	1.021194
	Q_r	57.47484	57.35331	133.6034	67.54092	56.87477
		53.24424	53.13247	190.7030	62.49008	52.69234
		57.26112	57.14048	170.8667	67.24729	56.66542
	Q_f	134.3107	134.0264	57.17245	0.000000	132.9072
		191.7119	191.3064	52.96613	225.2955	189.7098
		171.7665	171.4049	56.96093	201.6925	169.9809
	$D^*_i\left(p_{IV,i}\right)$	167.2636	167.2796	167.3035	0.000000	167.3428
		241.1826	241.2009	241.2281	239.6715	241.2730
		216.6217	216.6391	216.6649	215.1875	216.7075
	$\rho^*_i D^*_i\left(p_{IV,i}\right)$	172.6958	172.7007	172.7080	0.000000	172.7199
		247.4085	247.4140	247.4221	246.9557	247.4355
		221.2616	221.2695	221.2812	220.6101	221.3005
	PE	179.5999	178.8620	177.7666	164.9304	175.9706
	$\pi^*_{IV,r}$	26599.14	26598.60	26597.76	19812.42	26596.33
	$\pi^*_{IV,m}$	24200.61	23400.02	22199.18	13152.55	20197.91
	π^*_{IV}	50799.76	49998.62	48796.94	32964.97	46794.25

TABLE 13.8

Summary of Sensitivity Analysis of Table 13.7

Parameters	$T^*_{IV,i}$	$p^*_{IV,i}$	Q_r	Q_f	D^*_i	ρ^*_i	PE	$\pi^*_{IV,r}$	$\pi^*_{IV,m}$	π^*_{IV}
γ_i	Sensitive	Sensitive	Sensitive	Sensitive	MSensitive	ISensitive	Sensitive	MSensitive	MSensitive	MSensitive
τ_i	HSensitive	Sensitive	HSensitive	HSensitive	HSensitive	HSensitive	HSensitive	HSensitive	HSensitive	HSensitive
α_i	Sensitive	Sensitive	HSensitive	HSensitive	HSensitive	HSensitive	HSensitive	HSensitive	HSensitive	HSensitive
β_i	HSensitive	Sensitive	HSensitive	HSensitive	HSensitive	HSensitive	HSensitive	HSensitive	HSensitive	HSensitive
θ_i	Sensitive	HSensitive	Sensitive	Sensitive	Sensitive	Sensitive	HSensitive	HSensitive	HSensitive	HSensitive
A_r	Sensitive	MSensitive	Sensitive	Sensitive	HSensitive	MSensitive	Sensitive	Sensitive	Sensitive	Sensitive
$a_{r,i}$	Sensitive	MSensitive	HSensitive	HSensitive	HSensitive	HSensitive	HSensitive	HSensitive	HSensitive	HSensitive
$h_{r,i}$	ISensitive	Sensitive	HSensitive	HSensitive	HSensitive	HSensitive	HSensitive	HSensitive	HSensitive	HSensitive
$c_{r,i}$	HSensitive	Sensitive	HSensitive	HSensitive	HSensitive	HSensitive	HSensitive	HSensitive	HSensitive	HSensitive
μ_i	MSensitive	ISeensitive	MSensitive	MSensitive	MSensitive	ISensitive	MSensitive	MSensitive	MSensitive	Sensitive
$\theta_{m,i}$	Sensitive	Sensitive	Sensitive	Sensitive	Sensitive	Sensitive	Sensitive	Sensitive	Sensitive	Sensitive
u_i	HSensitive	Sensitive	Sensitive	Sensitive	Sensitive	HSensitive	Sensitive	Sensitive	Sensitive	Sensitive
A_m	Sensitive	Sensitive	HSensitive	HSensitive	HSensitive	HSensitive	HSensitive	HSensitive	HSensitive	HSensitive
$a_{m,i}$	HSensitive	HSensitive	HSensitive	HSensitive	HSensitive	HSensitive	HSensitive	HSensitive	HSensitive	HSensitive
$a_{mr,i}$	Sensitive	HSensitive	HSensitive	HSensitive	HSensitive	HSensitive	HSensitive	HSensitive	HSensitive	HSensitive
$h_{m,i}$	MSensitive	Sensitive	HSensitive	HSensitive	Sensitive	Sensitive	Sensitive	HSensitive	HSensitive	HSensitive
$h_{mr,i}$	ISensitive	ISensitive	HSensitive	HSensitive	ISensitive	ISensitive	MSensitive	ISensitive	ISensitive	ISensitive
$c_{mr,i}$	MSensitive	MSensitive	HSensitive	HSensitive	MSensitive	ISensitive	MSensitive	MSensitive	HSensitive	Sensitive

Note: M, Moderately; H, Highly; I, Insensitive.

TABLE 13.9

Total Profit with Publicity Effort with Effect of A_m and A_r

A_m/A_r	260	265	270	275	280
460	48,744	48,724.31	48,704.7	48,685.18	24,606.62
465	48,737.43	0.0001714	48,698.19	48,678.69	48,659.29
470	34,113.25	48,711.23	48,691.68	24,611.75	34.18558
475	48,724.31	33,555.11	139.1507	48,665.75	48,646.39
480	48,717.76	48,698.19	48,678.69	48,659.29	24,586.17

TABLE 13.10

ANOVA Test

Dependent Variable: Total Profit (TP)						
Sources	**Sum of Squares**		**Degree of Freedom**	**Mean Sum of Square**	***F***	***p*-Value**
A_m	Due to A_m effects	585287880.067	4	146321970.017	.436	0.781
A_r	Due to A_r effects	865286384.190	4	216321596.048	.644	0.639
Error	Due to errors	5371270244.183	16	335704390.261	–	–
Total	Due to total	6821844508	24	–	–	–

TABLE 13.11

Total Profit without Publicity Effort with Effect of A_m and A_r

A_m/A_r	260	265	270	275	280
460	48,080.81	48,058.29	48,035.86	48,013.53	47,991.3
465	48,073.29	48,050.8	48,028.41	48,006.11	47,983.91
470	48,065.79	48,043.33	48,020.96	47,998.7	47,976.53
475	48,058.29	48,035.86	48,013.53	47,991.3	47,969.16
480	48,050.8	48,028.41	48,006.11	47,983.91	47,961.81

TABLE 13.12

ANOVA Test

Target Variable: Total Profit (TP)						
Source	**Sum of Squares**		**Degree of Freedom**	**Mean Sum of Squares**	***F* value**	***p*-Value**
A_m	Due to the effect of A_m	2,765.897	4	691.474	105,932.469	0.000
A_r	Due to the effect of A_r	24,895.002	4	6,223.751	953,466.196	0.000
Error	Due to the errors	0.104	16	0.007	–	–
Total	Due to total	27,661.003	24	–	–	–

and mixed integer on the revenue growth of both the supply chain partners, namely manufacturers and retailers. From the sensitive analysis, it is concluded that the total profit per unit time with publicity effort is not significant for the different values of A_m and A_r. Under cooperative policies, the significant variables such as τ_i, α_i, β_i, θ_i, u_i, $a_{m,i}$, and $c_{mr,i}$ are highly sensitive for the multi-item SC model. So, selecting the values of these significant variables at a strategic level is a critical decision for optimizing supply chain profit. This chapter may be explored in some ways like applying stochastic deterioration rate, partial or full backlogging; partial or full trade credits system, and stochastic demand rate using real datasets for future research scope.

References

Abad, P.L., Jaggi, C.K., 2003. A joint approach for setting unit price and the length of the credit period for a seller when end demand is price sensitive. *International Journal of Production Economics*, 83, 115–122.

Ben-Khedher, N., Yano, C.A., 1994. The multi-item joint replenishment problem with transportation and container effects. *Transportation Science*, 28 (1), 37–54.

Chang, H.J., Hung, C.H., Dye, C.Y., 2001. An inventory model for deteriorating items with linear demand under condition of permissible delay in payments. *Production planning and control*, 12, 274–282.

Chen, J.M., Chen, T.H., 2005a. The multi-item replenishment problem in a two-echelon supply chain: The effect of centralization versus decentralization. *Computers and Operations Research*, 32, 3191–3207.

Chen, J.M., Chen, T.H., 2005b. Effects of joint replenishment and channel coordination for managing multiple deteriorating products in a supply chain. *Journal of the Operational Research Society*, 56 (10), 1224–1234.

Chen, J.M., Chen, T.H., 2005c. Effects of joint replenishment and channel coordination for managing multiple deteriorating products in a supply chain. *Journal of Operational Research Society*, 56 (10), 1224–1234.

Chen, T.H., Chen, J.M., 2005d. Optimizing supply chain collaboration based on joint replenishment and channel coordination. *Transportation Research Part E: Logistics and Transportation Review*, 41 (4), 261–285.

Chen, T.H., Chen, J.M., 2005e. Optimizing supply chain collaboration based on joint replenishment and channel coordination. *Transportation Research Part E: Logistics and Transportation Review*, 41, 261–285.

Chen, J.M., Chen, T.H., 2007. The profit maximization model for a multi-item distribution channel. *Transportation Research Part-E: Logistics and Transportation Review*, 43, 338–354.

Chung, K.J., 1998. A theorem on determination of economic order quantity under conditions of permissible in payments. *Computers and Operations Research*, 25 (1), 49–52.

Cohen, M.A., 1977. Joint pricing and ordering policy for exponentially decaying inventory with known demand. *Naval Research Logistics Quarterly*, 24, 257–268.

Goyal, S.K., 1988. Economic policy for jointly replenished items. *International Journal of Production Research*, 26 (7), 1237–1240.

Graves, S.C., 1979. On the deterministic demand multi-product single-machine lot scheduling problem. *Management Science*, 25 (3), 276–280.

Hahm, J., Yano, C.A., 1995a. The economic lot and delivery scheduling problem: The common cycle case. *IIE Transactions*, 27, 113–125.

Hahm, J., Yano, C.A., 1995b. The economic lot and delivery scheduling problem: Models for nested schedules. *IIIE Transactions*, 27, 126–139.

Hammer, M., 2001. The superefficient company. *Harvard Business Review*, 79, 82–91.

Hsu, P.H., Wee, H.M., 2005. Horizontal suppliers coordination with uncertain suppliers deliveries. *International Journal of Operations Research*, 2 (2), 17–30.

Kao, E.P.C., 1979. A multi-product dynamic lot-size model with individual and joint set-up costs. *Operations Research*, 27 (2), 279–289.

Khouja, M., 2003. Optimizing inventory decisions in a multi-stage multi-customer supply chain. *Transportation Research Part E: Logistics and Transportation Review*, 39 (3), 193–208.

Lee, H.L., 2004. The triple-a supply chain. *Harvard Business Review*, 82 (10), 102–112.

Li, X., Wang, Q., 2007. Coordination mechanisms of supply chain systems. *European Journal of Operational Research*, 179 (1), 1–16.

Narayanan, V.G., Raman, A., 2004. Aligning incentives in supply chains. *Harvard Business Review*, 82, 94–102.

Nielsen, C., Larsen, C., 2005. An analytical study of the Q(s,S) policy applied to the joint replenishment problem. European Journal of Operational Research, 163 (3), 721–732.

Rempala, R., 2003. Joint replenishment multiproduct inventory problem with continuous and discrete demands. *International Journal of Production Economics*, 81–82, 495–511.

Tsao, Y.C., Sheen, G.J., 2008. Dynamic pricing, promotion and replenishment policies for a deteriorating item under permissible delay in payments. *Computers & Operations Research*, 35, 3562–3580.

Tsao, Y.C., Sheen, G.J., 2012. A multi-item supply chain with credit periods and weight freight cost discounts. *International Journal Production Economics*, 135, 106–115.

vanEijs, M.J.G., 1994. Multi-item inventory systems with joint ordering and transportation decisions. *International Journal of Production Economics*, 35, 285–292.

Index